THE QUANTUM STORY

Praise for *The Quantum Story*

'Jim Baggott's survey of the history of the emergence of the twentieth century's most enigmatic but successful theory is a delight to read. It is clear, accessible, engaging, informative, and thorough. It illuminates an important, revolutionary era of modern science and the varied personalities behind it.'
Peter Atkins

'Jim Baggott's inspired – and inspiring – idea of presenting the history of quantum physics in terms of 40 key moments works both as an introduction for the uninitiated and as a refresher for anyone who thinks they know the story. Even familiar stories come up fresh in these juxtapositions. Great to dip in to!'
John Gribbin

'Intellectually gratifying.'
The Economist

'I have never come across a book quite like Jim Baggott's *The Quantum Story*. He has done something that I would have thought impossible in a popular book. He manages to present the full ambit of the theory, starting with the introduction of the quantum—the basic unit of energy—by the German physicist Max Planck in the beginning of the 20th century, and ending with the search for the Higgs particle at the collider at CERN in Geneva. In doing this Mr. Baggott navigates successfully between the Scylla of mathematical rigor and the Charybdis of popular nonsense. He also manages to get the people right. I know this because for many of the scenes he describes I was there.'
Jeremy Bernstein, *Wall Street Journal*

'An enjoyable addition to the overall quantum story.'
Chemistry World

'An accessible and informative history.'
Science magazine

'This is an engaging book, which situates some of the most important episodes in the history of quantum physics in their historical and social context. The author is equally at home with the classic work of Max Planck and with recent advances in the fields of superconducting electronic and string theory.'
Anthony J. Leggett, Winner of the 2003 Nobel Prize in Physics

'[A] wonderful history of the scientists and ideas behind quantum mechanics... The basic history behind the quantum revolution is well known, but no one has told it in quite such a compellingly human and thematically seamless way.'
Publishers Weekly

'Baggott shines... Those with a jones for physics will not be disappointed... Quantum theory may deny us the possibility of properly comprehending physical reality, but Baggott's account is smart and consoling.'
Kirkus Reviews

'... engrossing.'
Booklist

'Gripping story'
Flipside Magazine

THE Quantum Story

A HISTORY IN 40 MOMENTS

JIM BAGGOTT

OXFORD UNIVERSITY PRESS

OXFORD
UNIVERSITY PRESS

Great Clarendon Street, Oxford, OX2 6DP,
United Kingdom

Oxford University Press is a department of the University of Oxford.
It furthers the University's objective of excellence in research, scholarship,
and education by publishing worldwide. Oxford is a registered trade mark of
Oxford University Press in the UK and in certain other countries

First Edition published in 2011
First published in paperback 2013

Impression: 1

British Library Cataloguing in Publication Data
Data available

Library of Congress Cataloging in Publication Data
Data available

ISBN 978–0–19–956684–6 (hbk)
ISBN 978–0–19–965597–7 (pbk)

Printed in Great Britain
on acid-free paper by
Clays Ltd, St Ives plc

For MSS,

because every student deserves
at least one great teacher

CONTENTS

Preface xiii
Prologue: Stormclouds 1
London, April 1900

PART I: QUANTUM OF ACTION

1 The Most Strenuous Work of My Life 7
Berlin, December 1900
2 Annus Mirabilis 17
Bern, March 1905
3 A Little Bit of Reality 25
Manchester, April 1913
4 la Comédie Française 34
Paris, September 1923
5 A Strangely Beautiful Interior 43
Helgoland, June 1925
6 The Self-rotating Electron 51
Leiden, November 1925
7 A Late Erotic Outburst 60
Swiss Alps, Christmas 1925

PART II: QUANTUM INTERPRETATION

8 Ghost Field 71
Oxford, August 1926
9 All This Damned Quantum Jumping 79
Copenhagen, October 1926

10 The Uncertainty Principle — 87
Copenhagen, February 1927

11 The 'Kopenhagener Geist' — 95
Copenhagen, June 1927

12 There is No Quantum World — 103
Lake Como, September 1927

PART III: QUANTUM DEBATE

13 The Debate Commences — 115
Brussels, October 1927

14 An Absolute Wonder — 126
Cambridge, Christmas 1927

15 The Photon Box — 133
Brussels, October 1930

16 A Bolt from the Blue — 141
Princeton, May 1935

17 The Paradox of Schrödinger's Cat — 149
Oxford, August 1935

Interlude: The First War of Physics — 159
Christmas 1938–August 1945

PART IV: QUANTUM FIELDS

18 Shelter Island — 171
Long Island, June 1947

19 Pictorial Semi-vision Thing — 181
New York, January 1949

20 A Beautiful Idea — 193
Princeton, February 1954

21 Some Strangeness in the Proportion — 204
Rochester, August 1960

22 Three Quarks for Muster Mark! — 214
New York, March 1963

23 The 'God Particle' 225
Cambridge, Massachusetts, Autumn 1967

PART V: QUANTUM PARTICLES

24 Deep Inelastic Scattering 237
Stanford, August 1968

25 Of Charm and Weak Neutral Currents 247
Harvard, February 1970

26 The Magic of Colour 256
Princeton/Harvard, April 1973

27 The November Revolution 265
Long Island/Stanford, November 1974

28 Intermediate Vector Bosons 275
Geneva, January/June 1983

29 The Standard Model 285
Geneva, September 2003

PART VI: QUANTUM REALITY

30 Hidden Variables 297
Princeton, Spring 1951

31 Bertlmann's Socks 306
Boston, September 1964

32 The Aspect Experiments 318
Paris, September 1982

33 The Quantum Eraser 328
Baltimore, January 1999

34 Lab Cats 339
Stony Brook/Delft, July 2000

35 The Persistent Illusion 349
Vienna, December 2006

PART VII: QUANTUM COSMOLOGY

36 The Wavefunction of the Universe 361
Princeton, July 1966

37 Hawking Radiation 372
Oxford, February 1974

38 The First Superstring Revolution 381
Aspen, August 1984

39 Quanta of Space and Time 391
Santa Barbara, February 1986

40 Crisis? What Crisis? 399
Durham, Summer 1994

Epilogue: A Quantum of Solace? 407
Geneva, March 2010

Notes and Sources 411
Bibliography 441
Plate Acknowledgements 448
Index 451

ABOUT THE AUTHOR

Jim Baggott is an award-winning science writer. A former academic scientist, he now works as an independent business consultant but maintains a broad interest in science, philosophy, and history, and continues to write on these subjects in his spare time. His previous books have been widely acclaimed and include:

Atomic: The First War of Physics and the Secret History of the Atom Bomb 1939–49 (Icon Books, 2009);

A Beginner's Guide to Reality (Penguin, 2005);

Beyond Measure: Modern Physics, Philosophy and the Meaning of Quantum Theory (Oxford University Press, 2004);

Perfect Symmetry: The Accidental Discovery of Buckminsterfullerene (Oxford University Press, 1994); and

The Meaning of Quantum Theory: A Guide for Students of Chemistry and Physics (Oxford University Press, 1992).

PREFACE

The last century was defined by physics. From the minds of the world's leading physicists there flowed a river of ideas that would transport mankind to the very pinnacle of wonder and to the very depths of despair. This was a century that began with the certainties of absolute knowledge and ended with the knowledge of absolute uncertainty. It was a century in which physicists developed theories that would deny us the possibility that we can ever properly comprehend the nature of physical reality. It was also a century in which they built weapons with the capacity utterly to destroy this reality.

Almost everything we think we know about the nature of our world comes from one theory of physics. This theory was discovered and refined in the first thirty years of the twentieth century and went on to become quite simply the most successful theory of physics ever devised. Its concepts underpin much of the twenty-first century technology that we have learned to take for granted.

But this success has come at a price, for it has at the same time completely undermined our ability to make sense of the world at the level of its most fundamental constituents.

Rejecting the elements of uncertainty and chance implied by this new theory, Albert Einstein once famously declared that 'God does not play dice'. Niels Bohr claimed that anybody who is not shocked by the theory has not understood it. The charismatic American physicist Richard Feynman went further: he claimed that *nobody* understands it. To anyone tutored in the language and the logic of classical physics, this theory is at once mathematically challenging, maddeningly bizarre, and breathtakingly beautiful.

This is quantum theory, and this book tells its story.

If we fix on Max Planck's discovery of his 'quantum of action' in December 1900 as the historical origin of the quantum theory, then as I write this theory is 110 years old. Time enough, you would have thought, for physicists to get to grips with it and understand what it means. Time enough to come to terms with what quantum theory has to say about chance and causality and the nature of physical reality. And yet, if anything, the sense of shock has increased, not diminished, with the passing of time.

While nobody really understands how quantum theory actually works, the rules of its application are unquestioned and the accuracy and precision of its predictions are unsurpassed in the entire history of science. Although heated debate continues about how quantum theory should be interpreted, there can be no debate about whether or not the theory is fundamentally correct.

For more than four hundred years we nurtured the belief (should that, perhaps, be *faith*?) that evidence-based investigation meeting scientific standards of rigour would reveal the true mechanisms of nature. And yet when the mechanisms of nature were revealed to be quantum mechanisms, the worlds of science and philosophy were set on a collision course. Instead of truth and comprehension, we got deeply unsettling questions about what we can ever hope to know about the world. Quantum theory pushed us to the edge of an epistemological precipice. Since the mid-1920s we have lived in fear of stepping over the edge.

This book is a celebration of this wonderful yet wholly disconcerting theory, from its birth in the porcelain furnaces used to study black-body radiation in 1900 to the promise of stimulating new quantum phenomena to be revealed by CERN's Large Hadron Collider, over a century later. It is a history told in forty 'moments', significant moments of truth or turning points in the theory's development.

This history takes us on a long journey. Part I deals with Planck's discovery in 1900 and traces the development of early quantum theory through Einstein's light-quantum hypothesis, Bohr's quantum theory of the atom, Louis de Broglie's dual wave–particle hypothesis, Werner Heisenberg's matrix mechanics, the puzzling phenomenon of electron spin, and Wolfgang Pauli's exclusion principle. This section concludes with Erwin Schrödinger's 'late erotic outburst', which led him to wave mechanics in December 1925.

In Part II, the book traces the development of the Copenhagen interpretation of quantum theory. We move from Max Born's interpretation of the significance of Schrödinger's wavefunction in 1926, via the intense debates between Bohr, Heisenberg, and Schrödinger on the reality of quantum jumps, the development of Heisenberg's uncertainty principle, to Bohr's Como lecture in September 1927.

At this point, Einstein, an early champion of quantum theory, became one of the theory's most determined critics. Part III describes the Bohr–Einstein debate, one of the most profound debates in the history of science. We begin with Einstein's first thought experiments, outlined to a pensive audience during the fifth Solvay conference in October 1927, before moving on to the Einstein, Podolsky, Rosen argument and Schrödinger's famous cat paradox in 1935. The book pauses briefly in this section to catch the 'absolute wonder' that is Paul Dirac's relativistic quantum theory of the electron.

To study quantum theory is to study the physicists who made it. My original intention was to write a 'biography' of quantum theory based on the biographies of the physicists who forged it and refined it.[1] Many of these same physicists also played crucial roles in the development of the world's first atomic weapons, and I had intended to include a long section on their wartime exploits. In an already hopelessly ambitious book, this was an ambition too far. This section became a book in itself, titled *Atomic: The First War of Physics and the Secret History of the Atom Bomb, 1939–49*, published in 2009 by Icon Books. With the publisher's permission I have included here an interlude, based on extracts from this earlier book and focused primarily on the infamous meeting in September 1941 between Bohr and Heisenberg in Nazi-occupied Copenhagen.

As the physicists picked up the pieces of their academic careers after the war, quantum theory was in crisis. Part IV describes the series of crisis meetings that culminated in the development of quantum electrodynamics, by Julian Schwinger, Richard Feynman, Sin-itiro Tomonaga, and Freeman Dyson. This was followed in 1954 by the unheralded development of quantum field theory based on local gauge symmetry

[1] I wrote the original proposal for this book at a time when 'biographies' of inanimate subjects were still popular.

by Chen Ning Yang and Robert Mills. Sheldon Glashow, Abdus Salaam, and Stephen Weinberg went on to formulate early versions of a unified electro-weak theory in 1960, and predicted the existence of 'heavy photons', the W and Z particles. Although much of this effort was summarily dismissed by the physics community, this was a time of unprecedented fertility in theoretical physics. It culminated in Murray Gell-Mann's 1963 theory of quarks and the introduction of symmetry-breaking and the Higgs mechanism in 1967.

At this point, the history of quantum theory becomes synonymous with the history of particle physics. In Part V the book examines the roles of ever larger and more expensive particle accelerators and colliders in furnishing the evidence for the particular collection of quantum field theories that has become known as the Standard Model of particle physics. This section begins with the discovery, at the Stanford Linear Accelerator Center (SLAC) in 1968, that the proton possesses an internal structure. This is followed by the discoveries of a hypothetical charm quark and the colour force described by Gell-Mann and Harald Fritzsch's theory of quantum chromodynamics.

The experimental observation of the J/ψ meson (formed from a charm and anti-charm quark), simultaneously at SLAC and at Brookhaven National Laboratory in the 'November revolution' of 1974, and the subsequent observation of the W and Z particles at CERN in 1983, set the physicists on the path to the Standard Model. This model is constructed from three 'generations' of matter particles consisting of leptons (electrons and neutrinos) and quarks that interact through the exchange of force particles—photons, W and Z particles, and colour force gluons. There is as yet no room in the Standard Model for gravity. This section concludes by looking over the shoulders of the physicists gathered at CERN as they celebrate their successes in September 2003.

At this stage the book steps backwards, to 1951, and David Bohm's growing unease over the implications of the Copenhagen interpretation. Encouraged by Einstein, Bohm goes on to make further refinements to the Einstein, Podolsky, Rosen argument, bringing their thought experiment into the realms of practicality. Bohm goes on to develop an elaborate 'hidden variable' alternative to conventional quantum theory.

From these beginnings, Part VI traces the development of contemporary experimentation designed to probe the very nature of physical reality itself. The book recounts the development in 1964 of John Bell's theorem, and Bell's inequality, which exposed the true nature of Einstein's challenges and provided a straightforward test of local *versus* non-local reality. The first definitive experiments were performed by Alain Aspect and his colleagues in 1981 and 1982. The quantum world was shown to be determinedly non-local.

There followed a series of experiments demonstrating the truly incomprehensible nature of this world, leaving the committed realist grasping for straws. These included Marlan Scully and Kai Drühl's quantum eraser experiment and experiments demonstrating interference in macroscopic quantum objects, inanimate laboratory versions of Schrödinger's infamous cat. This section concludes with a description of some 2006 experiments by Anton Zeilinger and his colleagues, designed to test a further inequality devised by Anthony Leggett. The results strongly suggest that we can no longer assume that the particle properties we measure necessarily reflect or represent the properties of the particles as they really are.

These experiments tell us rather emphatically that we can never perceive reality 'as it really is'. We can only reveal aspects of an empirical reality that depend on the nature of the instruments we use and the questions we ask. Quantum physics, it seems, has completed its transformation into experimental philosophy.

The book closes with Part VII, which describes the efforts to forge together the two great physical theories of the twentieth century—quantum theory and general relativity—into a theory of quantum gravity or, alternatively, into a 'theory of everything', capable in principle of describing everything in the universe. This section begins with the development of canonical quantum gravity in the form of the Wheeler–DeWitt equation. Applying quantum field theory to the curved spacetime around a black hole, Stephen Hawking discovered in 1974 that black holes 'ain't so black'.

The first superstring revolution in August 1984 promised to provide a theory which could not only explain all the particles of the Standard Model but could also accommodate the graviton, the purported field particle of the gravitational force. This early promise faded, however, as different variants of superstring theory emerged and it lost any sense of uniqueness. Around the same time, the canonical approach was resurrected in the form of loop quantum gravity. Superstring theory experienced something of a renaissance in March 1995, in the second superstring revolution, and today dominates contemporary theoretical physics.

But there is growing impatience with the superstring programme's obsession with obscure hidden dimensions and its inability to make any kind of testable prediction. As has happened so often in its glorious 110-year history, quantum theory is once again in crisis. This section closes with an exploration of the role that *interpretation*, an obsession among physicists since the theory's inception, might still have yet to play.

The book concludes optimistically with, if you will, a quantum of solace. It has cost £3.5 billion and frustratingly blew up days after it was first switched on in September 2008, but the Large Hadron Collider (LHC) at CERN in Geneva gives some hope of resolving the current crisis. At worst, the LHC will simply confirm the existence of the Higgs boson, validating the mechanism of spontaneous symmetry-breaking, explaining how particles acquire mass, and putting the icing on the cake of the Standard Model. At worst, the LHC will provide *answers*.

At best, the LHC will turn up some bizarre new experimental facts; facts that simply can't be accommodated in the current quantum field theories that constitute the Standard Model, and the crisis will deepen. Physics will then truly come alive once again. Only from the depths of despair are we likely to see the breakthroughs needed to propel quantum theory on the next stage of its journey. At best, the LHC will beg *questions*.

Many of the 'moments' I have chosen to describe in this book suggest themselves as unambiguously key events in quantum theory's history. Others are a little less obvious and some have been chosen in an effort to maintain narrative consistency and flow. Whilst I make no apologies for my choices, I am very conscious of the risk that, taken together, these moments may be perceived to describe a smooth, irresistible progression towards some inexorable scientific truth.

This is just not how science works. It has not been possible to describe here all the blind alleys, the dead ends, the theories that dominated for a time only to be replaced by alternatives better able to explain the data. The reality of scientific endeavour is profoundly messy, often illogical, deeply emotional, and driven by the individual personalities involved as they 'sleepwalk' their way to a temporary scientific truth.

My thanks go to Latha Menon, my editor at Oxford University Press, for her patience, fortitude, and ability to channel my ambitions in more practical directions. I owe a debt of thanks to Anthony Leggett, Carlo Rovelli, and Peter Woit who read and commented on the draft manuscript. It goes without saying that all the errors, misconceptions, and misinterpretations that remain are entirely my fault.

I hope this book will stand as a testament to the intellectual and psychological challenge posed by a quantum theory so profoundly at odds with a common-sense conception of the world, and to those great physicists who were able to rise to this challenge. A testament to what can be achieved through the application of a theory that nobody understands.

Jim Baggott

Reading, July 2010

PROLOGUE

Stormclouds

London, April 1900

At the beginning of the twentieth century there were plenty of reasons for believing that the great journey that was physics was close to reaching its final destination.

The structure that was classical physics had been built on foundations laid by Isaac Newton's grand synthesis in the seventeenth century. A further two hundred years of scientific endeavour had created a seemingly unassailable model of the world. This model explained everything, from the interplay of force and motion in the dynamics of moving objects, thermodynamics, optics, electricity, magnetism, and gravitation. Its scope was vast. It described everything, from the familiar objects of everyday experience on earth to objects in the furthest reaches of the visible universe. There seemed little room for doubt about the basic correctness of classical physics, its essential truth.

But Newton had been obliged to compromise. He needed an absolute space and an absolute time to provide a framework against which all motion could be measured. Much more worrisome was the force of gravity. In all of Newton's mechanics, force is a physical phenomenon exerted through contact between one object and another. Newton's gravity was an influence felt through a curious, mutual action-at-a-distance between objects. He was accused of introducing 'occult agencies' into his otherwise rational, physical, and mathematically precise description of the world. In a *General Scholium*, added to the 1713 second edition of his most famous work *Philosophiæ Naturalis Principia Mathematica* (The Mathematical Principles of Natural Philosophy), he wrote:

> Hitherto we have explain'd the phænomena of the heavens and of our sea, by the power of Gravity, but have not yet assign'd the cause of this power...I have not been able to discover the cause of those properties of gravity from phænomena, and I frame no hypotheses.

Gravity was a force that was somehow exerted instantaneously, with no intervening medium other than a hypothetical, all-pervading, tenuous form of matter called the ether that was thought to fill the void.

Newton had also extended the scope of his mechanics to describe light, concluding that light consists of tiny particles, or corpuscles. Two of his contemporaries, English natural philosopher Robert Hooke and Dutch physicist Christiaan Huygens, had argued that light consists instead of waves. Such was Newton's standing and authority that the corpuscular theory held sway for a hundred years.

In a series of papers read to the Royal Society in London between 1801 and 1803, nearly eighty years after Newton's death, English physicist Thomas Young revived the wave theory as the only possible explanation of light diffraction and interference phenomena. In one experiment, commonly attributed to Young, it was shown that when passed through two narrow, closely spaced holes or slits, light produces a pattern of bright and dark fringes. These are readily explained in terms of a wave theory of light in which the peaks and troughs of the light waves from the two slits start out 'in step' (or in phase). Where a peak of one wave is coincident with a peak of another, the two waves add and reinforce. This is called constructive interference, and gives rise to a bright fringe. Where a peak of one wave is coincident with a trough of another, the two waves cancel. This is called destructive interference, and gives a dark fringe.

Despite the apparently inescapable logic of his explanation, Young's views were roundly rejected by the physics community at the time, with some condemning his explanation as 'destitute of every species of merit'.

The wave theory of light was to prove ultimately irresistible, however. In the 1860s Scottish physicist James Clerk Maxwell fused electricity and magnetism into a single theory of electromagnetism. The intimate connection between electricity and magnetism had been established for some years, most notably through the extraordinary experimental work of Michael Faraday at London's Royal Institution. Drawing on analogies with fluid mechanics, Maxwell proposed the existence of electromag-

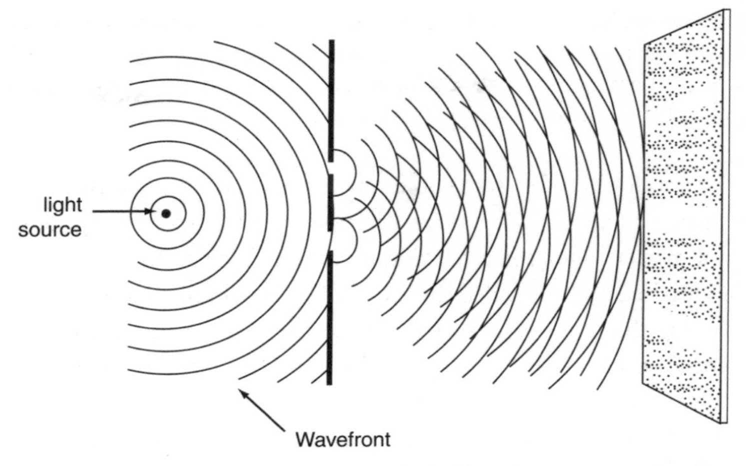

FIG 1 Thomas Young discovered that passing light through two narrow, closely-spaced holes or slits would produce a pattern of alternating light and dark fringes. These could only be explained, Young, believed, in terms of a wave theory of light in which overlapping wavefronts interfere constructively (bright fringe) and destructively (dark fringe).

netic fields whose properties are described by a set of complex differential equations.[2]

Maxwell had made no assumptions about how these fields were supposed to move through space. Nevertheless, when the equations are cast in a way that makes explicit the interdependence of the electric and magnetic fields as they move through free space, they point unambiguously to a wave-like motion. And, as Maxwell himself discovered, the speed of these electromagnetic waves is precisely the speed of light.

But waves are disturbances *in* something. Waves rippling on the surface of a pond are disturbances in water. The noise of a tree falling in a forest is derived from sound waves propagating through the air. All wave motion requires a medium to support it, so precisely what medium was meant to support light waves? The ether was once again called to duty. Faraday had rejected the notion of the ether, but Maxwell leaned heavily on the concept in developing his theory.

It seemed that neither gravity nor electromagnetism could be explained without the ether. But if the ether existed, then certain physical

[2] The historical development of Maxwell's equations is described in detail in Crease, *A Brief Guide to the Great Equations.*

consequences could be anticipated. The earth's motion through space could be expected to drag the ether along with it. And, just as a sound wave will *appear* to be travelling faster if a strong wind moves the air that carries it, so a light wave is expected to appear to travel faster if caught in an 'ether wind'. This meant that there should be measurable differences in the speed of light depending on the direction it is travelling relative to the earth's motion through space. This proposal was put to its most stringent test by American physicists Albert Michelson and Edward Morley in 1887. They could find no evidence for a drag effect and hence no evidence for relative motion between the earth and the ether.

As the century turned, Newton's grand mechanical design remained largely unassailable. Physicists were either willing to forgive gravitational action-at-a-distance or quietly forget that this was a problem, because the structure worked so wonderfully well and it was so obviously right. Newton had been shown to be fallible on the question of light, but it was now clear how a wave theory of light fitted into the equally wonderful structure created by Maxwell, despite some hesitancy on the question of the wave medium.

We could, perhaps, forgive physicists of the late nineteenth century their sense of triumph. In 1900, the great British physicist Lord Kelvin (William Thomson) famously declared to the British Association for the Advancement of Science that: 'There is nothing new to be discovered in physics now. All that remains is more and more precise measurement.'

It is a famous declaration, one that characterizes the mood of the time, although, it seems, one that may well be apocryphal.[3] The simple truth is that in the last decades of the nineteenth century the mechanical structure of classical physics was beginning to creak under the strain of accumulating contrary evidence. In April 1900, Kelvin delivered a lecture to the Royal Institution concerned with what he perceived to be nineteenth century 'clouds' over the dynamical theory of heat and light.[4]

Not triumphalist, but prescient.

The stormclouds were gathering. But nobody could tell precisely where the storm would break.

[3] In his 2007 biography of Einstein, Walter Isaacson explained that he could find no direct evidence that Kelvin had made this pronouncement.

[4] See Kragh, *Quantum Generations*, p. 9.

PART I

Quantum *of Action*

1

The Most Strenuous Work of My Life

Berlin, December 1900

Max Planck had once been counselled against a career in theoretical physics. His professor at the University of Munich had advised him that, with the discovery of the principles of thermodynamics, physics as a subject had been largely completed. There was, quite simply, nothing more to be discovered.

But, as the turn of the nineteenth century approached, there was nevertheless conflict between rival theories of physics. The principles of thermodynamics reinforced a vision of nature as one of harmonious flow. Energy, which could be neither created nor destroyed, flowed continuously between radiation and material substance, in themselves unbroken continua. Ranged against this view were the atomists, who offered a rather different perspective. Matter was not continuous, they argued, it was composed of discrete atoms or molecules. The thermodynamic properties of material substances could be calculated from the mechanical motions of the atoms or molecules from which they were composed, using statistics.

Planck was a master of classical thermodynamics. There were aspects of the atomists' statistical-mechanical models which served to undermine his world view, and his life's work. Although he accepted that the atomic conception of matter had scored some notable successes, he regarded it as a 'dangerous enemy of progress' which would ultimately 'have to be abandoned in favour of the assumption of continuous matter.'

In 1897 Planck had chosen the theory of cavity radiation, or so-called 'black-body' radiation, as the ground on which to engage his atomist rivals; as a place where he could finally reconcile mechanics with thermodynamics. But what he would discover just three years later would complete his slow conversion to the atomist doctrine. And, almost as a by-product, he would quietly sow the seeds for the most profound revolution in our scientific understanding of the world; a revolution whose repercussions can still be widely felt, more than a century later.

Planck's problem with the atomist doctrine was relatively simply stated. By reducing the calculation of thermodynamic quantities to the statistics of atomic or molecular motions, the atomists had opened the door to some discomforting consequences. What thermodynamics argued to be unquestionably irreversible and a matter of irresistible natural law, statistics argued was only the most probable of many different possible alternatives.

The conflict was most stark in the interpretation of the second law of thermodynamics. This was the subject of Planck's 1879 doctoral thesis, and it was a subject on which Planck prided himself as one of the world's leading experts. The second law claimed that for a substance—such as a gas—contained in a closed system, prevented from exchanging energy with the outside world, the thermodynamic quantity known as entropy would increase spontaneously and inexorably to a maximum as the gas reached equilibrium with its surroundings.

Entropy is a somewhat abstract quantity that we tend to interpret as the amount of 'disorder' in a system.[1] In 1895, with Planck's approval, his research assistant Ernst Zermelo took the argument directly to the atomists in the pages of the German scientific journal *Annalen der Physik*.

If, to take an example, we were to release two gases of different temperature in a closed container, the second law would predict that the gases would mix and the temperature would equilibrate, with the entropy of the mixture increasing to a maximum. However, according to the atomists, the behaviour of the gases is a consequence of the underlying mechanical motions of the atoms or molecules of each gas, and the equilibrium state of the mixture is simply its most probable state. This implies, argued Zermelo, that there is nothing in principle to rule out

[1] For example, as a block of ice melts, it transforms into a more disordered, liquid form. As liquid water is heated to steam, it transforms into an even more disordered, gaseous form. The measured entropy of water increases as water transforms from solid to liquid to gas.

a sequence of events in which the motions of the atoms or molecules are all reversed. If this were to happen, the gases would surely separate, returning to their initial temperatures and spontaneously reducing the entropy of the mixture, in complete contradiction to the second law.

The atomists' leading spokesman, Austrian physicist Ludwig Boltzmann, responded. Entropy does not always increase, he argued, in contradiction to the most commonly accepted interpretation of the second law. It just almost always increases. Statistically speaking, there are many, many more states of higher entropy than there are of lower entropy, with the result that the system spends much more time in higher entropy states. In effect, Boltzmann was saying that if we wait long enough,[2] we might eventually catch the system undergoing a spontaneous reduction in entropy.

This is as miraculous an event as a smashed cocktail glass spontaneously reassembling itself, to the astonishment of party guests.

To Planck, this stretched the interpretation of the second law to breaking point. In seeking to find a compelling refutation of Boltzmann's statistical argument, Planck chose as a battleground the physics of cavity radiation.

This seemed a perfectly safe choice. The theoretical physics of cavity radiation appeared to have no connection with atoms or molecules. It was a problem of continuous waves of electromagnetic radiation, as described by Maxwell's theory, and of thermodynamics, whose second law would drive the radiation to equilibrium. Planck had reasoned that if he could show how equilibrium was established without recourse to the statistical-mechanical models of the atomists, he could undermine the very basis of the mechanical description.

The behaviour of cavity radiation was by now well understood. Heat any object to a high temperature and it will gain energy and emit light. We say that the object is 'red hot' or 'white hot'. Increasing the temperature of the object increases the intensity of the light emitted and shifts it to a higher range of frequencies (shorter wavelengths). As it gets hotter, the object glows first red, then orange-yellow, then bright yellow, then brilliant white.

[2] Admittedly, we would have to wait for a time much longer than the present age of the universe.

Theoreticians had simplified the problem by invoking the idea of a 'black body', a hypothetical, completely non-reflecting (i.e. totally black) object that absorbs and emits light radiation without favouring any particular range of frequencies. The intensity of radiation emitted by a black body is directly related to the amount of energy in the body when it is in thermal equilibrium with its surroundings.

The theoreticians further realized that they could probe the properties of a black body by studying the radiation trapped inside a cavity consisting of perfectly absorbing walls, punctured with a small pinhole through which radiation can enter and leave. Early examples of such cavities included rather expensive closed cylinders made from porcelain and platinum.[3]

In the winter of 1859–60, the German physicist Gustav Kirchhoff had demonstrated that the ratio of emitted to absorbed energy depends only on the frequency of the radiation and the temperature inside the cavity. It does not depend in any way on the shape of the cavity, the shape of its walls, or the nature of the material from which the cavity is made. This implied that something quite fundamental concerning the physics of the radiation itself was being observed, and Kirchhoff challenged the scientific community to discover the origin of this behaviour.

Much progress had been made. Experimental studies of infrared (or heat) radiation emitted from a radiation cavity had led physicist Wilhelm Wien in 1896 to devise a relatively simple mathematical relationship between the radiation frequency and cavity temperature. Wien's law seemed to be quite acceptable, and was supported by further experiments carried out by Friedrich Paschen at the Technical Academy in Hanover in 1897. But new experimental results reported in 1900 by Otto Lummer and Ernst Pringsheim at the Reich Physical-Technical Institute in Berlin showed that Wien's law failed at lower frequencies. Wien's law was clearly not the answer.

Planck had succeeded Kirchhoff at the University of Berlin in 1889, rising to full professor in 1892. He was a most unlikely scientific revolutionary. Descended from a line of pastors and professors of theology

[3] The study of cavity radiation was not just about establishing theoretical principles, however. It was also of interest to the German Bureau of Standards as a reference for rating electric lamps.

and jurisprudence, at school Planck was diligent and personable but not especially gifted. Physics was a subject for which Planck himself felt he had no particular talent, but he had risen through the academic ranks and established a solid international reputation. Now in his early forties, he worked at a slow, steady, and conservative pace. He preferred the stability and predictability of a science which reflected the character of the bourgeois German society of which he was a part. By his own subsequent admission he was 'peacefully inclined', and rejected 'all doubtful adventures'.

During a visit to Planck's villa in the Berlin suburb of the Grünewald on 7 October 1900, experimental physicist Heinrich Rubens had told him about some new experimental results he had obtained with his associate Ferdinand Kurlbaum. They had studied cavity radiation at even lower frequencies. Ruben's description of the behaviour of the radiation at these frequencies set Planck thinking. After Rubens had left, Planck continued to work alone in his study. He adapted Wien's earlier law and arrived, largely through some inspired guesswork, at an expression which fit all the available experimental data.

Planck had discovered his radiation law.

The law required two fundamental constants, one relating to temperature and a second relating to radiation frequency. This second constant would eventually gain the label *h* and become known as Planck's constant. When combined with the speed of light and Newton's gravitational constant, the two constants in Planck's radiation law promised a fundamental underpinning for all physical quantities. Planck wrote that the constants offered: 'the possibility of establishing units of length, mass, time and temperature which are independent of specific bodies or materials and which necessarily maintain their meaning for all time and for all civilizations, even those which are extraterrestrial and nonhuman, constants which therefore can be called "fundamental physical units of measurement".'

He sent Rubens a postcard on which he had written details of the new radiation law, and he presented a crude derivation of the law at a meeting of the German Physical Society on 19 October. He declared: 'I therefore feel justified in directing attention to this new formula, which, from the standpoint of electromagnetic radiation theory, I take to be the simplest excepting Wien's.' The next day, Rubens advised Planck that he had

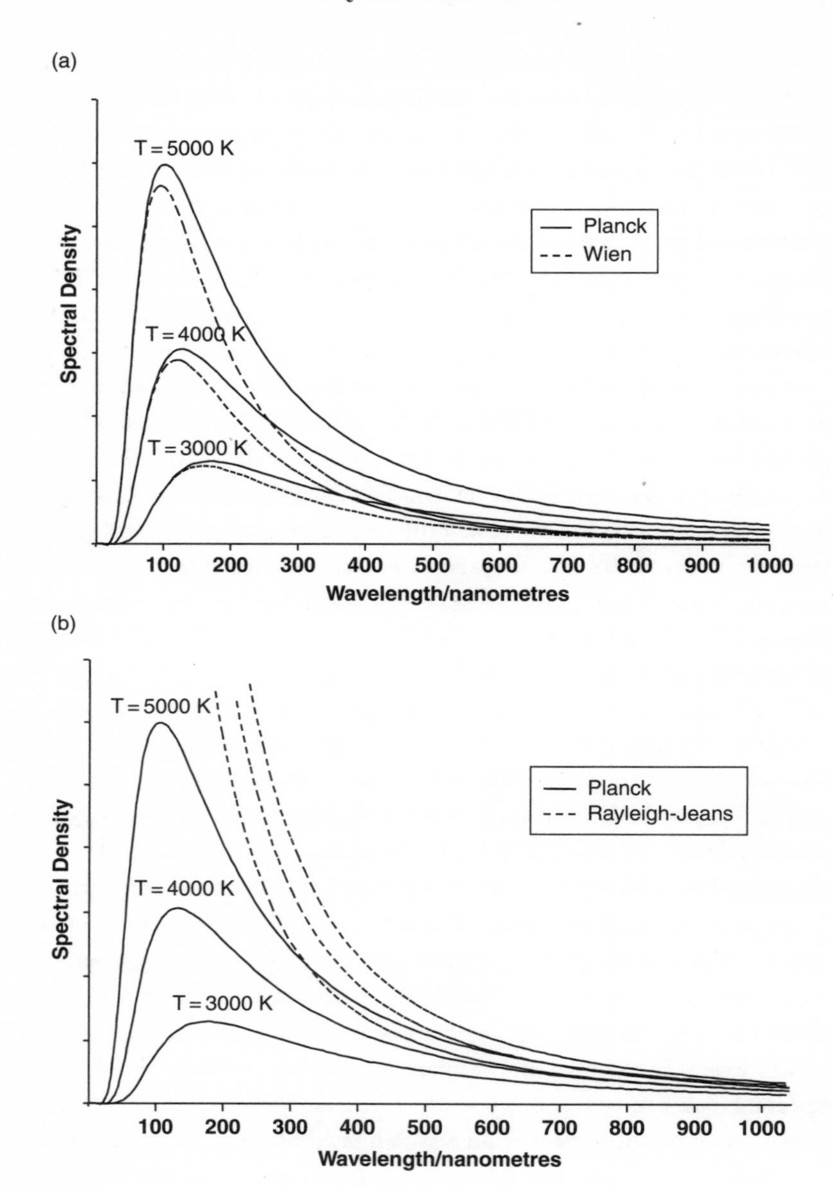

FIG 2 a) Comparison of the predictions of Planck's radiation law and Wien's law for three different temperatures. Wien's law accurately reproduces the behaviour of black-body radiation at very high frequencies (short wavelengths) but fails at lower frequencies (longer wavelengths). The discrepancy is most noticeable at higher temperatures. b) Planck's radiation law compared to the Rayleigh-Jeans law for the same three temperatures. The Rayleigh-Jeans law approaches the behaviour of black-body radiation at very low frequencies (very long wavelengths) but gives catastrophic results in the ultraviolet.

compared the experimental results with the new law and found 'completely satisfactory agreement in all cases'.

It seemed that Planck's radiation law was *the* answer, at least as far as experiment was concerned. Planck now turned his attention to finding a proper theoretical basis for the law, a task which was to lead to 'some weeks of the most strenuous work of my life.'

Planck set out the problem in terms of the interaction between the electromagnetic field and a set of vibrating 'oscillators' in the cavity material. The primary purpose of these oscillators was to ensure that the energy was properly equilibrated among the possible radiation frequencies through a continual, dynamic process of absorption and emission.[4] Being an expert in entropy and the second law, he began by using the radiation law to derive an expression for the entropy of an individual oscillator in terms of its internal energy and its frequency of oscillation, which would give rise to the same frequency of radiation inside the cavity. This gave him an expression for the oscillator entropy known to be consistent with experiment. His challenge now was to derive a similar expression from 'first principles', compare the two, and draw appropriate conclusions.

It was this task that proved to be 'strenuous'. Planck may have tried several different approaches, but he found that he would always be compelled to return to an expression strongly reminiscent of the statistical methods of his rival Boltzmann. The mathematics was leading him in a direction he had not wanted to go.

Boltzmann's approach to calculating the entropy of a gas was to assume that its total available energy could be thought of as being organized into a series of 'buckets'. The lowest energy bucket was assigned an energy ε, the next an energy 2ε, the next 3ε, and so on. The gas molecules would then be distributed among the buckets and the number of different possible permutations of molecules in the buckets calculated. In this analysis the energy itself remains continuously variable. All that

[4] Planck originally referred to these as 'resonators', but by 1909 he had accepted that the special properties of resonators were not required. Today we would identify these oscillators as highly excited electrons within the atoms of the cavity material (but recall that the very existence of electrons had been confirmed only three years earlier, in 1897).

Boltzmann had done was parcel it up so that he could count the number of molecules *in the energy range* zero to ε, the range ε to 2ε, and so on, and thus calculate the number of different possible permutations.

For example, consider a gas consisting of just three molecules, which we label *a*, *b*, and *c*. Let's assume this gas has a total energy of 4ε. We can achieve this by putting two molecules in the lowest, ε, energy bucket, and one in the 2ε bucket. How many permutations are possible? There are just three. We can put molecules *a* and *b* into the lowest energy bucket and *c* in the next, a permutation we can label as [*ab*,*c*]. We can also put molecules *a* and *c* in the lowest energy bucket, and *b* in the next, labelled [*ac*,*b*]. The third possible permutation is [*bc*,*a*].

Boltzmann reasoned that the most probable state of the gas would be the one with the highest number of possible permutations for the available energy, representing maximum entropy at that energy. By equating the maximum number of possible permutations with the most probable distribution of energy it was a relatively simple step to the calculation of the entropy itself.

Planck had been fighting a losing battle against Boltzmann's logic for at least three years. He now succumbed to the inevitable. As he later explained: 'I busied myself, from then on, that is, from the day of its establishment, with the task of elucidating a true physical character for the [new distribution law], and this problem led me automatically to a consideration of the connection between entropy and probability, that is, Boltzmann's trend of ideas.'

Even though the problem of cavity radiation appeared to be totally unrelated to the question of whether or not a gas was composed of atoms or molecules, Planck now reached for the statistical methods of the atomists. But there was a catch. Because he was working backwards from the result he was aiming for, the statistical methods he needed were actually far removed from those used by Boltzmann.

Planck's statistical distribution was of a subtly different kind. Boltzmann had examined the permutations arising from the distribution of *distinguishable* molecules over the various possible energy buckets. Planck, however, examined instead the permutations of *indistinguishable* energy elements (which we continue to label as ε) over the various oscillators in the cavity material. For example, if we use Planck's methodology to distribute four energy elements (4ε) over three oscillators, then we find there are now fifteen possible permutations. We can put all the elements in the first oscillator, and none in the other two, a permutation which we can write as $(4\varepsilon,0,0)$. Other permutations are $(3\varepsilon,\varepsilon,0)$, $(2\varepsilon,\varepsilon,\varepsilon)$, $(\varepsilon,2\varepsilon,\varepsilon)$, and so on.

Moreover, to get the result he wanted Planck found that the energy elements had to be directly related to the frequency of the oscillators (and hence the frequency of the radiation) according to his now famous relation: $\varepsilon = h\nu$—energy element equals Planck's constant multiplied by frequency. He further discovered that the energy elements had to be fixed in size as integer multiples of $h\nu$. In making these choices he was following a very different path from Boltzmann.

Many years later Planck described his state of mind as follows:

> Briefly summarised, what I did can be described as simply an act of desperation. By nature I am peacefully inclined and reject all doubtful adventures. But by then I had been wrestling unsuccessfully for six years (since 1894) with the problem of equilibrium between radiation and matter and I knew that this problem was of fundamental importance to physics... A theoretical interpretation therefore had to be found at any cost, no matter how high.

Planck, by now a willing and enthusiastic convert to the atomist view, presented this new derivation of his radiation law to a regular fortnightly meeting of the German Physical Society on 14 December 1900 shortly after 5 pm. As he explained to the assembled audience: 'We therefore regard—and this is the most essential point of the entire calculation—energy to be composed of a very definite number of equal finite packages.' He submitted a paper to the journal *Annalen der Physik* in January 1901. About the physical constant that was to carry his name he had this to say:

> ...since it has the dimensions of a product of energy and time, I called it the elementary quantum of action or element of action in contrast with the energy element $h\nu$.

The 14 December 1900 is widely acknowledged to be date on which the quantum revolution began. But, in truth, Planck did not yet recognize that his equation $\varepsilon = h\nu$ represented a fundamental unraveling of the structure of classical physics.

In a possibly apocryphal tale, during a walk in the Grünewald, Planck is reported to have told his seven-year-old son Erwin that he: 'felt that he had possibly made a discovery of the first rank, comparable perhaps only to the discoveries of Newton.' If true, it is likely that Planck was referring to the discovery of the properties of the second constant in his radiation law—which he had called Boltzmann's constant, *k*—and *not* the discovery of the quantum of action and the fixed energy elements of electromagnetic radiation.

Planck had used a statistical procedure, distributing the fixed energy elements over the oscillators, without giving much thought to the physical significance of this step. If atoms and molecules were real entities, something Planck was now ready to accept, then in his own mind energy itself remained determinedly continuous, to flow uninterrupted back and forth between radiation and matter. But in deriving his radiation law, Planck had inadvertently introduced the idea that energy itself could be 'quantized'. There the idea sat, in Planck's lectures and writings, unarticulated, unnoticed, and unremarked.

It would take true genius to see what everyone else could not.

2

Annus Mirabilis

Bern, March 1905

Planck used his radiation law to score some notable successes. In 1901 he used the available experimental data to obtain estimates for both the Planck and Boltzmann constants. He went on to use his estimate for Boltzmann's constant to calculate Avogadro's number (the number of atoms or molecules present in a characteristic amount of pure substance called a mole).[1] *He then used his estimate for Avogadro's number to determine the charge of the electron. His estimates were all accurate to within one to three per cent of the currently accepted values for these fundamental constants.*

The convert became an ardent atomist: Planck began referring to atoms and molecules as though they were real.

But whilst Planck's results appeared to be well-founded, there remained doubts about his derivation; about the somewhat devious manner in which he had arrived at his result. Many were puzzled. With the benefit of hindsight, this should not be surprising. Planck had sprung a profoundly non-classical concept from within an otherwise entirely classical formulation. This was not something that could be done without some violence to the classical prescription.

One young physicist who remained sceptical about Planck's derivation was in 1905 working in the Swiss Patent Office in Bern, as a 'technical expert, third class'. His name was Albert Einstein.

[1] A mole is the atomic or molecular weight of the substance in grams.

Einstein had joined the Swiss Patent Office on 16 June 1902. This had been something of a relief. After completing his graduate studies in physics at the Zurich Polytechnic in August 1900 he had tried, and failed, to find an academic position at universities in Germany, Holland, and Switzerland. He had been unemployed for a time before finding temporary work as a high-school teacher.

Growing increasingly desperate about his employment prospects, he had sought help from his fellow student and good friend Marcel Grossmann. Grossmann had become aware of an impending vacancy at the Patent Office and his father, who knew the director personally, was happy to suggest Einstein's name. Einstein had moved to Bern in anticipation of his appointment to the post and, as he waited on a decision from the Swiss Council, he had supported himself by offering private tuition in mathematics and physics.

Later that year Einstein's father, Hermann, died. On his deathbed Hermann finally yielded to his son's requests and gave permission for him to marry his student sweetheart, Mileva Marić. Einstein had first met Mileva when they both enrolled as students at the Polytechnic in 1896. She had become his muse; in their tender love letters he called her 'Dollie', she called him her 'Johnnie'. However, it was not initially a match that had met with approval from Einstein's parents.

They were married in Bern on 6 January 1903, in a civil ceremony witnessed by Einstein's friends Maurice Solovine and Conrad Habicht. Both had responded to the advert Einstein had placed in a newspaper on his arrival in Bern, offering tutorials (with trial lessons offered free). The three had formed a close friendship. They had held frequent discussion meetings on a wide variety of subjects, proclaiming themselves the 'Olympia Academy'.

Theirs was to be a long and enduring friendship. But there was one secret that Einstein did not choose to share with even his closest friends. He had fathered a child with Mileva. Their daughter, Lieserl, was born a few days after Einstein had first arrived in Bern. Mileva had returned to her parents' home in the Serbian city of Novi Sad for the latter stages of her pregnancy and the birth. Einstein had written of his plans to bring Mileva and the baby to Bern once things were settled, but then their plans changed.

Whilst Einstein's job at the Patent Office gave him the financial means to support Mileva and their child, his position in Swiss officialdom demanded a respectability incompatible with a child born out of wedlock. Mileva returned to Zurich alone, leaving Lieserl with relatives and friends eventually, it seems, to be given up for adoption. Einstein never saw his first child, or held her in his arms. All mention of Lieserl ceased; her very existence became a close family secret.[2] Her ultimate fate is unknown.[3]

Mileva bore Einstein a second, 'replacement' child, Hans Albert, on 14 May 1904. By this time Einstein was settled into his job at the Patent Office. He found the work interesting, and used it to sharpen his critical intellect and ground his thinking about physical theories in terms of their direct practical consequences. His discussions with his friends in the Olympia Academy provided a rich diet of empiricist philosophy and the determinism of Dutch philosopher Baruch Spinoza's impersonal *Deus sive Natura*, God or Nature. These early intellectual influences were to remain very powerful for the rest of Einstein's life.

If Einstein had succeeded in his attempts to find an academic position, it is quite possible that the challenge of climbing the academic career ladder would have demanded a certain conformity, a tendency to choose 'safe' research topics and churn out admirable, but hardly revolutionary, research papers. But working quietly outside the strictures of academe left Einstein free to be rebellious, to think—even suggest—the unthinkable.

Habicht had moved away from Bern in the spring of 1905. Einstein wrote to him in late May to tell him of his recent work:

> I promise you four papers in return. The first deals with radiation and the energy properties of light and is very revolutionary... The second paper is a determination of the true sizes of atoms... The third proves that bodies on the order of magnitude 1/1000 mm, suspended in liquids, must already

[2] This secret was revealed only in 1986 when researchers discovered a few, scant references to her in Einstein's private correspondence.

[3] It has been suggested that Lieserl may have died from scarlet fever in September 1903. Another hypothesis suggests that Lieserl was adopted by Mileva's friend Helene Savić. See Isaacson, p. 87.

> perform an observable random motion that is produced by thermal motion. Such movement of suspended bodies has actually been observed by physiologists who call it Brownian motion. The fourth paper is only a rough draft at this point, and is an electrodynamics of moving bodies which employs a modification of the theory of space and time.

Any one of these four papers would grant Einstein lasting recognition, even fame. The first of these, written in March 1905, contains the bold assertion that Planck's radiation law only makes sense if radiation itself is thought to be composed of discrete packets of energy which he called light-quanta. The second and third, written in April and May, deal with the physical reality of molecules and some observable consequences of molecular motions.[4] In the fourth, which Einstein would complete in June, he derived his special theory of relativity. In a subsequent fifth paper, a supplement to the June special relativity paper which was published in September 1905, he would go on to derive his most famous equation, $E = mc^2$.

This was Einstein's *annus mirabilis,* his miracle year. He was just 26 years old.

The problem with Planck's derivation of his radiation law was that it was inconsistent. Planck had first assumed energy to be continuously variable, as the principles of classical physics demanded, and then he had assumed precisely the opposite by distributing the fixed energy quanta over the oscillators of the cavity material.

In June 1900 the English physicist Lord Rayleigh (William Strutt) had published details of another theoretical model for cavity radiation and had derived a radiation law that was somewhat different from both Wien's and Planck's laws.[5] Rayleigh had applied the principles of classical physics in ways that Planck had not. An error in Rayleigh's calculation was corrected by fellow Englishman James Jeans in June 1905. The result is now known as the Rayleigh–Jeans law.

[4] He submitted the second paper as a doctoral dissertation to the University of Zurich in July 1905 (at that time the Zurich Polytechnic was unable to confer doctoral degrees). Einstein maintained that his dissertation was first rejected because it was too short. Einstein added a single sentence and resubmitted it, and it was accepted. Einstein became 'Herr Doctor'. See Isaacson, p. 103.

[5] Rayleigh's paper was published some months before Planck's October paper on his radiation law, but Planck had been unaware of Rayleigh's result at the time.

Rayleigh's reasoning and use of thermodynamic principles was both logical and convincing, but the result was little short of disastrous. The Rayleigh–Jeans law implies that the intensity of cavity radiation should increase in proportion to the square of the radiation frequency without limit. The law predicts that the total emitted energy should quickly increase to absurd levels at high radiation frequencies, or short wavelengths.[6] The Rayleigh–Jeans law provided a relatively close fit to the experimental data at low frequencies, where Wien's law failed, and Wien's law provided a reasonable fit to the data at high frequencies, where the Rayleigh–Jeans law failed. Both were clearly limiting cases of Planck's law, which fit the data at all frequencies.

It seemed that a truly classical derivation produced laws that failed. Only Planck's mysterious non-classical derivation could produce a law that worked.

Einstein proved this much to himself, independently deriving the correct form for the Rayleigh–Jeans law. This was puzzling enough, but Einstein was in pursuit of a bigger prize. He opened his March 1905 paper with the statement:

> A profound formal difference exists between the theoretical concepts that physicists have formed about gases and other ponderable bodies, and Maxwell's theory of electromagnetic processes in so-called empty space. While we consider the state of a body to be completely determined by the positions and velocities of an indeed very large yet finite number of atoms and electrons, we make use of continuous spatial functions to determine the electromagnetic state of a volume of space, so that a finite number of variables cannot be considered as sufficient for the complete determination of the electromagnetic state of space.

The evidence for atoms and molecules was becoming overwhelming and this particulate view of matter was now in the ascendancy. Einstein was pointing to the dichotomy created by models of *particulate* matter on the one hand and *waves* of radiation described by Maxwell's electromagnetic field theory on the other.

[6] In 1911 the Austrian physicist Paul Ehrenfest called this problem the 'Rayleigh–Jeans catastrophe in the ultraviolet', now commonly known as the *ultraviolet catastrophe*.

Einstein was about to make a very bold move, one that he later told his friend Habicht was 'very revolutionary'. His problem was that this was a move that could not be justified from within the framework of the classical theory, from within the constraints that led to the Rayleigh–Jeans law.

He solved this problem by introducing what he called the 'heuristic principle'. He suggested, purely as an unproven hypothesis, that:

> If monochromatic radiation (of sufficiently low density) behaves, as concerns the dependence of its entropy on volume, as though the radiation were a discontinuous medium consisting of energy quanta of magnitude [$h\nu$], then it seems reasonable to investigate whether the laws governing the emission and transformation of light are also constructed such as if light consisted of such energy quanta.

Two hundred years after Newton, Einstein was ready to return to a 'corpuscular' theory of light. He was ready to assume that: '...in the propagation of a light ray emitted from a point source, the energy is not distributed continuously over ever-increasing volumes of space, but consists of a finite number of energy quanta localised at points of space that move without dividing, and can be absorbed or generated only as complete units.'

Einstein was not proposing to abandon the wave theory of radiation completely. There was simply too much experimental evidence for the wave properties of light—phenomena such as light diffraction and interference could only be explained in terms of a wave model. Einstein proposed to reconcile these two contrasting descriptions by recognizing that wave phenomena were the result of time-averaged observations. So, wave interference reflects not the instantaneous 'snapshot' of the motion of individual, localized light-quanta, but rather the collective motions of many light-quanta statistically averaged over time.

Although unproven, Einstein proposed to use his light-quantum hypothesis to solve other problems in physics unconnected with cavity radiation and Planck's law. He now turned his attention to the photoelectric effect.

This was another phenomenon that had been puzzling physicists for some time. It was known that shining light on metal surfaces could result in the ejection of electrons from these surfaces. In the wave model of light, the light energy is proportional to the amplitude (or intensity) of the light wave. So, if the intensity of the light is increased, it seemed sensible to assume that the energy of the ejected electrons would increase as a consequence. But this is not what was found in the laboratory.

Only the *number* of ejected electrons was found to increase with increasing light intensity, not their individual energies. Instead, the energy of the electrons was found to increase with increasing light frequency, completely at odds with accepted theories of electromagnetic radiation.

Einstein solved this problem by suggesting that a light-quantum incident on the surface transfers all its energy to a single electron. That electron is ejected with an energy equal to the energy of the light-quantum less an amount expended by escaping the surface, and which is therefore characteristic of the metal (a property now known as the work function).

Einstein now took Planck's equation, $\varepsilon = h\nu$, and invested it with a completely new meaning. Instead of representing a convenient mathematical relationship between the fixed energy quanta and the frequency of the oscillators, as Planck had done, Einstein assumed it to represent the relationship between the energy of the light-quantum and its (radiation) frequency. In doing this Einstein divorced Planck's relationship completely from the problem of cavity radiation. He made it a fundamental property of light relevant in all situations.

By imparting an energy $h\nu$ to an electron in the metal, each light-quantum would produce an ejected electron whose energy could be expected to increase with increasing radiation frequency, ν. Increasing the intensity of the radiation would increase the number of light quanta incident on the metal surface, increasing the number of ejected electrons, but not their energies.

Einstein's theory was very simple, and yet it made a number of important, testable predictions. He verified that the voltage of photo-electricity produced in such an experiment was of the same order of magnitude as voltages already observed in experiments carried out by German physicist Phillipp Lenard in 1902. He then went on to predict that a graph of the voltage of photo-ejected electrons *versus* the frequency of the incident light would be a straight line, with a slope that would be independent of the nature of the metal used in the experiments.[7]

[7] The slope of such a graph actually gives the value of Planck's constant.

The physics community was not at all sure what to make of this. Einstein's genius, revealed in his five 1905 papers, was clear. Planck became an enthusiastic supporter of the special theory of relativity, but baulked at the idea that light energy could be quantized. As far as he was concerned, the counting procedure he had used in his derivation of the radiation law was a mathematical convenience, with no grounding in reality. Einstein had gone too far.

When Einstein was subsequently recommended for membership of the prestigious Prussian Academy of Sciences, its leading members—Planck among them—acknowledged his remarkable contributions to physics. But they also acknowledged what they took to be errors in his judgement, errors that could be forgiven. 'That he may sometimes have missed the target in his speculations, as, for example, in his hypothesis of light-quanta, cannot be really held against him, for it is not possible to introduce really new ideas even in the most exact sciences without sometimes taking a risk.'

Planck and Einstein entered into a lively correspondence. Planck was prepared to concede that the appearance of fixed energy quanta might be the result of some as yet unknown internal properties of the atoms in the cavity material. In other words, the fixed energy elements did not arise because light is composed of such elements, they arise because only fixed energy elements could be emitted by the atoms that formed the cavity itself.

Planck had lit a slow-burning fuse in December 1900. Five years later the quantum revolution was begun, but it had begun with the merest whisper. Reaction to the light-quantum hypothesis had been strongly negative. Einstein was very aware of the potential for conflict between the light-quantum and the wave theory of light, and this made him cautious, although he remained unrepentant.

In 1906 Einstein was promoted, to 'technical expert, second class'.

3

A Little Bit of Reality

Manchester, April 1913

Einstein's vision was broader even than the light-quantum. He anticipated that the quantization of energy might be a much more universal phenomenon and in 1907 went on to develop a quantum theory of the heat capacities of crystalline solids. Although nobody took this work very seriously at the time, by 1909 new experimental results were forcing the physics community to sit up and take notice.

In 1910 the eminent German scientist Walther Nernst chose to pay Einstein a visit (by this time Einstein was back in Zurich, at the ETH,[1] but was still a relative unknown). The visit by Nernst encouraged greater respect for Einstein and his work, with one Zurich colleague remarking: 'This Einstein must be a clever fellow, if the great Nernst comes so far from Berlin to Zurich to talk to him.' Nernst was instrumental in drawing attention to Einstein's quantum approach and from early 1911 a growing number of scientists began to cite Einstein's papers and embrace quantum ideas.

In the meantime, atoms had evolved from hypothetical entities, dismissed by some physicists as the result of wildly speculative metaphysics, into the objects of detailed laboratory studies. The discovery of the negatively charged electron by English physicist Joseph John Thomson in 1897 implied that atoms, indivisible for more than 2000 years, now had to be recognized as having some kind of internal structure.

[1] The Eidgenossische Technische Hochschule. This was the same Zurich Polytechnic where Einstein had completed his graduate studies, by now restructured as a university and able to confer PhD degrees. It was named the ETH in 1911.

Further secrets of this structure were revealed by New Zealander Ernest Rutherford in 1909–11. Together with his research associates Hans Geiger and Ernest Marsden in Manchester, he had been conducting experiments in which energetic alpha-particles,[2] emitted with high velocity through the breakdown of certain radioactive elements, were fired through thin gold foil. To their astonishment, they found that about one in 8000 alpha-particles was deflected by the foil, sometimes by more than ninety degrees. This was no less startling than observing high velocity machine-gun bullets deflected by tissue paper.

These results were subsequently interpreted by Rutherford to mean that most of the atom's mass is concentrated in a small central nucleus, with the much lighter electrons orbiting the nucleus much like the planets orbit the sun. According to this model, the atom is largely empty space.[3] As a visual image of the internal structure of the atom, Rutherford's planetary model remains compelling to this day.

It was the theory of the electron that occupied the thoughts of the 25-year old Danish physicist Niels Bohr in September 1911 as he left Denmark on the ferry through the *Storebælt*, the strait between the islands of Zealand and Fyn. He was bound for England. He had left behind Margrethe Nørlund, whom he had first met in 1909 and to whom he had become engaged in the summer of 1910. Clutching a poor English translation of his PhD thesis and a stipend from the Carlsberg Foundation, Bohr was heading for Cambridge to work in J.J. Thomson's laboratory.

In the early years of the twentieth century, Cambridge was one of the leading centres of theoretical and experimental physics and Thomson was universally admired not only for his contributions to science but also for his irrepressible enthusiasm. He had won the 1906 Nobel Prize for physics for his discovery of the electron and had since immersed himself in the theory of atomic structure. His often disastrous encounters with experimental apparatus effectively ruled him out of any form of 'hands-on' experimentation (he once declared all the glass in his laboratory to be bewitched).

[2] Alpha particles are the result of a certain type of radioactivity. They are the positively charged nuclei of helium atoms, consisting of two protons and two neutrons (although this was not understood at the time of Rutherford's experiments).

[3] If the dimensions of the atom are compared to those of our solar system, and Pluto (admittedly, no longer classified as a planet) were to be thought of as the atom's outermost electron, then the atomic nucleus would have a radius about a tenth of the radius of the sun.

Thomson was determined to explain the behaviour of atoms and molecules in terms of the particle he had discovered. The result of his endeavours was a theoretical model of atoms consisting of uniform sphere of weightless positive charge in which hundreds of negatively charged electrons are embedded, like currants in a cake. In this model, the electrons are largely responsible for the mass of the atom.

The model was not without its problems. Whilst he could devise some stable configurations of motionless electrons embedded in the positively charged medium, he suspected that it was the motion of electrons inside atoms that was responsible for the properties of magnetic materials. However, any model involving moving electrons was predicted to be inherently unstable.

Thomson had been forced to re-think this model when in 1910 experiments at the Cambridge laboratory showed that he had greatly overestimated the number of electrons present in each atom. There were not hundreds, as he had originally proposed. There were considerably fewer.

Bohr went to Cambridge because he judged it to be the centre of physics and Thomson a wonderful man. Unfortunately, their relationship got off to a poor start from which it seems never to have recovered. As a young postdoctoral student, Bohr's grasp of the English language was poor. Though always polite and courteous, Bohr's manner could sometimes appear brusque and was open to misinterpretation. His first meeting with Thomson was not auspicious. He entered Thomson's office with a copy of one of Thomson's books on atomic structure, pointed to a particular section and declared: 'This is wrong.' It is hardly surprising that Thomson did not immediately warm to him.

Bohr struggled on, growing increasingly frustrated. He attended some lectures, carried out some experiments under Thomson's direction that he thought were fairly pointless and tried hard to learn English. In Denmark, Bohr had enjoyed some success on the football field, as a goalkeeper (his brother Harald had played a few games for the Danish national team and had won a silver medal at the Olympic Games in London in 1908). Bohr joined a local football club, but his physics was going nowhere. In October 1911 he wrote to his brother complaining that Thomson was very difficult to talk to, and seemed unable to accept criticism.

Bohr met Rutherford for the first time in early November 1911, during a visit to Manchester. He resolved to transfer to Rutherford's research team for the final months of his postdoctoral year and learn about radioactivity. Bohr was aware of Rutherford's planetary model of the atom, but at this time his principal interest was radioactivity and the Manchester laboratory was the world's leading centre for experimental studies of the phenomenon.

In truth, Rutherford's planetary model had not been given much consideration by the physics community. Although the image might be compelling, the planetary structure was also quite impossible, for the same reason that moving electrons in Thomson's model were impossible. Unlike the sun and planets, electrons and atomic nuclei carry electrical charge. It was known from Maxwell's theory that electrical charges moving in an electromagnetic field would radiate energy in the form of waves. These waves could be expected to carry energy away from the orbiting electrons, slowing them down and leaving them exposed to the irresistible pull of the positively charged nucleus. The electrons in such a planetary model would spiral down towards the nucleus as they lost their energy and the atoms would collapse in on themselves within about one hundred-millionth of a second.

Thomson, for one, didn't believe a word of it.

Rutherford agreed to accept Bohr into his team for the remainder of his postdoctoral year, provided Bohr could get Thomson to agree. Thomson did not object, and arrangements for Bohr's transfer to Manchester were made in December 1911. He started work there the following March. 'The whole thing was very interesting in Cambridge,' Bohr said years later, 'but it was absolutely useless.'

Bohr initially set to work on experiments on the absorption of alpha-particles by aluminium. But experimental physics was not his forte, and after a few weeks he asked Rutherford if he could instead work on theoretical problems. Although Rutherford took a keen interest in the various research programmes and the work of his associates, he was busy writing a book on the physics of radioactive substances and had little time for extensive discussions. Bohr learnt what he needed about radioactivity from two Manchester colleagues, George von Hevesy and Charles Darwin (whom Bohr would always introduce to others as the grandson of the 'real' Darwin).

It was whilst working on some problems raised by Darwin that he was led from radioactivity to atomic structure. He still had to deal with the fact that a system of negatively charged electrons orbiting a positively charged nucleus is inherently unstable in classical terms. Perhaps, he reasoned, some progress could be made by employing quantum ideas. He had become convinced that the inner electronic structure of the Rutherford model was governed in some way by Planck's quantum of action.

Like Einstein before him, Bohr now realized that to expect to resolve the paradoxes of the atom from within an entirely classical description was to expect too much. The classical picture said that atoms should not exist. Yet atoms clearly do exist and so it was likely to be impossible to deduce a theoretical description using only the mathematical methods of classical mechanics.

Something else was needed.

Faced with this kind of impasse, in 1905 Einstein had invoked his 'heuristic principle'. Bohr now did something very similar. He simply hypothesised that, if atoms are stable, then this surely means that there must exist certain stable configurations of the electrons orbiting a central nucleus. These stable orbits would be dependent in some, as yet unspecified, way on Planck's constant. 'This seems to be nothing else than what was to be expected as it seems rigorously proved that the mechanics cannot explain the facts in problems dealing with single atoms.'

There was more. Planck's derivation of his radiation law had suggested that the cavity oscillators could absorb or emit energy only in integral multiples of the energy element $h\nu$. Bohr now assumed that the electron orbits would be organized similarly, the orbital energies increasing as $nh\nu$, where $n = 1, 2, 3, \ldots$ The lowest energy, innermost stable orbit would correspond to $n = 1$.

In June 1912 he wrote a letter to his brother Harald:

> Perhaps I have found out a little about the structure of atoms. Don't talk about it to anybody, for otherwise I couldn't write to you about it so soon. If I should be right it wouldn't be a suggestion of the nature of a possibility (i.e., an impossibility, as J.J. Thomson's theory) but perhaps a little bit of reality... You understand that I may yet be wrong; for it hasn't been worked out fully yet (but I don't think so); also, I do not believe that Rutherford thinks that it is completely wild... Believe me, I am eager to

> finish it in a hurry, and to do that I have taken off a couple of days from the laboratory (this is also a secret).

The model was still nevertheless rife with contradictions. He summarized his work in a manuscript he submitted to Rutherford on 6 July, but which was never published. He left Manchester and returned to Copenhagen a couple of weeks later, carrying the problems back with him in his briefcase. On 1 August he and Margrethe were married. They honeymooned in England, Bohr returning to the Cambridge and Manchester laboratories for brief visits before setting off for a holiday in Scotland and then back to Copenhagen, where an academic position now awaited him.

Bohr continued to work on the problem of atomic structure through the rest of 1912 and into early 1913. For his hypothesis concerning the stable orbits to have any value, he needed to use it to explain the results of recent experiments or predict the results of experiments yet to be done. His next breakthrough came in February 1913, when he was given a clue that would unlock the entire mystery. Hans Hansen, a young professor of physics who had done some experimental work on atomic spectroscopy at the University of Göttingen in Germany, drew Bohr's attention to something called the Balmer formula.

Spectroscopy is the study of the absorption and emission of electromagnetic radiation by atoms and molecules. The simplest atoms exhibit the simplest spectra and the hydrogen atom, consisting of a nucleus and just one electron, has the simplest spectrum of all. Whilst it might have been anticipated from classical physics that atoms such as hydrogen should absorb and emit energy continuously, favouring no particular radiation frequencies, the opposite is actually found. The hydrogen emission spectrum is a 'line' spectrum of discrete, narrowly defined frequencies.

In 1885, the Swiss mathematician Johann Jakob Balmer had studied the measurements of one series of hydrogen emission lines and found them to follow a relatively simple pattern. He found that the frequencies of the lines are proportional to the differences between the inverse squares of pairs of integer numbers. In other words, the frequencies depend on the numbers m and n where, for the hydrogen series that Balmer had investigated, $m = 2$ and n takes the values 3, 4, 5, and so on.

Balmer's formula was generalized in 1888 by Swedish physicist Johannes Rydberg. He found that other spectral series follow a similar relation, with the value of m taking different integer values. In themselves the Balmer and Rydberg formulas were purely empirical and their origin in terms of the underlying physics of the atom was quite obscure. Bohr, however, immediately understood where the integer numbers had come from.

He realized that an electron moving from an outer, high-energy orbit to a lower-energy inner orbit causes the release of energy as emitted radiation. If, as he had hypothesized, the orbits are fixed, with energies that depend on integer numbers that can be counted outwards from the nucleus in a linear sequence, then the energy differences between the orbits are also therefore fixed.

For example, in the case where an electron circling the nucleus in orbits characterized by the integer numbers $n = 3, 4, 5, \ldots$ drops down into a lower-energy orbit characterized by $m = 2$ the result is the series of emission lines studied by Balmer, which had become known as the Balmer series. Putting $m = 3$ and $n = 4, 5, 6, \ldots$ gives another series that had been observed in 1908 by German physicist Friedrich Paschen. Bohr predicted the existence of a further series in the ultraviolet with $m = 1$, and series in the infrared characterized by $m = 4$ and 5.

Bohr was further able to show that the constant of proportionality that appeared in Rydberg's formula (known as the Rydberg constant) can be calculated directly by combining a number of fundamental physical constants, including Planck's constant, the electron charge, and the electron mass. The Rydberg constant was well known at the time from spectroscopic measurements. Bohr's calculation was within six per cent of the experimental value, a margin well within the experimental uncertainty of the values he had used for the fundamental constants.

A further series of emission lines named for American astronomer and physicist Edward Charles Pickering was thought by experimentalists also to belong to the hydrogen atom. However, at the time, the Pickering series was characterized by half-integer numbers which are not admissible in Bohr's theory. Instead, Bohr proposed that the formula be rewritten in terms of integer numbers, suggesting that the Pickering series belongs not to hydrogen atoms but to ionized helium atoms. An awkward mismatch was later resolved by Bohr himself. The correction gave a Rydberg constant for ionized helium some 4.00163 times greater than that for hydrogen. The experimentalists had found this ratio to be 4.0016. This kind of agreement between theory and experiment was simply unprecedented.

Bohr's idea of stable electron orbits had a further consequence. The electrons had to possess a fixed angular momentum—a fixed momentum associated with their 'rotation' around the central nucleus—in units of h divided by 2π. Transitions between the orbits had to occur in instantaneous 'jumps', because if the electron were to move gradually from one orbit to another, it would again be expected to radiate energy continuously during the process. Transitions between inherently non-classical stable orbits must themselves involve non-classical discontinuous jumps. Bohr wrote that:

> ...the dynamical equilibrium of the systems in the [stable orbits] can be discussed by help of the ordinary mechanics, while the passing of the systems between the different [stable orbits] cannot be treated on that basis.

The integer numbers that characterize the electron orbits would come to be known as quantum numbers, and the transitions between different orbits would be called quantum jumps.

Bohr wrote to Rutherford on 6 March 1913, enclosing with his letter a manuscript with the title: 'On the Constitution of Atoms and Molecules'. In his reply Rutherford reacted favourably but raised some difficult questions. He was particularly puzzled by the fact that, in Bohr's model, an electron in a high-energy orbit would somehow need to 'know' beforehand the energy of the final, destination, orbit in order to emit radiation of just the right frequency. Rutherford was already ringing alarm bells about the implications of the new quantum theory for our understanding of cause-and-effect, bells that would continue to peal loudly for another century.

He also warned that Bohr's manuscript was rather long: 'I do not know if you appreciate the fact that long papers have a way of frightening readers,' he wrote. Bohr was nonplussed. The day before he received Rutherford's letter he had sent a revised draft of his manuscript. It was even longer.

Bohr resolved to go at once to Manchester and discuss the paper directly with Rutherford. He wrote back on 26 March and declared his intention to visit Rutherford in the first days of the following week. Rutherford was patient. After discussions through several long evenings, during which he declared he had never thought Bohr should prove to be so obstinate, he agreed to leave all the detail in the final paper and communicate it to the journal *Philosophical Magazine* on Bohr's behalf.

The paper was published in July 1913, and was followed by two further papers published in the same journal in September and November that year.

Bohr's model of atomic structure was a triumph. But, much like Planck's achievement in 1900, it was also rather mysterious. There were many unanswered questions. The most pressing of these concerned the quantum numbers. What did they mean? From where, exactly, had they come?

4

la Comédie Française

Paris, September 1923

Experiment forced the pace on the new quantum theory of the atom. It turned out that the spectrum of the hydrogen atom was not quite so simple, after all. Some of the lines in the spectrum had already been found some years before actually to be two, closely spaced lines. Furthermore, lines in the spectrum were found to be split in low-intensity electric and magnetic fields. In addition to Bohr's quantum number n, *German physicist Arnold Sommerfeld introduced two additional quantum numbers,* k *and* m, *to explain the experimental results. These new quantum numbers were thought to be related in some way to the quantization of the geometry of the stable electron orbits.*

As the various lines in the spectrum were identified with different quantum jumps between different orbits, it was soon discovered that not all the possible jumps were appearing. Some lines were missing. For some reason certain jumps were forbidden. An elaborate scheme of 'selection rules' was established by Bohr and Sommerfeld to account for those jumps that were allowed and those that were forbidden.

In the meantime, evidence for Einstein's light-quantum hypothesis was also accumulating. Einstein had used his hypothesis to make predictions about the photoelectric effect, predictions that were duly borne out in experiments by American physicist Robert Millikan in 1915.[1] *The status of light 'particles' was further enhanced when, in 1923,*

[1] Einstein was awarded the 1921 Nobel Prize for physics for his work on the photoelectric effect (not relativity). Bohr was awarded the 1922 Nobel Prize for his work on atomic structure. Planck had been awarded the Prize in 1918 for his discovery of the quantum.

American physicist Arthur Compton and Dutch theorist Pieter Debye showed that light quanta could be 'bounced' off electrons, with a consequent (and predictable) change in their frequencies. These experiments appeared to demonstrate that light consists of particles with directed momenta, like small projectiles.

Despite this evidence, many physicists, including Planck and Bohr, dismissed the light-quantum. They preferred to think of quantization as having its origin in atomic structure, retaining Maxwell's classical wave description for electromagnetic radiation. Whatever theory was going to replace classical physics had to confront the difficult task of somehow reconciling the wave-like and particle-like aspects of light in a single theory. And, somehow, this theory also had to account for the inner structure of the atom. It had to account for the stable electron orbits characterized by their quantum numbers.

An important clue would come from Einstein's special theory of relativity.

Einstein had introduced his special theory in the fourth paper he published in 1905. He had struggled, and failed, to find a way to reconcile one of the most disturbing observations in late nineteenth-century physics with the body of accepted classical theory.

When in 1887 Michelson and Morley could find no evidence for relative motion between the earth and the hypothetical, all-pervading ether they concluded that the speed of light is constant and independent of the motion of the light source. This was one of the most important 'negative' experiments ever performed, and led to the award of the 1907 Nobel Prize for physics to Michelson.

The absence of an ether and an apparently universal speed of light could not be accommodated in any kind of Newtonian conception of absolute space and time. As he had done some months earlier for the light-quantum, now Einstein decided to build a new theory out of the bare minimum of assumptions in which the observations made by Michelson and Morley would result. He found that he needed only two.

He assumed that the laws of physics should be identical for all observers. In particular, they should not depend in any way on how an observer is moving in uniform motion, relative to an observed object. In practical terms, this means that the laws of physics should appear to be identical in any so-called inertial frame of reference, and so all such frames of reference are equivalent. An observer stationary in one frame of reference (standing on the ground, say) should be able to draw the same conclusions from some set of physical measurements as another observer moving uniformly relative to the first, or stationary in his own moving frame of reference, such as a moving train or a spaceship.

Einstein also assumed that the speed of light should be regarded as a fundamental, universal constant, representing an ultimate speed which cannot be exceeded.

From these assumptions there flowed a number of bizarre consequences. Out went any idea of an absolute frame of reference (and hence the idea of a stationary ether), together with absolute space, time, and simultaneity. In came all sorts of strange effects predicted for moving objects and clocks within what came to be recognized as a four-dimensional spacetime. The theory was 'special' only insofar as it treated the special case of observers moving in constant uniform motion in a straight line.

As Einstein explained it to his Olympia Academy friend Maurice Solovine:

> All movements of bodies were supposed to be relative to the light-carrying ether, which was the incarnation of absolute rest. But after efforts to discover the privileged state of movement of this hypothetical ether through experiments had failed, it seemed that the problem should be restated. That is what the theory of relativity did. It assumed that there are no privileged physical states of movement and asked what consequences can be drawn from this.

There was a further, fundamentally important consequence of the principle of relativity, which Einstein had explored in a short, fifth paper published in 1905. Suppose a body emits light of total energy E equally in two, opposite directions. When seen from the perspective of an observer in an inertial frame of reference moving uniformly with respect to the body, there is a perceived difference in the amount of energy lost by the body, such that:

> ...its mass decreases by [E/c^2]. Here it is obviously inessential that the energy taken from the body turns into radiant energy, so we are led to the more general conclusion: The mass of a body is a measure of its energy-content...

The theory demanded that energy and mass be recognized as equivalent and interchangeable, through the approximate relation $E = mc^2$.

This relation suggests that energy can be considered equivalent to mass, and that all mass represents energy. Einstein's earlier light-quantum hypothesis makes the connection between the energy of a light-quantum and its frequency. Now there were two simple, yet fundamental equations connecting energy to mass and energy to frequency. This seems to beg an obvious question. Could these equations be combined?

French physicist Prince Louis de Broglie certainly thought so.

Thirty-one year old de Broglie[2] was the younger son of Victor, fifth duc de Broglie. He had originally intended a career in the humanities, and had studied medieval history and law at the Sorbonne, receiving a degree in 1910. But he gained a passion for physics through the influence of his older brother Maurice and his experiences during World War I serving in the French Army, in field radio communications, stationed at the Eiffel Tower.

After the war, de Broglie joined a private physics laboratory headed by his brother which specialized in the study of X-rays. It was whilst he was working at this laboratory in 1923 that he thought to combine the two most iconic equations of special relativity and quantum theory:

> The notion of a quantum makes little sense, seemingly, if energy is to be continuously distributed through space; but, we shall see that this is not so. One may imagine that, by cause of a meta law of Nature, to each portion of energy with a proper mass [m], one may associate a periodic phenomenon of frequency [ν], such that one finds... [$h\nu = mc^2$].

This seemed to imply that a light-quantum with frequency ν would possess a mass, and hence a momentum, calculable as mass times velocity. Such directed momentum had been revealed in Compton's experiments, so this appeared to make sense.

But it was de Broglie's next step that was breathtaking. He later wrote:

> After long reflection in solitude and meditation, I suddenly had the idea, during the year 1923, that the discovery made by Einstein in 1905 should be generalized by extending it to all material particles and notably to electrons.

If electromagnetic waves, characterized by a frequency ν, possessed associated particle-like properties such as momentum then, de Broglie

[2] Pronounced 'de Broy'.

reasoned, perhaps particles such as electrons, characterized by a mass m, might possess associated wave-like properties. He went on:

> An electron is for us the archetype of [an] isolated parcel of energy, which we believe, perhaps incorrectly, to know well; but, by received wisdom, the energy of an electron is spread over all space with a strong concentration in a very small region, but otherwise whose properties are very poorly known. That which makes an electron an atom of energy is not its small volume that it occupies in space, I repeat: it occupies all space, but the fact that it is undividable, that it constitutes a unit.

De Broglie was suggesting that the electron could be thought of as a wave. A beam of electrons could be diffracted, just like a beam of light. The connection could be reduced to a simple equation, $\lambda = h/p$, the wavelength of the wave is equal to Planck's constant divided by the momentum, p (mass times velocity), of the particle. That this wave nature of particles is not apparent in everyday macroscopic objects is due to the very small size of Planck's constant h.[3] The dual wave–particle nature of matter is apparent only in the microscopic world of fundamental particles, atoms, and molecules.

Whatever they were, these 'matter waves' could not be considered to be in any way equivalent to more familiar wave phenomena, such as sound waves or ripples on the surface of a pond. De Broglie was able to show that the velocity of such matter waves would be greater than the speed of light—forbidden in Einstein's special theory of relativity—and could not therefore be waves carrying energy. He therefore concluded that the matter wave: 'represents a spatial distribution of *phase*, that is to say, it is a "*phase wave*".'

The concept of phase was therefore critical to de Broglie's work right from the beginning. We can think of phase in terms of a simple sinusoidal oscillation, with 'peaks' and 'troughs' in the height or amplitude of the wave. A series of waves in which the peaks and troughs coincide in space and in time are said to be 'in phase'.

De Broglie's interest in chamber music now led him to a major breakthrough. Musical notes produced by string or wind instruments are the result of so-called standing waves, vibrational patterns which 'fit'

[3] If Planck's constant were very much larger, the macroscopic world would be an even more peculiar place than it is.

within the stopped length of the string or the length of the pipe. A variety of standing wave patterns is possible provided the patterns meet the requirement that they fit between the string's secured ends or the ends of the pipe. This means that they must have zero amplitude at each end. This is possible only for vibrational patterns which contain an integral number of half-wavelengths.

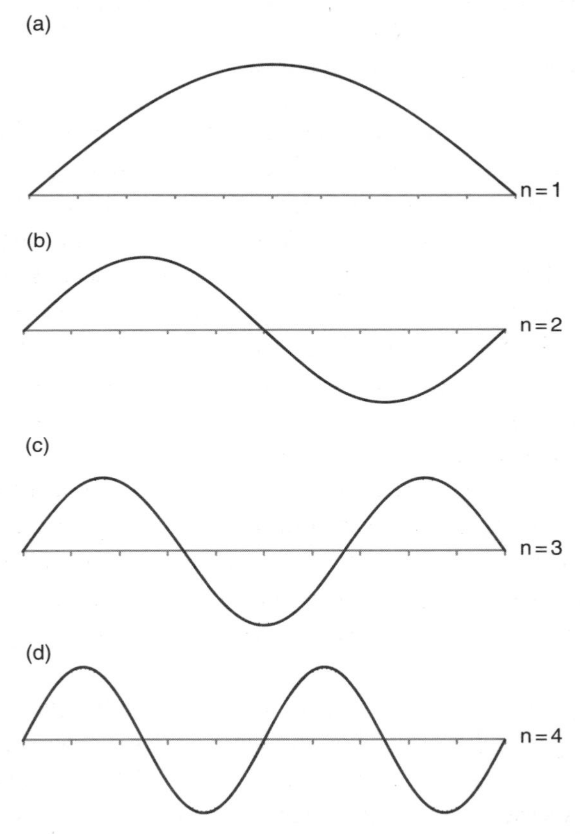

FIG 3 De Broglie made a connection between the physics of musical notes and quantum numbers. Musical notes generated by string or wind instruments are derived from standing wave patterns. These wave patterns must have zero amplitude at each end of the string or pipe, and so the only possible patterns contain an integral number of half-wavelengths. In this figure, pattern (a) has no nodes (points where the wave passes through zero), (b) has one node, (c) has two and (d) has three. If we define the quantum number n as the number of nodes +1 (or the number of half-wavelengths) then these patterns correspond to n = 1, 2, 3 and 4.

The longest wavelength standing wave is therefore equal to twice the length of the string or pipe. Such a wave has no 'nodes'—points where the wave amplitude passes through zero—between the ends. The next wave is characterized by a wavelength equal to the length of the string or pipe, with one node between the ends. The third has one-and-a-half wavelengths and two nodes, and so on.

De Broglie saw that whole-number requirement in Bohr's theory of atomic structure would emerge naturally from a model in which an electron wave is confined to a circular orbit around the nucleus. Perhaps, he reasoned, the stable electron orbits of Bohr's theory represented standing electron waves, just like the standing waves in strings and pipes responsible for musical notes. The same kinds of arguments would apply. For a standing wave to be produced in a circular orbit, the electron wavelengths must fit exactly within the orbit circumference.

Put simply, by the time the electron wave has performed one complete orbit and returned to the starting point, the value of the associated wave amplitude and the phase of the wave (its position in its peak–trough cycle) must be the same as at the starting point. If this is not the case, the wave will not 'join up' with itself: it will interfere destructively and no standing wave will be produced.

To satisfy the requirement, the wavelength of the electron wave must be such that an *integral* number of wavelengths will fit into the circumference of the orbit. This is called the resonance or phase condition. Bohr's quantum numbers could therefore be thought of as the number of wavelengths of the electron wave present in each orbit. 'Thus,' wrote de Broglie, 'the resonance condition can be identified with the stability condition from quantum theory.'

De Broglie published his ideas in a series of three short papers in the *Comptes Rendus* (proceedings) of the Paris Academy in September and October 1923. He collected these papers together, extended his ideas and presented them as a PhD thesis to the Faculty of Science at Paris University. He produced three typed copies and passed one of these to the prominent French physicist Paul Langevin, who was to act as an examiner.[4]

[4] De Broglie's other examiners were Jean Perrin, Élie Cartan and Charles Mauguin.

Langevin didn't know quite what to make of de Broglie's bold (and, possibly, crazy) ideas and sought guidance from Einstein, now professor at the University of Berlin. Einstein asked for a further typed copy to be sent to him.

Einstein was by now famous for his work on both the special and general theories of relativity, propelled into the public consciousness by the observations of the bending of starlight by the sun during the total solar eclipse of 1919 that had spectacularly confirmed the predictions of the general theory. The peculiar action-at-a-distance demanded by Newton's theory of gravity, denounced at the time as an 'occult agency' and made even more peculiar through the non-existence of the ether, had now been replaced by the motions of gravitational bodies in a curved space–time.

Einstein may have recognized a little of his former rebelliousness in de Broglie's work. He wrote back to Langevin offering encouragement: 'He [de Broglie] has lifted a corner of the great veil'. This was enough for Langevin. He accepted the thesis and de Broglie was duly awarded his doctorate in November 1924. De Broglie's thesis was published in its entirety in the French scientific journal *Annales de Physique* in 1925.

In a December 1924 letter to Dutch physicist Hendrik Lorentz, Einstein remarked that:

> ...de Broglie has undertaken a very interesting attempt to interpret the Bohr–Sommerfeld quantum rules (Paris dissertation 1924). I believe it is a first feeble ray of light on this worst of our physics enigmas.

Not everyone was convinced. Most physicists were rather sceptical: 'some wit, in fact, dubbed de Broglie's theory "la Comédie Française".'

De Broglie's ideas were illuminating but they were far from representing a solution. Assuming that material particles like electrons could possess associated wave-like properties had led de Broglie to identify the quantum numbers of Bohr's theory with the resonance condition for standing waves. But this was nothing more than a theoretical connection. De Broglie had not derived the quantum numbers from some kind of wave theory of the atom. And he had as yet no explanation for quantum jumps:

> From this we see why certain orbits are stable; but, we have ignored passage from one to another stable orbit. A theory for such a transition can't be studied without a modified version of electrodynamics, which so far we do not have.

Einstein was already concerned about the implications of quantum jumps for the principles of causality, the notion that every effect in the physical universe is traceable to a direct physical cause. It was clear that an electron in a high-energy atomic orbit will fall to a more stable, lower-energy orbit. It is caused to do so by the mechanics of the atom. But the theory allows only the calculation of the *probability* that a spontaneous jump will take place. The exact moment of the jump and the direction of the consequently emitted light-quantum appeared to be left entirely to chance. These were things that could not be predicted.

Einstein was not at all comfortable with this. In 1920 he had written to German physicist Max Born on the subject, noting that he: 'would be very unhappy to renounce complete causality'. From the very beginning, Einstein had viewed the quantum hypothesis as provisional, to be replaced eventually by a new, more complete theory that would properly explain quantum phenomena.

After pioneering quantum theory through one of its most testing early periods, Einstein was beginning to have grave doubts about what the theory implied.

One thing was clear. The quantum jumps between stable atomic orbits simply could not be accommodated in accepted theories of the dynamics of electrically charged particles. As de Broglie had acknowledged in his thesis, a modified electrodynamics was required. Familiar models were no longer of any use when applied to the mysterious interior of the atom. A radical new vision was needed.

5

A Strangely Beautiful Interior

Helgoland, June 1925

Despite the success of the Bohr–Sommerfeld rules, the 'old' quantum theory (as it came to be known) was creaking under the strain of a growing body of inexplicable experimental results. Whilst the spectrum of the simplest atom—that of hydrogen—could be 'explained' using these rather ad hoc *rules with some degree of confidence, the spectrum of the next simplest atom—helium—could not. The spectra of certain other types of atoms, such as sodium and the atoms of rare-earth elements, showed 'anomalous' splitting when placed in a magnetic field.*[1] *The old quantum theory, created by shoe-horning quantum rules into a classical mechanical structure, simply could not account for this effect.*

The problems with theory spilled over into confrontation. Bohr's general reluctance to embrace Einstein's light-quantum led him to formulate an alternative approach which did not invoke point-like particles to explain phenomena such as the photoelectric effect. He found, however, that there was a high price to be paid for this reluctance. In early 1924, *together with Dutch physicist Hendrik Kramers and American John C. Slater, Bohr had proposed a theory in which the principles of conservation of momentum and energy within individual atomic events had been abandoned, replaced instead by the appearance of conservation as a result of statistical averaging. Einstein was furious.*

*The Bohr–Kramers–Slater (BKS) proposal was subsequently shot down by further experiments in early/mid-*1925. *Bohr, whose personality and power of persuasion had*

[1] In fact, this 'anomalous' Zeeman effect (named for Dutch physicist Pieter Zeeman) is actually the normal behaviour. It is the general behaviour of simpler atoms, regarded by the physicists of the time as normal, that is actually anomalous.

convinced his junior colleagues of the merits of the proposal against their better judgement, agreed to give the effort an 'honourable funeral'. There was now little alternative but to accept the reality of light-quanta.

The BKS proposal was symptomatic of the physicists' growing sense of desperation. It was becoming increasingly clear that a new quantum theory was needed, constructed somehow from the 'bottom up', from a new set of principles that did not depend on contradictory classical pictures of charged, material particles in planetary-like orbits around a central nucleus.

There was revolution in the air.

It had become apparent in 1917 that the cramped working conditions at the University in Copenhagen could not accommodate Bohr's vision for Danish physics. He had set off on the long and rather tortuous journey that would lead eventually to the construction of a new Institute for Theoretical Physics. The Institute was completed in 1920 and inaugurated in March 1921. Bohr had involved himself directly in the construction planning. He secured further funding in the early 1920s from the Carlsberg Foundation and the International Education Board (founded by John D. Rockefeller in 1923) to purchase scientific equipment, provide student grants and funds for visiting scholars, and to extend the Institute further to accommodate this burgeoning activity.

There followed a grand procession of aspiring young physicists, beating a path to the neo-classical style doors of Bohr's Institute, to study with the master of atomic theory. One young German physicist recently arrived during the Easter vacation in 1924, experiencing problems with Danish customs and unable to speak the language, realized that all he had to do was mention Bohr's name:

> ...as soon as it became clear that I was about to work in Professor Bohr's Institute, all difficulties were swept out of the way and all doors were opened to me. And so from the very outset I felt safe under the protection of one of the greatest personalities in this small but friendly country.

Werner Heisenberg had first met Bohr in Göttingen in June 1922, during a scientific festival held in Bohr's honour. Just 20 years old, he had been studying for his doctorate in physics under Sommerfeld at the University of Munich, and had rather impudently stood at the end of the third of

Bohr's lectures to make a critical remark. Bohr had replied hesitantly, and after the lecture had invited Heisenberg to join him on a walk over the nearby Hain Mountain. Bohr explained that he fully understood the nature of Heisenberg's doubts, and invited the young physicist to join him in Copenhagen for a term's study leave.

In July 1923 Heisenberg completed his doctorate in Munich, youthful arrogance and disdain for aspects of experimental physics causing him some problems during his oral examination. Unable to derive the equations for the resolving power of a microscope to the satisfaction of the aged Wien, Heisenberg achieved a poor pass result. It was a salutary experience for the young doctor, dismissed by Wien as an upstart.

From Munich Heisenberg had travelled to Göttingen to work with Max Born with the intention of qualifying as a university lecturer. 'He looked like a simple peasant boy,' Born later recalled, 'with short, fair hair, clear bright eyes and a charming expression.' In Göttingen he had his first experiences wrestling with the problems that beset the quantum theory of the atom. When in January 1924 Bohr repeated his invitation for Heisenberg to come to Copenhagen, he accepted and made arrangements to travel to Denmark during the Easter break.

Heisenberg was received warmly at Bohr's Institute, but was overwhelmed and somewhat intimidated by what he found there. The Institute was a bustling centre for atomic physics and he encountered: 'a large number of brilliant young men from every part of the world, all of them greatly superior to me, not only in linguistic prowess and worldliness, but also in their knowledge of physics.'

He saw little of Bohr in those first few days of his stay in Copenhagen. But Bohr eventually found some time and asked Heisenberg to join him on another lengthy walk, through the island of Zealand to the Kronborg Castle, long associated with Denmark's most famous, but possibly mythical, prince Hamlet.

In coming to Copenhagen Heisenberg had, for a short time at least, escaped the vicissitudes of a Germany in social and political turmoil. In February 1924 Adolf Hitler had been sentenced to five years' imprisonment for his part in the failed beer-hall putsch of the previous November, but his trial had been a major propaganda victory for the National

Socialist party.[2] Bohr was interested in what the young German had to say about events in his country.

'But now and then our papers also tell us about more ominous, anti-Semitic, trends in Germany, obviously fostered by demagogues. Have you come across any of that yourself?' he asked Heisenberg. Bohr was part Jewish.

Indeed he had. Heisenberg told Bohr of his experiences during a lecture on general relativity delivered by Einstein in Leipzig in the summer of 1922. He had thought that science would be above political strife of the kind that had led to civil war in Munich, and was greatly saddened to witness student demonstrations against Einstein's 'Jewish physics', sponsored by Philipp Lenard.

Heisenberg's first visit to Copenhagen was short, but it was followed by many longer visits. He returned to Bohr's Institute in September 1924, supported by grants from the Carlsberg Foundation and the International Education Board. During this second visit he worked closely with Hendrik Kramers, a talented Dutch theoretician who by this time had become Bohr's right-hand man.

It was this work with Kramers that helped to convince Heisenberg that progress in atomic theory could only be made by abandoning any attempt to 'understand' the interior workings of the atom. He came to realize that the planetary model, with its images of material particles circulating around a central nucleus in a series of stable orbits, was visually rich but analytically empty. As a classical mechanical model, its usefulness had been prolonged by the addition of rather arbitrary quantum rules, but it was surely doomed to failure.

Back in Göttingen, Born had argued for a new 'quantum mechanics' to replace the classical theory as applied to the internal structure of the atom. Heisenberg had elected to focus his attention on what could be seen, rather than what could only be guessed at. The secrets of the atom where revealed in atomic spectra, in the precise patterns of frequencies and intensities (or 'brightness') of individual spectral lines. Heisenberg now realized that a new quantum theory of the atom should deal only

[2] Hitler served only eight months of his sentence. Whilst in prison he wrote *Mein Kampf* with Rudolph Hess.

in these observable quantities, not unobservable mechanical 'orbits' conforming to arbitrary quantum rules.

On Hain Mountain, Bohr had spoken of the challenges they faced. 'These models,' he had said, 'have been deduced, or if you prefer guessed, from experiments, not from theoretical calculations. I hope that they describe the structure of atoms as well, but only as well, as is possible in the descriptive language of classical physics. We must be clear that, when it comes to atoms, language can be used only as in poetry. The poet, too, is not nearly so concerned with describing facts as with creating images and establishing mental connections.'

Heisenberg decided it was now time for a new language.

He returned once more to Göttingen in April 1925. Towards the end of May he succumbed to a severe bout of hay fever, and asked Born for 14 days' leave of absence to recuperate. On 7 June he arrived on the small island of Helgoland, just off the north coast of Germany, hoping that the clear North Sea air would facilitate a speedy recovery. His face was so badly swollen that the landlady of the boarding house where he stayed was convinced he had been in a fight.

Free from distractions, he now made swift progress. He had been working on an approach to atomic theory in which the parameters of an unobservable interior mechanics were replaced by terms corresponding to atomic events that could be observed—the jumps or transitions between orbits which were manifested as spectral lines. He had constructed a rather abstract model consisting of an infinite series (called a Fourier series) of harmonic oscillators, each characterized by an amplitude and a frequency.[3] He had identified each oscillator—each term in the Fourier series—with a quantum jump from some stable orbit characterized by a quantum number *n* to an orbit characterized by a quantum number *m*. The result was an infinite table of symbols or terms in the Fourier series, organized into columns and rows, each term representing a quantum transition from some initial to some final state.

He now assumed that he could calculate the intensities of the resulting spectral lines as the squares of the amplitudes of the terms that appeared

[3] More familiar examples of harmonic oscillators include a swinging pendulum and a ball moving up and down in regular motion attached to a spring.

in the table. In the case of a jump from state n to $n - 2$, for example, he found it necessary to multiply the amplitude of the term corresponding to the jump n to $n - 1$ with the amplitude of the term corresponding to the jump from $n - 1$ to $n - 2$. More generally, he calculated the intensity of a spectral line resulting from any quantum jump as the sum of the products of the amplitudes for all possible intermediate jumps.

Whilst this multiplication rule seemed straightforward and satisfactory, Heisenberg was aware of a potential paradox. If this same rule is used to calculate the product of two different physical quantities (x and y, say) then it suggests that there may arise situations in which the product x multiplied by y is not equal to the product y multiplied by x. Heisenberg was quite unfamiliar with this kind of result and greatly unsettled by it.

He wasn't quite sure where all this was leading when he arrived on Helgoland. He had become aware that the scheme was not guaranteed to be free of contradictions and, in particular, may not respect the need for the conservation of energy. The lesson from the failure of the BKS proposal was that conservation of energy was an imperative for any purported new quantum mechanics of the atom. Heisenberg now set about the task of calculating the energies to check that everything was in order. He worked on it long into the night:

> When the first terms seemed to accord with the energy principle, I became rather excited, and I began to make countless arithmetical errors. As a result, it was almost three o'clock in the morning before the final result of my computations lay before me. The energy principle had held for all the terms, and I could no longer doubt the mathematical consistency and coherence of the kind of quantum mechanics to which my calculations pointed. At first, I was deeply alarmed. I had the feeling that, through the surface of atomic phenomena, I was looking at a strangely beautiful interior, and felt almost giddy at the thought that I now had to probe this wealth of mathematical structures nature had so generously spread out before me.

He was too excited to sleep. He quietly left his lodgings and walked in the darkness, climbing a rock jutting out to sea on the southern tip of the island. He watched as the sun rose.

Heisenberg hastily wrote up his calculations and passed a copy of the manuscript to Born. Despite his experience on Helgoland, he still wasn't certain if this rather intuitive approach made any real sense. Born recalled Heisenberg telling him that: 'he had written a crazy paper and did not dare send it in for publication; I should read it and, if I liked it, send it to *Zeitschrift für Physik*.' Born was enthusiastic about the paper but puzzled by the multiplication rule that Heisenberg had used. It seemed familiar. On 10 July 1925 he finally remembered. Born had been taught this multiplication rule as a student. It was the rule for multiplying matrices.

A matrix is a square or rectangular array of numbers organized in columns and rows. Like ordinary numbers, matrices can be added, subtracted, multiplied, and divided. Specific rules are necessary to guide matrix multiplication, showing how each element in each matrix must be combined to give the corresponding elements in the final, product matrix. Unlike ordinary numbers, it is possible to find matrices (for example, $\mathbf{x}$ and $\mathbf{y}$) that do not commute; that is, the product of $\mathbf{x}$ multiplied by $\mathbf{y}$ is not equal to $\mathbf{y}$ multiplied by $\mathbf{x}$.

Born now worked with his student Pascual Jordan to recast Heisenberg's calculations into the language of matrix multiplication. They discovered that the matrix for the energy of the system is diagonal—all the elements in the matrix are zero but for those along the diagonal of the array, which are time-independent and represent the stable quantum states (the 'orbits') of the system.

They also discovered that Heisenberg had been right to be disconcerted. They found that the matrix equivalents of classical physical quantities such as position and momentum do not commute. Denoting the position matrix as $\mathbf{q}$ and the momentum matrix as $\mathbf{p}$, they discovered that the difference between the product $\mathbf{pq}$ and the product $\mathbf{qp}$ (called the commutation relation) is equal to $-ih/2\pi$, where i is the square-root of minus 1 and h is Planck's constant.[4] Classical mechanics (in which the values of position and momentum are represented by ordinary numbers which always commute) was once again seen as a limiting approximation of quantum mechanics, in which Planck's constant h is assumed to be zero.

[4] Strictly speaking, $\mathbf{pq}-\mathbf{qp} = -ih1/2\pi$, where 1 is the unit or identity matrix.

Heisenberg was at once both pleased and relieved: 'Only much later did I learn from Born that it was simply a matter here of multiplying matrices, a branch of mathematics that had hitherto remained unknown to me.' He acquired some textbooks on the subject and quickly caught up. He was soon collaborating with Born and Jordan and together they published a further paper on the new quantum mechanics in November 1925.

A proof copy of Heisenberg's paper found its way to respected English physicist Ralph Fowler in Cambridge. Unimpressed, Fowler passed the copy to his young student, 23-year-old Paul Dirac.

It was only when Dirac studied the proofs properly, about a week or so after he had first been given them, that he realized precisely what Heisenberg had done. He now worked intensely on the problem, identifying analogies between Heisenberg's multiplication rule and mathematics developed by French theoretician Siméon Poisson in 1809. When he had finished he had independently arrived at the position–momentum commutation relation, proved the principle of energy conservation, and had derived the relationship between the energies of the stable orbits and frequency of emitted radiation, the relation originally deduced by Bohr.

Dirac sent a handwritten version of his paper describing these results to Heisenberg, and was subsequently disappointed to learn that his work had been anticipated by Born and Jordan. Heisenberg was nevertheless full of praise: '...on the one hand, your results, especially concerning the general definition of the differential quotient and the connection of the quantum conditions with the Poisson brackets, go considerably further than [Born and Jordan's] work; on the other hand, your paper is also written really better and more concisely than our formulations given here.'

The new quantum revolution was now firmly underway. The beginnings of a true quantum mechanics had been formulated, but with new, unprecedented levels of mathematical abstraction. Heisenberg's prescription, as elaborated by Born and Jordan, came to be known as matrix mechanics (a name that Heisenberg hated, because it sounded too mathematical).

But the new theory was not to everybody's taste. If it was to displace the 'old' quantum theory, the new quantum mechanics had to show that it could do everything the old theory could do. Not least it had to predict the emission spectrum of the hydrogen atom. There was no shortage of challenges to be faced.

6

The Self-rotating Electron

Leiden, November 1925

There was no great rush to embrace the new quantum mechanics, with its abstract, infinite-dimensional tables of numbers and obscure multiplication rules. Bohr was enthusiastic. Einstein was not.

As work began to deploy the new methods in the calculation of the emission spectrum of hydrogen, the problem of the 'anomalous' Zeeman effect continued to tease the intellects of the new quantum physicists. The 'normal' Zeeman effect had been nicely accommodated in the old quantum theory through the introduction of a third (magnetic) quantum number, m. *But the anomalous effect remained stubbornly intransigent. When young Austrian physicist Wolfgang Pauli was stopped in the street in Copenhagen and asked why he was unhappy, he replied: 'How can one look happy when he is thinking about the anomalous Zeeman effect?'*

In an attempt to classify the nature and magnitude of the splitting of the spectral lines of multi-electron atoms in 1920 Sommerfeld had introduced a fourth quantum number, which he called the 'inner quantum number', j, *and a new selection rule. Where the first three quantum numbers had evolved from reference to classical conceptual models of the inner workings of the atom, this fourth quantum number was entirely* ad hoc. *Sommerfeld simply assumed that the motions of multi-electron atoms are complex and characterized somehow by a 'hidden rotation'. In 1921 German spectroscopist Alfred Landé at the University of Tübingen suggested that the quantum numbers* m *and* j *could take half-integral values. Two years later he proposed that for atoms with a many-electron 'core' and a single outlying, 'valence' electron, the hidden rotation should be associated with the core electrons.*

Landé's approach was relatively successful, but it was still all quite confusing and barely comprehensible. To add insult to injury, in 1922 German physicists Otto Stern and Walther Gerlach had studied the effects of a magnetic field on a beam of silver atoms, produced by effusion from a heated oven. Stern had been looking for evidence for the kind of space quantization predicted by the Bohr–Sommerfeld theory and when they found that the beam was indeed split into two components by the magnetic field, they thought they had found it. Others, including Einstein, were convinced they had not, and were greatly puzzled by these results.

The explanation, when it came, would be the last hurrah of the old quantum theory.

Pauli had already established a reputation as a prodigy of Einstein's relativity when he arrived in Munich in the autumn of 1918, at the tender age of eighteen, to join Sommerfeld's research team. He wrote a review article on general relativity which was published a few years later and of which Einstein wrote: 'Whoever studies this mature and grandly conceived work might not believe that its author is a twenty-one year-old man.' The renowned physicist and philosopher Ernst Mach was Pauli's godfather.

At Munich, Sommerfeld introduced Pauli to the intricacies of atomic theory and the number-games and selection rules of the old quantum theory. Pauli received his doctorate in 1921, moving to Göttingen to work with Born for a short period before moving to the University of Hamburg. Like Heisenberg, Pauli first met Bohr during the festival in Göttingen in 1922, and was deeply impressed by Bohr's personality, wisdom, and insight. On Bohr's invitation, he spent a year in Copenhagen before returning to Hamburg in late 1923.

It was in Copenhagen that Pauli had begun to wrestle with the problem of the anomalous Zeeman effect. It caused him considerable difficulties and regular periods of anguish and despair. Whilst he was able to improve somewhat on the proposals of Sommerfeld and Landé, he disliked the *ad hoc* nature of the theorizing that was involved. He was looking for something more fundamental.

Back in Hamburg towards the end of 1924, Pauli's attention was drawn to the work of Cambridge physicist Edmund Stoner, described in the preface to the fourth edition of Sommerfeld's book *Atombau und Spektrallinien*. In Stoner's original paper, published in October 1924 in the

British journal *Philosophical Magazine*, he had set out a scheme describing the relationship between the quantum numbers and the idea of electron 'shells', surrounding the nucleus and imagined to nest one inside the other like a Russian *matryoshka* doll. The energy of each shell is determined by the principal quantum number, n. The number of possible states or 'orbits' within each shell is then determined by the values that the quantum numbers k and m can take for a given value of n.

The rules dictated that k must be an integral number greater than zero and less than or equal to n,[1] and m could take a total of $2k - 1$ values over the range $-(k-1), -(k-2), \ldots 0, \ldots, (k-2), (k-1)$. So, for $n = 1$, the only values of k and m that are possible are $k = 1$ and $m = 0$. This implies a single state, or orbit. For $n = 2$, the rules imply four different orbits, for $n = 3$ the number of orbits is nine. This implies that the number of possible orbits increases as n^2.

But the pattern reflected in the periodic table of the elements tells a slightly different story. German physicist Walther Kossel had earlier argued that the striking stability and inertness of the noble gases (such as helium, neon, argon, krypton) could be understood in terms of Bohr's atomic theory if these atoms were assumed to have filled, or 'closed', shells. The periodic table could then be understood as a progression of occupancy of the electron shells, forming a pattern in which first two (hydrogen, helium), then eight (lithium through to neon), then another eight (sodium to argon), then eighteen electrons (potassium to krypton) are added in sequence until each successive shell is filled, or closed.

Stoner had gone one step further in his prescription. Instead of assigning a single electron to each orbit he had chosen to assign two: 'In the classification adopted, the remarkable feature emerges that the number of electrons in each completed level is equal to double the sum of the inner quantum numbers as assigned…'. In n^2 orbits, Stoner suggested, there should be $2n^2$ electrons. For $n = 1$ there is only one orbit, implying occupancy up to a total of two electrons, for $n = 2$ there are four orbits,

[1] The restriction that k must be greater than zero was impossible to prove in the old quantum theory. In modern atomic physics, k has been replaced by the orbital angular momentum (or azimuthal) quantum number l, which takes values $l = 0, 1, 2, \ldots, n$. Consequently, the magnetic quantum number m takes the values $-l, \ldots, 0, \ldots, l$.

implying up to eight electrons, for $n = 3$ there are nine orbits and up to eighteen electrons.[2]

Pauli put two and two together. He believed that Landé's model in which a 'hidden rotation' was ascribed to the core of electrons in a multi-electron atom was wrong. And yet the model appeared to work quite well. Stoner was suggesting a doubling of the electron count ascribed to each orbit, and hence each shell. The answer, Pauli reasoned, was to ascribe the fourth quantum number not to the core of electrons, but to *each individual electron*. He was led to the inspired conclusion that the electron must have a curious, non-classical 'two-valuedness' characterized by a half-integral quantum number. There was more. The shell structure of atoms and the periodic table of the elements implied that each orbit could accommodate two and only two electrons. Pauli wrote:

> There can never be two or more equivalent electrons in the atom, for which, in strong fields, the values of all quantum numbers... coincide. If one electron is present in the atom, for which these quantum numbers (in the external field) have definite values, then this state is 'occupied'.

This is Pauli's *exclusion principle*. It says that no electron in an atom can have the same set of four quantum numbers. An electron in a state characterized by $n = 1$, $k = 1$, $m = 0$, and $j = +½$ can enter the same orbit (with the same values of n, k, and m) only by taking the quantum number $j = -½$. No other values of j are possible so each orbit can contain only two electrons.

Pauli was unable to provide any formal explanation for this rule and had no alternative but to argue that it seemed to 'offer itself automatically in a natural way'. When in December 1924 he sent a copy of the

[2] To make explicit the connection with the 2, 8, 8, 18,... pattern of the periodic table it is necessary to jump ahead to our current understanding of the atomic orbits and their relative energies. The orbit with $n = 1$, $k = 1$ ($l = 0$) is spherical and named by convention as the 1s orbit. This can accommodate up to two electrons (accounting for hydrogen and helium). For $n = 2$ the possible orbits are 2s and three different 2p orbits, accommodating up to a total of eight electrons (lithium to neon). For $n = 3$ the orbits are 3s, 3p (up to eight electrons—sodium to argon) and five different 3d orbits (ten electrons). However, the 4s orbit actually lies somewhat lower in energy than 3d and is filled first. Therefore, the combination 4s, 3d, and 4p (accommodating up to eighteen electrons in total) together account for the next row of the periodic table, from potassium to krypton.

manuscript of a paper describing his results to Bohr and Heisenberg in Copenhagen, he received an exuberant, though barbed, response from Heisenberg:

> I am the one who *rejoices most* about it, not only because you push the swindle to an unimagined, giddy height (by introducing *individual* electrons with 4 degrees of freedom) and thereby have broken all hitherto existing records of which you have insulted me, but quite generally, I triumph that you too (*et tu, Brute!*) have returned with lowered head to the land of the formalism pedants...

Although its origin in the as yet undefined quantum mechanics of the atom remained vague, Pauli's exclusion principle accounted for an important feature of multi-electron atoms—the fact that they exist at all. There had been nothing in the earlier atomic theory to provide a reason why all the electrons in a multi-electron atom should not simply collapse down into the lowest-energy orbit.

For neutral atoms, increasing the number of electrons implies an increase in the total positive charge of the nucleus. The greater the nuclear charge, the smaller the radius of the innermost orbit, as the electrons are pulled more tightly towards the nucleus. Now we would expect that the repulsion between increasingly closely packed electrons would tend to resist collapse of the atom into ever-smaller orbits and hence ever-smaller volumes, but it is relatively straightforward to show that electron–electron repulsion cannot prevent heavier atoms from shrinking dramatically in size as the charge of the central nucleus is increased. The repulsion between neighbouring electrons simply isn't strong enough to overcome the force of attraction. Atomic volumes, easily calculated from atomic weights and densities of the elements, follow a complex variation with atomic charge but they do not systematically shrink with increasing charge.

By preventing the electrons from collapsing or condensing into the lowest-energy orbit, the exclusion principle allows complex multi-electron atoms to exist in the pattern described by the periodic table. It enables the existence of a marvellous variety of elements, the multitude of possible chemical combinations, and hence all material substance, living and non-living. This was a fantastic achievement.

But, still, why only *two* electrons per orbit?

Perhaps, argued a young American physicist named Ralph de Laer Kronig, this is because the electron's 'two-valuedness' is associated with *self-rotation*. Kronig was born in Dresden in 1904 to American parents. In January 1925 he had travelled back to Germany from America, where he had recently completed a PhD at Columbia University in New York. He had returned to Germany to work with Landé.

In late December 1924 Pauli advised Landé by postcard that he intended to visit Tübingen on 9 January. He was particularly interested to discuss the exclusion principle and to try to find any spectroscopic data among Landé's extensive collection that might tell against it.

Landé showed his new American assistant the correspondence he had received from Pauli concerning the exclusion principle. Kronig was struck by the application of the fourth quantum number to individual electrons and immediately saw a potential explanation. Sommerfeld had ascribed the fourth quantum number to a 'hidden rotation'. What if this rotation were, in fact, a real self-rotation of individual electrons? If the electron rotates about its own axis, in much the same way that the earth rotates on its axis as it orbits the sun, then this would generate a small, local magnetic field. The electron in an atom would behave like a tiny bar magnet. It would possess a magnetic moment that can become aligned with or aligned against the lines of force of an applied external magnetic field, giving two states of different energy which would appear as a splitting of spectral lines.

In the absence of this splitting, there is only one state of the electron and hence only one line in the spectrum. Kronig calculated that the angular momentum of electron self-rotation was required to have the fixed value of $\frac{1}{2}\hbar$, where $\hbar$, is a shorthand for Planck's constant h divided by 2π. In addition, he was able to show that the ratio of the magnetic moment and the angular momentum due to self-rotation, a characteristic factor known as the Landé 'g-factor' for the electron, had to have the value 2. This was rather curious, as the ratio of the electron orbital magnetic moment to the orbital angular momentum is 1, as predicted by classical mechanics.

There was a further problem. From Kronig's hastily derived expressions he calculated a splitting of spectral lines that was a factor of two larger than the splitting observed experimentally. This was not connected with the assumption of $g = 2$, as this was needed to explain other experimental observations. Electron self-rotation was able to resolve two out of three of the problems with Landé's core model, but could not yet resolve them all.

The next day Kronig went to collect Pauli from the railway station:

> For some reason I had imagined him [Pauli] as being much older and as having a beard. He looked quite different from what I had expected, but I felt immediately the field of force emanating from his personality, an effect fascinating and disquieting at the same time. At Landé's institute a discussion was soon started, and I also had occasion to put forward my ideas. Pauli remarked: *'Das ist ja ein ganz witziger Einfall,'* [this is indeed quite a witty idea], but did not believe that the suggestion had any connection with reality.

Pauli, noted as much for his biting wit as for his talents as a theoretical physicist, completely dismissed Kronig's suggestion. Kronig discussed it further with Bohr, Kramers, and Heisenberg, who were similarly dismissive. He was in any case troubled by the factor of two discrepancy between prediction and experimental observation, and concerned also that the equator of a spinning sphere of charged matter would be required to move ten times faster than the speed of light, forbidden by Einstein's special theory of relativity. He subsequently dropped the idea.

Ten months can be a long time in physics. When two young Dutch physicists Samuel Goudsmit and George Uhlenbeck, based in Leiden in the Netherlands, independently reached the same conclusion, the climate in the physics community was to prove more temperate. They summarized their arguments in favour of electron self-rotation in a paper they submitted to the journal *Naturwissenschaften*. After submitting the paper, they talked to the esteemed physicist Hendrik Lorentz. He advised them that their proposal was almost impossible in classical electron theory. Fearing that they had made a significant error, they scrambled to withdraw their paper before it could be published. But it was too late.

The paper was published in November 1925. Initially it provoked the same concerns that Kronig's proposal, now almost forgotten, had attracted. Travelling to Leiden in December 1925, Bohr was met at the railway station by Pauli and Stern and asked for his opinion about the Dutch proposal. Bohr may have said that he found it 'very interesting', Bohr-speak implying that it was probably wrong. On arrival in Leiden he was met by Einstein and Austrian physicist Paul Ehrenfest, who asked him the same question. When Einstein went on to explain how some of his objections to the proposal could be overcome, Bohr began to have a change of heart.

From Leiden Bohr travelled to Göttingen, where he was met by Heisenberg and Jordan, who asked the question again. This time Bohr was enthusiastic, although Heisenberg vaguely recalled having heard a similar proposal some time before. On his way back to Copenhagen Bohr's train stopped in Berlin, where he was met by Pauli who had travelled from Hamburg expressly to ask Bohr once again what he thought about electron self-rotation. Bohr now said it represented a great advance. Pauli, still unable to get past the fact that the classical picture of a spinning bit of charged matter made absolutely no sense in the context of atomic physics, called it 'a new Copenhagen heresy'.

Bohr became a strong advocate of electron self-rotation and may have been the first to use the term 'electron spin'. This is a term that has stuck, despite the fact that its meaning in quantum theory is considerably far removed from its classical interpretation. As the idea gained wider acceptance, it became clear that Pauli (and Bohr, Kramers, and Heisenberg) had discouraged Kronig from developing his own proposal further and from therefore becoming the 'discoverer' of electron spin.[3] Kronig tended to play down the matter, but could not help feeling some bitterness: 'I should not have mentioned the matter [to Kramers] at all,' he wrote later to Bohr, 'if it were not to take a fling at the physicists of the preaching variety, who are always so damned sure of, and inflated with, the correctness of their own opinion.'

As Bohr had anticipated, the problem of the mysterious factor of two discrepancy between the predicted and observed splitting of spec-

[3] A verse penned some time later summarized the situation: *'Der Kronig hätt' den Spin entdeckt, hätt' Pauli ihn nicht abgeschreckt.'* (Kronig would have discovered the spin if Pauli had not discouraged him.) See Enz, p. 117.

tral lines was resolved satisfactorily. English physicist Llewellyn Hilleth Thomas subsequently showed that re-casting the problem in the proper rest frame of the electron changed the expression for the splitting, replacing the appearance of g in the expression with $g-1$. Assuming $g = 2$ effectively reduces the predicted splitting by half.

It was Paul Dirac who subsequently suggested that if the electron can be considered to possess two possible 'spin' orientations then this, perhaps, explains why each atomic orbit can accommodate only two electrons. The two electrons must be of opposite spin to 'fit' in the same orbit. An orbit can hold a maximum of two electrons provided their spins are *paired*.

This was great progress, but there were still many, many puzzles. A classical spinning object is not constrained in principle to only two positions of alignment of its magnetic moment. It was reasoned that this restriction must be somehow due to the quantum nature of the electron.

It was just not clear how.

7

A Late Erotic Outburst

Swiss Alps, Christmas 1925

Both Pauli and Dirac moved quickly to use the new matrix mechanics to derive key features of the hydrogen emission spectrum, most notably the Balmer formula. Pauli submitted a paper describing his results to the journal Zeitschrift für Physik *on 17 January 1926, beating Dirac, who submitted his paper to the British journal* Proceedings of the Royal Society, *by just five days.*

Pauli also used the new mechanics to account for the Stark effect, named for German physicist Johannes Stark, in which spectral lines are split by a static electric field. However, certain relationships still had to be assumed, in the guise of an ansatz, *an 'educated guess' or hypothesis to be verified through agreement with experimental results. Furthermore, the theory was not yet fully relativistic—that is, it did not conform to the demands of Einstein's special theory of relativity. And it did not yet accommodate Pauli's own exclusion principle, electron spin, and the anomalous Zeeman effect.*

Whilst older physicists may have struggled with the theory's mathematical complexity and lack of 'visualizability', matrix mechanics looked set to become a romping ground for a new generation of quantum physicists. In his paper Pauli wrote: 'Heisenberg's form of quantum theory completely avoids a mechanical-kinematic visualization of the motions of electrons in the stationary states [i.e. the stable orbits] of an atom.' The message was clear. To make progress we first must rid ourselves of the conceptual baggage of our classical physical heritage and focus our attention only on what can be observed and measured in the laboratory.

The direction for future development of the theory seemed set. Yet the young turks of the quantum revolution were about to be upstaged by a 38-year-old Austrian *physicist named Erwin Schrödinger. By the time he was finished, the landscape of the new quantum theory had been almost completely transformed.*

Schrödinger had secured his doctorate in physics at the University of Vienna in May 1910. After a little more than a year of obligatory military service, he returned to the University and qualified as a lecturer (a *Privatdozent*) in 1914. Four years of military service in the First World War then intervened. He returned to Vienna in 1917, becoming an associate professor in 1920. He then moved first to Jena, then Stuttgart, then Breslau before securing the professorship in theoretical physics at the University of Zurich in 1921.

He arrived in Zurich with his wife, Annemarie (known to him affectionately as Anny), in October 1921. Diagnosed with suspected pulmonary tuberculosis just a few months later, he was ordered to take a complete rest cure. He and Anny retreated to a villa in the Alpine resort of Arosa, near the fashionable ski resort of Davos, where they stayed for nine months as Anny nursed him back to health. The altitude and the relative isolation provided valuable opportunities for Schrödinger to think, and he wrote two scientific papers whilst on his road to recovery. He returned to Zurich and resumed his heavy teaching schedule in November 1922, delivering his delayed inaugural lecture on 9 December.

His teaching commitments left little time for research and, though cured of his respiratory illness, he remained weak and tended to tire easily. As he settled into his new life in Zurich he may have pondered on his place in the firmament of physics. He had drawn praise for his versatility as a scientist and the breadth of his knowledge and accomplishments, but he had yet to make a noteworthy contribution to any branch of physical science with which he was familiar. As he grew older, he had little choice but to watch as a younger generation of physicists overtook him. It seemed that he would be sidelined, worthy of little more than a footnote in the history of physics. When, in 1924, he was invited to attend the fourth in a series of conferences on physics established by the wealthy Belgian industrialist Ernest Solvay, he was not asked to present a paper.

His marriage to Anny was also in trouble. Theirs had been a relatively 'open' marriage, with both Erwin and Anny indulging in extra-marital affairs. Their infidelity caused friction but had thus far generated insufficient heat for either of them to sue for divorce. Schrödinger was keen to father a child which, given their sexual incompatibilities, seemed unlikely to happen. Yet Anny was in favour of continuing the marriage, with all its openness. Anny was a Catholic, divorce was expensive, and the arrangement may have appealed to Schrödinger's anti-bourgeois sensibilities.[1] Despite their incompatibilities, Anny remained enraptured by Erwin's looks and intellect: 'You know it would be easier to live with a canary than a racehorse, but I prefer the racehorse.' The heat was building, however.

Schrödinger would sometimes seek refuge in the bohemian nightlife of the city, in the company of colleagues from the neighbouring ETH; Dutch physicist Pieter Debye and mathematician Hermann Weyl. They formed a complex social set. Anny fell in love with Weyl, and they began an affair. Meanwhile Weyl's wife Helene Joseph, who had studied philosophy under phenomenologist Edmund Husserl, had fallen for Swiss ETH physicist Paul Scherrer.

The turmoil in Schrödinger's private life inevitably affected his work. He published no papers in 1923, but managed to regain his appetite for research the following year. In October 1925 his attention was drawn to a footnote in one of Einstein's recent papers. It mentioned 'a very notable contribution' by de Broglie.

Intrigued, Schrödinger acquired a copy of de Broglie's thesis. What he read would have deep resonance. Whilst recovering from illness in Arosa in 1922 he had written a paper, 'On a Remarkable Property of the Quantized Orbits of a Single Electron', which he had submitted to the *Zeitschrift für Physik*. Building on ideas he had derived from his close friend Weyl, he had observed a relationship between the stable electron orbits and a so-called 'gauge factor', a characteristic of the orbits themselves.[2] He had found that he could understand the relationship between the different electron orbits—and the origin of the principal quantum number n—by interpreting the gauge factor as a *phase factor*.

[1] Schrödinger kept a notebook detailing his sexual encounters. This was not merely a log of sexual conquests; it was rather intended as a personal exploration of female sensuality.

[2] We will return to the idea of gauge theories in Chapter 20.

Although at the time he had recognized this result as important, he had chosen not to pursue it further. As he now read de Broglie's thesis he understood that the relationship he had observed was simply de Broglie's phase condition for a standing electron wave confined to a circular orbit around the nucleus.

Physicists at the University of Zurich and the ETH had established joint bi-weekly colloquia at which they would discuss current topics of mutual interest. Debye asked Schrödinger if he would be prepared to present a colloquium on de Broglie's thesis, which had recently been published in the French journal *Annales de Physique*. Schrödinger agreed, and wrote to Einstein on 3 November, drawing parallels with his earlier 1922 paper on electron orbits but noting that: 'Naturally de Broglie's consideration in the framework of his comprehensive theory is altogether of far greater value than my single statement, which I did not know what to make of at first.'

The seminar was held on 23 November. In the audience was a young Swiss student called Felix Bloch, who had started at the ETH the previous year with the intention to study engineering. But he found physics held the greater fascination, and he learned more from one of Debye's introductory classes than all his other courses combined. Schrödinger's colloquium on de Broglie's thesis was to prove particularly memorable, as he recalled more than fifty years later:

> So in one of the next colloquia, Schrödinger gave a beautifully clear account of how de Broglie associated a wave with a particle and how he could obtain the quantization rules of Niels Bohr and Sommerfeld by demanding that an integer number of waves should be fitted along a stationary orbit. When he had finished, Debye casually remarked that this way of talking was rather childish. As a student of Sommerfeld he had learned that, to deal properly with waves, one had to have a wave equation. It sounded quite trivial and did not seem to make a great impression, but Schrödinger evidently thought a bit more about the idea afterwards.

A few days before Christmas 1925, Schrödinger left Zurich for a short vacation in the Swiss Alps, returning once more to the villa in Arosa where he had stayed in 1922 and at which he and Anny had vacationed during Christmas 1923 and 1924. But his relationship with his wife was at an all-time low. He therefore chose to invite an old girlfriend from Vienna to join him, leaving Anny in Zurich. He also took with him his notes on de Broglie's thesis.

We do not know who the girlfriend was or what influence she might have had on him, but when he returned on 8 January 1926, he had discovered *wave mechanics*.

It is impossible to provide a physically rigorous derivation of the quantum mechanical Schrödinger wave equation starting from classical physics. It is possible, however, to follow Schrödinger's reasoning from notebooks he kept at the time. His starting point was the well-known equation of classical wave motion, which interrelates the space and time dependences of any wave form described by a wavefunction, typically given the symbol ψ (psi). He chose to work initially with the time-independent version of the wave equation, suitable for describing standing (rather than travelling) waves.

Into this classical wave equation he then substituted the relationships between the particle mass (more specifically, particle momentum) and wave frequency that had been deduced by de Broglie. He realized that in order to obtain meaningful solutions of the wave equation it would be necessary to make some assumptions about the form of the electron wavefunction—the mathematical description of the electron as a wave. He needed to restrict the range of functions that could be accepted. Specifically, the functions had to be single-valued (only one value for a given set of spatial coordinates), finite (no infinities), and continuous (no sudden 'breaks' or discontinuities).

Like de Broglie, he adopted a fully relativistic approach, seeking to derive a wave equation for the electron in a hydrogen atom that would necessarily conform to the requirements of Einstein's special relativity. He cast the wavefunction of the electron in a coordinate system more suited to the problem of motion in a spherical electrostatic field. This is a system of so-called spherical polar coordinates, which consist of the distance, r, between the electron and the nucleus, an angle θ (theta—a co-latitude), and a second angle φ (phi—a longitude). He found that he could factor the wavefunction into two separate functions, a 'radial' function depending only on r, and an angular, or 'spherical harmonic' function depending only on the angles θ and φ. The result was a series of highly complex differential equations.

It was here that he reached the limit of his mathematical competence. He could solve the differential equation for the spherical harmonics but struggled to solve the equation for the radial function. Nevertheless, he had seen enough. On 27 December he wrote a letter to Wien:

> At the moment I am struggling with a new atomic theory. If only I knew more mathematics! I am very optimistic about this thing and expect that if I can only... solve it, it will be *very* beautiful.... I hope I can report soon in a little more detailed and understandable way about the matter. At present, I must learn a little more mathematics...

He returned to Zurich on 8 January and immediately sought help from Weyl. Within about a week he had solved the differential equation for the radial function. Perturbed by the fact that he now had to deal with half-integral quantum numbers, he reverted to the very similar non-relativistic wave equation which yielded only integer values. The solution to all the mysteries now lay before him.

By constraining the properties of the three-dimensional electron wavefunction fitted around the central nucleus, a very specific pattern of possible solutions of the wave equation emerged naturally and directly. Schrödinger found that solutions for the radial function depend on two integer numbers, equivalent to the principal quantum number n and the quantum number k. Solutions for the angular function depend on k and m. The energies of the various solutions depend on the square of the principal quantum number, thereby reproducing the Balmer formula.

In Schrödinger's derivation, the only *ansatz* concerned the constraints he had placed on the form of the electron wavefunction. All the rest—the quantum numbers, their interrelationships, the hydrogen energy levels, the Balmer formula—was the result of irresistible mathematical logic. It was indeed *very* beautiful.

Schrödinger submitted a paper describing his results to *Annalen der Physik*. It was received on 27 January 1926, barely three weeks after he had made his initial discovery and just ten days after Pauli's paper had been received by the *Zeitschrift*. In his paper Schrödinger showed that the quantum numbers, introduced in *ad hoc* fashion by Bohr and Sommerfeld, emerged 'in the same natural way as the integers specifying the number of nodes in a vibrating string'.

The differential equations arising in Schrödinger's wave mechanics have a special property. A differential operator[3] operates on a function to yield the same function multiplied by some resultant quantity. In the case of the differential operator for the total energy of a dynamical system (called the Hamiltonian operator[4]), the quantity returned after the operation is the total energy of the system. The functions satisfying such equations are given the special name *eigenfunctions*, and the quantities produced as a result of the operation are called *eigenvalues*. Consequently, Schrödinger chose to title his paper 'Quantization as an Eigenvalue Problem'.

For Schrödinger, the triumph of wave mechanics meant an opportunity to put significant elements of classical mechanics back into the quantum picture. He now sought to replace discontinuous quantum jumps between unvisualizable electron orbits with a more classical, visually appealing picture of smooth, continuous transitions between the stationary wavefunctions of the system.

In the six months from January to June 1926, Schrödinger produced a series of six highly creative papers on the new wave mechanics, born from what Weyl called a 'late, erotic outburst in his life'. Among them was a paper in which Schrödinger demonstrated the mathematical equivalence between matrix and wave mechanics. In classical mechanics, the momentum (p) of a particle is given by its mass multiplied by velocity. In wave mechanics this expression for momentum is replaced by a differential operator. It is now accepted practice to distinguish between the classical p and the wave mechanical operator equivalent by drawing a caret,^, above the letter to indicate that it is now an operator, $\hat{p}$

Needless to say, the effect on a wavefunction of a series of operations depends on the order in which the operations are applied. In Schrödinger's wave mechanics, the effect of multiplying the wavefunction by the value

[3] An operator is simply an instruction to do something to a mathematical function, such as multiply it, differentiate it, etc.

[4] Named for the nineteenth century Irish mathematician William Rowan Hamilton.

of the position, q, and then applying the momentum operator, $\hat{p}$, is different from applying the momentum operator and then multiplying the result by the position. In other words $\hat{p}\, q\psi$ is not equal to $q\, \hat{p}\psi$. In wave mechanics the momentum operator is structured such that the position–momentum commutation relation, $\hat{p}\, q - q\, \hat{p}$, is equal to $-ih/2\pi$.

'I was absolutely unaware of any generic relationship with Heisenberg,' Schrödinger wrote in a footnote to his paper, 'I naturally knew about his theory, but because of the to me very difficult-appearing methods of transcendental algebra and because of the lack of *Anschaulichkeit* [visualizability], I felt deterred by it, if not to say repelled.' The feelings were mutual. As Heisenberg wrote to Pauli shortly after the publication of Schrödinger's equivalence paper: 'The more I think about the physical portion of the Schrödinger theory, the more repulsive I find it... What Schrödinger writes about the visualizability of his theory "is [paraphrasing Bohr] probably not quite right", in other words, it's crap.'

But, despite Heisenberg's misgivings, Schrödinger's *tour de force* began to win many adherents. Attention now turned to the electron wavefunctions themselves. What were they and how were they meant to be interpreted?

PART II

Quantum *Interpretation*

8

Ghost Field

Oxford, August 1926

As far as Schrödinger was concerned, the interpretation of the wavefunctions of his new wave mechanics was simple and straightforward. Although de Broglie had developed his ideas around the core concept of a duality of wave-like and particle-like behaviour, Schrödinger was happy to dispense with particles altogether. He viewed the wavefunctions as the very real manifestations of a completely undulatory material world. In his description, particle-like behaviour was an illusion created by the overlapping and reinforcing of collections of 'matter waves'.

What Schrödinger had in mind was a so-called 'wave packet' state. If a series of waves with high amplitude gathered around a specific point in space and time are added together, one on top of the other, the resultant wave may have a large amplitude peak concentrated at that point and little amplitude elsewhere. Such a wave packet would appear, to all intents and purposes, like a concentrated bit of matter—a particle. If such a collection, or 'superposition' of waves, were then to move collectively through space and time, it would carve out a path that would look like a particle trajectory.

Simple, perhaps, but it was in fact not so straightforward. Except for the special case that Schrödinger had considered, wave packet states are not generally sustainable unless their dimensions are much greater than the wavelengths of the waves that comprise them. This was not behaviour that could be expected of wave packets constructed on atomic scales. Dutch physicist Hendrik Lorentz argued that an electron wave packet would not hold together. It would disperse, dissolving rapidly into nothingness as its constituent waves went separate ways.

Schrödinger began to have doubts. There were also troublesome phenomena, such as the photoelectric effect, which he struggled to reconcile with his model of smooth, continuous transitions between stable wave states.

Meanwhile, in Göttingen, Max Born took a distinctly different view of the meaning of the wavefunctions. He came to reject Schrödinger's attempts to bring a classical physical perspective back into the heart of the atom. Although at the time Born's interpretation was not perceived as particularly radical by physicists of the Copenhagen and Göttingen schools, it was to provoke a debate about the nature of reality at the quantum level that continues to this day.

Max Born had studied mathematics at universities in Breslau, Heidelberg, and Zurich before completing his doctorate in theoretical physics at the University of Göttingen. At Göttingen he had come into contact with the 'mandarins' of mathematics: Felix Klein, David Hilbert, and Hermann Minkowski. He was quickly appointed as the note-taker for Hilbert's lectures, eventually becoming Hilbert's unpaid assistant.

Born went on to become professor of theoretical physics at the University of Berlin in 1915. It was here that he met Einstein, and they became close friends. Einstein had dragged Born from his sickbed in November 1918 to help him release the university's rector and deans, who had been imprisoned by student revolutionaries. In 1919 Born moved to Frankfurt, before taking up an appointment as Director of the newly established Institute for Theoretical Physics in Göttingen. The Institute became an important centre for theoretical physics which, like Bohr's Institute in Copenhagen, attracted a continual stream of esteemed visitors and bright students from around the world.

It had been Born who, in July 1925, had realized that Heisenberg's odd multiplication rule was, in fact, the rule for multiplying matrices. He was therefore one of the founders of quantum mechanics. And yet when he saw the new wave mechanics, he immediately recognized the utility of Schrödinger's approach and was initially enamoured of his attempt to restore a classical space–time description to quantum mechanics. But he was appalled by Schrödinger's attempt to eliminate quantum jumps.

Born had left Göttingen in November 1925 for a lecture tour of America. During a lecture at the Massachusetts Institute of Technology (MIT) in Boston, he emphasized the need for caution regarding the matrix approach:

> Only a further extension of the theory; which in all likelihood will be very laborious, will show whether the principles [of matrix mechanics] are really sufficient to explain atomic structure. Even if we are inclined to put faith in this possibility, it must be remembered that this is only the first step towards the solution of the riddles of the quantum theory.

On his return, he quickly reached for Schrödinger's wave approach to tackle problems concerning the nature of the interactions between quantum particles such as electrons and atoms. Both matrix and wave mechanics had been shown to be at least partially successful[1] in providing a framework within which the stable orbits of electrons bound in atoms could be understood and the positions and intensities of spectral lines could be predicted. These new quantum theories addressed the issue of structure, but they did not address issues concerning *transitions* (quantum jumps) between structures.

Born hoped that by formulating a quantum theory of collisions between electrons and atoms he would be well on the way to formulating a theory of the interaction between radiation (light quanta) and matter. In other words, he would have a theory that would incorporate quantum jumps into Schrödinger's wave mechanics.

This was the reason he had chosen to abandon matrix mechanics. Heisenberg's theory had been designed to describe the stationary states (the stable orbits) of electrons in atoms and allow predictions of the positions and intensities of spectral lines. It was not a theory that could be easily stretched to include collisions. Born had tried to apply matrix methods, in vain. Wave mechanics, however, proved to be far more amenable.

He quickly finalized a paper 'Quantum Mechanics of Collision Phenomena' which he submitted to *Zeitschrift für Physik* in June 1926. Although he had employed the methods of wave mechanics the paper contained the seeds of a radical re-interpretation of the wavefunctions.

Born had interpreted the collision between an electron and an atom as an interaction between a plane electron wave and an atom 'vibrating' at

[1] I say partially successful, as neither matrix nor wave mechanics had yet been cast in forms that were consistent with Einstein's special relativity and there was still no accounting for Pauli's exclusion principle or the anomalous Zeeman effect.

a characteristic frequency determined by its state. The result would be a complex vibration consisting of a superposition of these waves which would then separate, the electron wave 'scattering' as a consequence of the interaction. In the case of colliding billiard balls, the direction of scattering of one from the other can be predicted from the masses, speeds, and directions of the balls immediately prior to the collision. Born now saw that the wave interpretation eliminated this kind of predictability. He reasoned that the direct causal connection between the state of the electron and atom before and after the collision is lost.

In the wave theory of light, the connection between the square of the wave amplitude and the intensity of the light was well understood. In his papers Schrödinger had tried to establish a connection—through a 'heuristic hypothesis'—between the modulus-square of the amplitude of the wavefunction for a single electron and the density of electric charge.[2] Now Born was saying that the wavefunctions represented the *probabilities* that the electron wave will be scattered in certain directions: '...only one interpretation is possible, [the wavefunction] gives the probability for the electron, arriving from [a specific initial] direction, to be thrown out into [a final] direction.' In the proofs to this hastily written paper Born added a footnote: 'A more precise consideration shows that the probability is proportional to the square of [the wavefunction].'

Born subsequently claimed that he had been influenced by a remark that Einstein had made in one of his unpublished papers. In the context of light quanta interpreted using de Broglie's wave–particle ideas, Einstein had suggested that the waves represented a *Gespensterfeld*, a kind of 'ghost field', which determines the probability for the light quantum to follow a specific path. Born had therefore chosen to reject Schrödinger's attempts to provide a literal interpretation of the wavefunction as a real wave disturbance and, following Einstein's logic, instead regarded the wavefunction as a measure of the probability of realizing specific outcomes in a quantum transition, such as a collision.

But the reason that Einstein had not published his speculations is that this probabilistic interpretation has profound implications for the

[2] The modulus-square of the amplitude is the wavefunction amplitude multiplied by its complex conjugate, written $|\psi|^2$. If the wavefunction is not complex (i.e. if it does not contain *i*, the square-root of −1), then the modulus-square of the wavefunction is simply its square, ψ^2.

notions of causality and determinism, notions that Einstein held very dear. Born understood these implications only too well. In his June 1926 paper he wrote:

> Schrödinger's quantum mechanics therefore gives quite a definite answer to the question of the effect of the collision; but there is no question of any causal description. One gets no answer to the question: 'what is the state after the collision,' but only to the question, 'how probable is a specified outcome of the collision'...
>
> Here the whole problem of determinism comes up. From the standpoint of our quantum mechanics there is no quantity which in any individual case causally fixes the consequence of the collision; but also experimentally we have so far no reason to believe that there are some inner properties of the atom which conditions a definite outcome for the collision. Ought we to hope later to discover such properties... and determine them in individual cases? Or ought we to believe that the agreement of theory and experiment—as to the impossibility of prescribing conditions for a causal evolution—is a pre-established harmony founded on the nonexistence of such conditions? I myself am inclined to give up determinism in the world of atoms. But that is a philosophical question for which physical arguments alone are not decisive.

These paragraphs frame a debate that would rage for decades. If the wavefunctions carry only information about probabilities then they are not 'real' in the sense that Schrödinger imagined them to be real. If the only information available from quantum mechanics concerns the probabilities of certain outcomes, then causality and determinism are abandoned. In the realm of quantum transitions, we can no longer say: 'if we do this, then that will happen,' we can only say: 'if we do this, then that will happen with a certain probability.' It might be possible to restore causality and determinism if some new, but currently 'hidden', properties of quantum systems were to be uncovered that would allow a cause to be traced directly through to an effect. Born did not feel the need to invoke such hidden properties.

If the wavefunctions are not real then they are no longer obliged to behave as real systems would be expected to behave. At a stroke Born resolved many of the conceptual issues with Schrödinger's wave mechanics. There was now no need to invoke untenable wave packet states.

Schrödinger had also been troubled by the fact that his wavefunctions could be complex, i.e. they contained 'imaginary' numbers based on the quantity *i*, the square-root of minus one. For complicated systems containing two or more electrons, the wavefunctions had to be written not in three spatial coordinates representing three dimensions but in many coordinates representing many dimensions. The wavefunction of a system containing *N* particles depends on 3*N* position coordinates and is a function in a 3*N*-dimensional configuration space or 'phase space'. It is difficult to visualize a reality comprising imaginary functions in an abstract, multi-dimensional space. No difficulty arises, however, if the imaginary functions are not to be given a real interpretation.

In his hastily prepared June paper Born promised a more considered perspective. This duly arrived a month later. In this second paper he considerably sharpened and deepened his interpretation and acknowledged the inspiration he had drawn from Einstein's work. He considered a system whose wavefunction, as a result of a transition of some kind, can be represented as a superposition of two or more discrete eigenfunctions of the system, each mixed together in specific proportions. The proportion or amplitude of each eigenfunction ψ_n in the sum is determined by a 'mixing' factor c_n. Born now argued that the probability of the system to be in the state characterized by ψ_n after the transition is given by the modulus-square of its amplitude, $|c_n|^2$, by definition a number between zero and one.

In his June paper he had talked about the probabilities of transitions between states involving collisions between electrons and atoms. Now he was talking about the probabilities of the specific quantum states themselves.

By the time Born came to deliver a lecture to a meeting of the British Association at Oxford, England, in August 1926, his understanding was complete. His lecture was translated by American physicist J. Robert Oppenheimer, who at that time was working with Born's colleague James Franck in Göttingen, and subsequently published in the British journal *Nature* in 1927. In this lecture, for the first time Born drew a clear distinction between the statistical probabilities of classical physics and the quantum probabilities associated with the wavefunctions. He wrote:

> In classical dynamics the knowledge of the state of a closed system (the position and velocity of all its particles) at any instant determines unambiguously the future motion of the system; that is the form that the principle of causality takes in physics... But besides these causal laws, classical physics always made use of certain statistical considerations. As a matter of fact, the occurrence of probabilities was justified by the fact that the initial state was never exactly known; so long as this was the case, statistical methods might be, more or less provisionally, adopted.

We deal with probabilities in classical physics because we are often *ignorant* of the states of large, complex systems. A good example is Boltzmann's use of statistical methods to describe the properties of atomic and molecular gases. In such a situation, we can perhaps be confident that the principles of causality and determinism hold at the microscopic level for each individual particle undergoing a sequence of collisions, but we are not in a position to follow these motions experimentally. Instead, we deal with statistical averages.

Born now contrasted this situation with probability in quantum mechanics:

> The classical theory introduced the microscopic co-ordinates which determine the individual process, only to eliminate them because of ignorance by averaging over their values; whereas the new [quantum] theory gets the same results without introducing them at all. Of course, it is not forbidden to believe in the existence of these co-ordinates; but they will only be of physical significance when methods have been devised for their experimental observation.

Born concluded his lecture with the observation: '... the fundamental idea of waves of probability will probably persist in one form or another.'

This distinction between classical and quantum probability might seem trivial, or irrelevant. To the physicists of the Göttingen and Copenhagen schools, Born's interpretation seemed intuitive and obvious. It was no big deal. Consequently, when Heisenberg adopted the quantum probability interpretation in a paper submitted in November 1926, he did not feel the need to cite Born's June or July papers.

But this use of probability in quantum mechanics had removed a fundamental building-block of physical theory. Quantum mechanics appeared to provide a recipe for identifying the different possible outcomes of a transition and for establishing their relative probabilities. Applying the recipe was equivalent in many ways to stating the cause of the transition. There was, however, nothing in the theory to enable prediction of which of the various possible outcomes would be realized in actuality. Having established the cause, and the range of possible outcomes, the actual effect appeared to be left entirely to chance.

Some physicists were deeply troubled. Schrödinger was not persuaded by Born's arguments, as he explained in a letter to Wien:

> From an offprint of Born's last work in the *Zeitsch. f. Phys.* I know more or less how he thinks of things: the *waves* must be strictly causally determined through field laws, the *wavefunctions* on the other hand have only the meaning of probabilities for the actual motions of light- or material-particles. I believe that Born thereby overlooks that... it would depend on the taste of the observer which he now wishes to regard as real, the particle or the guiding field. There is certainly no criterion for reality if one does not want to say: the *real* is only the complex of sense impressions, all the rest are only pictures.

Most importantly, Schrödinger's opposition to the idea of quantum jumps remained unshaken.

Born had acknowledged his debt of inspiration to Einstein's *Gespensterfeld* in a letter to Einstein dated 30 November 1926. In his reply, Einstein summarized the nature and extent of his doubts:

> Quantum mechanics is very impressive. But an inner voice tells me that it is not yet the real thing. The theory produces a good deal but hardly brings us closer to the secret of the Old One. I am at all events convinced that *He* does not play dice.

Einstein, whose genius and insight had established the foundations for the construction of the new quantum theory, was rapidly transforming into one of the theory's most determined critics.

Born was dismayed by Einstein's reaction. An intense debate about the nature of reality at the quantum level now loomed.

9

All This Damned Quantum Jumping

Copenhagen, October 1926

Heisenberg was absolutely furious. The foundations of his new quantum mechanics were at risk of being undermined by Schrödinger's intuitively appealing and mathematically more familiar and accessible wave mechanics. As more and more members of the physics community began to defect—with great relief—to the wave approach, Heisenberg grew increasingly dismayed. Matrix mechanics was losing out.

There was more at stake here than the question of interpretation; of Schrödinger's stubborn attempts to eliminate quantum jumps and return to a classical space–time perspective. If wave mechanics came to be adopted as the 'deepest formulation of the quantum laws',[1] this not only tended to invalidate Heisenberg's personal experience of discovery on Helgoland but also thwarted his ambition to dominate the new field which he had helped to create. Heisenberg's future career was potentially at stake.

In April 1926 he had turned down the offer of an associate professorship in theoretical physics at the University of Leipzig in favour of a return to Bohr's Institute in Copenhagen, as a replacement for Hendrik Kramers, who had taken a professorial position in Utrecht. The tradition in Germany was for a young, aspiring academic always to accept the first offer of a professorial position. His father had advised him to accept. Virtually all of the esteemed senior physicists he had consulted had advised him to go to Copenhagen. He arrived back in Copenhagen in May 1926.

[1] In his June 1926 paper, Born had written: 'Of the different forms of the theory only Schrödinger's has proved suitable for [the description of a collision] and for this reason I might regard it as the deepest formulation of the quantum laws.'

Whilst he acknowledged the familiarity and utility of Schrödinger's mathematical formalism, he argued in a letter to Pauli in June 1926 that these methods should be deployed in the service of matrix mechanics, confined to the calculation of individual matrix elements. At the same time, he also sensed that the wave approach could provide some elements of physical interpretation that had thus far been lacking in matrix mechanics.

All was in flux. Heisenberg had to find a way to regain the initiative. He resolved to force the issue in a direct debate with Schrödinger.

The first opportunity for such a debate arose towards the end of July 1926. Schrödinger had been invited by Wien and Sommerfeld to give two lectures in Munich, one to the Bavarian section of the German Physical Society and the other to a select group of quantum physicists the next day. Heisenberg, settling into his new post as university lector and assistant to Niels Bohr in Copenhagen, had been busily familiarizing himself with wave mechanics. He took a brief holiday in Norway before arriving in Munich in July to stay with his parents. He attended both of Schrödinger's lectures armed with a raft of objections.

Heisenberg held his tongue as Schrödinger delivered his first lecture to a packed auditorium. But, as Schrödinger concluded his second lecture on the morning of Saturday, 24 July, Heisenberg stood to voice his objections.

Wave mechanics as conceived by Schrödinger, he argued, could not explain some relatively fundamental physical phenomena. It could not explain Planck's radiation law, or the Compton effect, or the intensities of spectral lines. These were all phenomena that could only be explained using the ideas of discontinuity and quantum jumps. And these were ideas that could not be accommodated in Schrödinger's continuous wave mechanics.

The response from the audience was not the one he might have hoped for. Wien motioned for him to shut up and sit down. Heisenberg later wrote that Wien:

> ...told me rather sharply that while he understood my regrets that [matrix] mechanics was finished, and with it all such nonsense as quantum jumps, etc., the difficulties I had mentioned would undoubtedly be solved by Schrödinger in the very near future. Schrödinger himself was not quite so certain in his own reply, but he, too, remained convinced that it was only a question of time before my objections would be removed.

These sentiments appeared to be shared by the majority of physicists in the audience. The elderly experimentalist Wien may never have forgiven the young theoretician Heisenberg's shirking of experiment and the theory of experimental instruments as he had studied for his doctorate in Munich three years before. Heisenberg's poor doctoral grade was testimony to the depth of Wien's anger with the upstart theoretician. But even Sommerfeld, who had defended Heisenberg's 'extraordinary abilities' in the ensuing row with Wien over the relative importance of theory and experiment, now seemed to have succumbed to the simple elegance of Schrödinger's mathematics.

Heisenberg was bitterly disappointed. Shortly after the lecture he wrote to Bohr about the unhappy outcome of his intervention. Bohr shared Heisenberg's misgivings, and decided to invite Schrödinger to Copenhagen to debate the issues further. Schrödinger agreed to visit Bohr towards the end of September 1926. Heisenberg sped back to Denmark.

The stage was set for the first round in a dialogue that would shape the future development of quantum theory, and with it our conception of physical reality.

Towards the end of August 1926, Schrödinger took Anny on a three-week holiday in the South Tirol. Anny then went to visit her parents in Salzburg while Schrödinger spent a few days with the Wiens before making the journey to Copenhagen towards the end of September.

The debate began even as Bohr and Heisenberg collected Schrödinger from the railway station in Copenhagen. Bohr and Schrödinger had not met before, and Schrödinger must now have wondered just what he had let himself in for. Bohr, normally considerate and friendly in his dealings with visitors, confronted Schrödinger like some kind of remorseless fanatic. Bohr would concede nothing. He and Heisenberg were right and Schrödinger was wrong, and it was now Bohr's mission to make Schrödinger see the light.

But Schrödinger, too, had deeply held convictions and an entrenched position on how he believed his wave mechanics should be interpreted. Moreover, he had been greatly encouraged by the warmth with which the physics community had received his approach. He was in no mood to back down.

'Surely you realise,' Schrödinger pleaded, 'that the whole idea of quantum jumps is bound to end in nonsense.'

He pointed to the difficulties of understanding precisely how an electron moving inside an atom was meant to make such discontinuous jumps. 'Is this jump supposed to be gradual or sudden?' he asked. 'If it is gradual, the orbital frequency and energy of the electron must change gradually as well. But in that case, how do you explain the persistence of fine spectral lines? On the other hand, if the jump is sudden ... then we must ask ourselves precisely how the electron behaves during the jump. Why does it not emit a continuous spectrum, as electromagnetic theory demands? And what laws govern its motion during the jump? In other words, the whole idea of quantum jumps is sheer fantasy.'

Bohr's response established the grounds for the real argument, which concerned the visualizability and understandability of physical theories.

'What you say is absolutely correct,' Bohr replied, 'But it does not prove that there are no quantum jumps. It only proves that we cannot imagine them, that the representational concepts with which we describe events in daily life and experiments in classical physics are inadequate when it comes to describing quantum jumps. Nor should we be surprised to find it so, seeing that the processes involved are not the objects of direct experience.'

But Schrödinger did not want to be drawn into a philosophical debate about the development and meaning of concepts. 'I wish only to know what happens inside an atom. I don't really mind what language you choose to discuss it,' he said.

If electrons were assumed to be tiny particles orbiting the nucleus, then Schrödinger expected that a physical theory should describe how these particles move in their orbits and make transitions from one orbit to another. But neither wave nor matrix mechanics could provide such a rational description. Change the picture to one of matter waves, Schrödinger argued, then everything suddenly looks very different, and the unresolved contradictions disappear.

'I beg to disagree,' Bohr insisted. 'The contradictions do not disappear; they are simply pushed to one side.' Bohr reiterated the problem of explaining Planck's radiation law using wave mechanics, and the fact that Planck's law demanded that an atom possess discrete energy values which change discontinuously. 'You can't seriously be trying to cast doubt on the whole basis of quantum theory?' he said.

Schrödinger had no choice but to concede the point. 'I don't for a moment claim that all these relationships have been fully explained,' he said. He suggested that the application of thermodynamics to wave mechanics may eventually provide some answers.

No, Bohr countered, one cannot hope for that. 'We have known what Planck's formula means for twenty-five years. And, quite apart from that, we can see the inconstancies, the sudden jumps in atomic phenomena quite directly, for instance when we watch sudden flashes of light on a scintillation screen or the sudden rush of an electron through a cloud chamber. You cannot simply ignore these observations and behave as if they did not exist at all.'[2]

'If all this damned quantum jumping were really here to stay,' an exasperated Schrödinger exclaimed, 'I should be sorry I ever got involved with quantum theory.'

Bohr tried to calm the discussion. 'But the rest of us are extremely grateful that you did,' he said, 'Your wave mechanics has contributed so much to mathematical clarity and simplicity that it represents a gigantic advance over all previous forms of quantum mechanics.'

Schrödinger stayed at the Bohrs' home and the debate raged day and night. Perhaps as a result of the enormous intellectual effort demanded of him, Schrödinger succumbed to a feverish cold. Bohr pursued him even to his sick-bed. As Bohr's wife Margrethe attempted to nurse him back to health with tea and cake, Bohr sat at the edge of the bed and continued his harangue: 'But you must surely admit that...'

But Schrödinger would admit nothing.

He was, nevertheless, deeply affected by the intensity of the debate and recognized that it was going to be necessary somehow to accommodate

[2] The cloud chamber was invented by Charles Wilson, a student of J.J. Thomson. As an energetic particle passes through such a chamber, it dislodges electrons from atoms in the vapour contained in the chamber, leaving charged ions in its wake. Water droplets condense around the ions, revealing the particle trajectory.

both the wave and particle perspectives. Wave mechanics alone, he now realized, couldn't be the answer.

In a letter to Wien he described his first encounter with Bohr:

> Bohr's...approach to atomic problems...is really remarkable. He is completely convinced that any understanding in the usual sense of the word is impossible. Therefore the conversation is almost immediately driven into philosophical questions, and soon you no longer know whether you really take the position he is attacking, or whether you really must attack the position he is defending.

But there were no hard feelings. This had been a stimulating, if somewhat exhausting, intellectual debate about matters of profound importance to the future development of quantum theory. In the same letter, Schrödinger continued:

> In spite of all [our] theoretical points of dispute, the relationship with Bohr, and especially Heisenberg, both of whom behaved towards me in a touchingly kind, nice, caring and attentive manner, was totally, cloudlessly, amiable and cordial.

The simple truth was that nobody could yet offer a coherent interpretation of quantum mechanics. However, Bohr and Heisenberg were convinced it was they who were on the right track.

The debate with Schrödinger had a singularly powerful effect on Bohr. Schrödinger's grasping for the kinds of wave concepts associated with classical physics and his relentless pursuit of a theory that provided some kind of space–time visualizability brought home to Bohr the precise nature of the problems they were all grappling with.

Schrödinger cherished the continuity and visualizability that was afforded by wave concepts, and simply could not see how discontinuous quantum jumps could fit with such a scheme. Bohr had argued that '... the pictorial concepts we use to describe the events of everyday life and the experiments of the old physics do not suffice to represent also the process of a quantum jump.'

Bohr and Heisenberg had abandoned any attempt to use classical space–time concepts to describe an essentially unvisualizable physics. But the very fact that wave mechanics had struck such a chord with physicists now suggested to Bohr that the use of classical concepts to describe decidedly non-classical quantum phenomena was not going to be so readily abandoned.

Perhaps, Bohr reasoned, there was a sense in which both classical wave and particle concepts were in some way equally valid; each helpful in describing some aspect of an otherwise unfathomable quantum reality depending on the specific phenomena that were being observed. Each was mutually exclusive to the other, yet both were needed for a complete description of the inner workings of the atom.

Heisenberg, too, paused to reflect. Inevitably, whilst he was content to embrace Born's probability interpretation, he was rather unhappy that Born had reached for wave mechanics as the 'deepest formulation of the quantum laws'. He felt that Born's conclusion still left too much room for alternative interpretations. Born, in the meantime, was beginning to regret the 'victory' of Schrödinger's approach over matrix mechanics, as the approach continued to drag Schrödinger's interpretation around closely behind it.

Heisenberg was now living in a small attic flat in Bohr's Institute, with slanting walls and windows which overlooked the entrance to Fælled Park. Bohr would come to the flat late at night, and the two would discuss their scientific problems into the small hours.

But Heisenberg disliked the approach that Bohr was developing. It seemed too vague, too philosophical, too arbitrary. He was uncomfortable with a theory that could be this or that, one thing then another. What he wanted was a theory that imposed a unique interpretation, if not of physical, sub-atomic entities such as electrons, then at least of the magnitudes of their observable properties, such as energy and momentum.

Although they disagreed on many points, Heisenberg had good reason to believe they were heading for the same result. He was astonished at how simple phenomena—such as the observed trajectory of an electron revealed in a cloud chamber—were proving so intractable. Indeed, the very concept of a trajectory was absent from matrix mechanics. Wave mechanics would have the electron matter wave spreading out, dispersing or dissolving as it moved through the chamber. And yet it was difficult for anyone who had looked at the track left by an electron in a cloud chamber not to be convinced of the reality of the electron's particle-like trajectory.

Heisenberg summarized this period of intense debate as follows:

> I remember discussions with Bohr which went through many hours till very late at night and ended almost in despair; and when at the end of the discussion I went alone for a walk in the neighbouring park I repeated to myself again and again the question: Can nature possibly be as absurd as it seemed...?

They reached no real conclusions. Their protracted debate left them exhausted and somewhat tense. When Bohr decided to go on a skiing holiday to Norway in February 1927, Heisenberg was greatly relieved.

He was left alone in Copenhagen, '...where I could think about these hopelessly complicated problems undisturbed.'

10

The Uncertainty Principle

Copenhagen, February 1927

Born had eagerly embraced Schrödinger's wave mechanics in June 1926, but by November his opposition had hardened. He told Schrödinger: 'It would have been beautiful if you were right. Something that beautiful happens, unfortunately, seldom in this world.' In a letter to Einstein he explained that: 'Schrödinger's achievement reduces itself to something purely mathematical; his physics is quite wretched.' Born would go on to champion the matrix approach long after it had ceased to be fashionable.[1]

But the Göttingen school remained a lone voice. Elsewhere, physicists began to use wave mechanics to duplicate results that had already been obtained using matrix mechanics. And, as Born had done, they also applied wave mechanics where matrix methods were inapplicable or regarded as too cumbersome.

Schrödinger's intransigence during their debate in Copenhagen in October 1926 left Bohr and Heisenberg more determined than ever to find a satisfactory resolution. Heisenberg may have wanted simply to retreat to matrix mechanics, as his colleagues in Göttingen had done, but his experiences in Munich and his long discussions with Bohr convinced him that a fresh approach was needed. It was time to go back to the drawing-board. It was time to think the unthinkable.

[1] Born and Pascual Jordan wrote a book on elementary quantum mechanics which was published in 1930. In it they made no reference at all to wave mechanics. Pauli wrote a withering review, finding only one feature about which to be positive: 'The setup of the book as far as printing and paper are concerned is splendid,' he wrote. Quoted in Beller, p. 38.

From the outset, Heisenberg had resolved to eliminate classical space–time pictures involving particles and waves from the quantum mechanics of the atom. He had wanted to focus instead on the properties that were actually observed and recorded in laboratory experiments, such as the positions and intensities of spectral lines. Left alone in Copenhagen in February 1927, he now pondered on the significance and meaning of such experimental observables. Somehow, he needed to introduce at least some form of 'visualizability' into matrix mechanics.

This train of thought was to lead him to another fundamental discovery.

In explaining his opposition to wave mechanics many years later, Born claimed that he: '...was witnessing the fertility of the particle concept every day in [James] Franck's brilliant experiments on atomic and molecular collisions and was convinced that particles could not simply be abolished.'

Indeed, the track left by an electron in a cloud chamber seemed to provide unambiguous evidence of the electron's particle-like properties. Yet this was something that Schrödinger's wave mechanics seemed unable to rationalize except in terms of completely untenable 'wave packets'. It was here that wave mechanics was most vulnerable.

The situation was very confusing, however. What Heisenberg needed to do was regain the 'high ground'. He needed to find a way to use matrix mechanics to formulate a specifically particulate description, with the attendant quantum jumping, thereby refuting Schrödinger's exclusively wave interpretation. This was not specifically about matrix *versus* wave mechanics, or about particles *versus* waves, as the matrix approach in any case sought to eliminate such classical concepts. This was about making a connection; it was about using the underlying quantum behaviour as detailed in matrix mechanics to describe large-scale particle-like properties that could in principle be observed in a laboratory. It was about doing what Schrödinger's wave mechanics could not.

He drew inspiration from dialogues with Pascual Jordan in Göttingen and with Dirac, who had arrived in Copenhagen in September 1926 for a six-month period as a visiting fellow. Pauli, too, provided some significant clues. In a letter Heisenberg had received in October 1926, Pauli had proposed to re-interpret the modulus-square of the wavefunction not as a transition probability or as the probability for the system to be in a specific state, as Born had done, but as the probability of 'finding' the electron at a specific position in its orbit inside an atom.[2] This gives rise to an image of a particulate electron which is 'smeared' through space with, at any one time, different probabilities of being found in different positions around the nucleus.[3]

In essence, Pauli was re-admitting a space–time description, by fusing together Schrödinger's wavefunction and an adaptation of Born's probability interpretation. Pauli, at least, was convinced of the need for such a connection. He wrote: 'I am now however, convinced with all the fervour of my heart that *the matrix elements must be connected with principally observable kinematic (perhaps statistical) data of the particles in question in the stationary states*.' This was a complete reversal of the position he had taken in his January 1926 paper on the matrix mechanics of the hydrogen atom.

Pauli had gone on in his letter to make an important observation concerning the relationship between position and momentum in quantum mechanics. He had considered the situation where two electrons collide. When the electrons are far apart they can be conveniently treated as plane waves each with a clearly defined position (q) and momentum (p). But as they come together they manifest what Pauli called a 'dark point', at which things become fuzzy. If the positions are assumed to be controlled, then the momenta are uncontrolled, and *vice versa*. 'One may view the world with the p-eye and one may view it with the q-eye, but if one opens both eyes at the same time one becomes crazy,' he wrote.

This was behaviour that could be traced back to the position–momentum commutation relation.

The matrix elements must be connected with principally observable kinematic data, Pauli had written, nearly four months earlier. Heisenberg now tried to use matrix mechanics to do precisely this; to describe the clearly visible path of an electron in a cloud chamber. He quickly ran into trouble:

[2] Strictly speaking, Pauli suggested that the probability of finding the electron between positions q and $q + dq$, where dq is an infinitesimal increment along the q-coordinate, is given by $|\psi(q)|^2dq$.

[3] This image remains with us today in the form of atomic 'orbitals', drawn as maps of electron density or as boundary surfaces within which there is a high probability (typically 90 per cent or more) of finding the electron.

> ... when I realised fairly soon that the obstacles before me were insurmountable, I began to wonder whether we might not have been asking the wrong sort of question all along. But where had we gone wrong? The path of the electron through the cloud chamber obviously existed; one could easily observe it. The mathematical framework of quantum mechanics existed as well, and was much too convincing to allow for any changes. Hence it ought to be possible to establish a connection between the two, hard though it appeared to be.

He then remembered a conversation he had once had with Einstein, following a lecture he had delivered on the new matrix mechanics in Berlin. Einstein had challenged the philosophical basis of the new theory, and particularly Heisenberg's insistence that it should deal only with the observable properties of atomic systems.

'But you don't seriously believe that none but observable magnitudes must go into a physical theory?' Einstein had protested.

Heisenberg charged that Einstein himself had done just this in formulating his theories of relativity. Einstein admitted that he might have done, but it was all still nonsense. 'It is the theory that decides what we can observe,' he continued, 'You must appreciate that observation is a very complicated process. The phenomenon under observation produces certain events in our measuring apparatus. As a result, further processes take place in the apparatus, which eventually and by complicated paths produce sense impressions and help us to fix the effects in our consciousness. Along this whole path—from the phenomenon to its fixation in our consciousness—we must be able to tell how nature functions, must know the natural laws at least in practical terms, before we can claim to have observed anything at all.'

The phenomenon under observation produces certain events in our measuring apparatus, Einstein had said. It was after midnight, but Heisenberg set off for a walk in Fælled Park. As he walked in the darkness he asked himself some fairly searching, fundamental questions, such as: what do we actually *mean* when we speak about the *position* of an electron? The track caused by the passage of an electron through a cloud chamber seems real enough—surely it provides an unambiguous measure of the electron's trajectory through space?

But wait. The track is made visible by the condensation of water droplets around atoms that have been ionized by the electron as it passes through the chamber. The droplets are much larger than the electron they have been used to 'detect', suggesting that the instantaneous position and velocity of the electron through the cloud chamber can, in truth, be known only approximately.

He had found the right question: 'Can quantum mechanics represent the fact that an electron finds itself approximately in a given place and that it moves approximately with a given velocity, and can we make these approximations so close that they do not cause experimental difficulties?' He returned to his room and quickly demonstrated to himself that he could, indeed, represent such approximations mathematically using matrix mechanics. He had found the connection he was looking for.

Heisenberg had discovered the uncertainty principle: the product of the 'uncertainties' in position and momentum cannot be smaller than Planck's constant h.[4] In other words, quantum mechanics places a fundamental limit on the precision with which both position and momentum can be jointly determined in any laboratory experiment. In an entirely classical world, where h is assumed to be zero, no such constraint exists.

Having deduced the uncertainty principle, Heisenberg now had to show that this was a fundamental principle which could not be violated. He developed a number of hypothetical 'thought' experiments, or *gedankenexperiments,* designed to illustrate how the principle is physically manifested. These were not necessarily meant to be taken seriously as proposals for real experiments; they were imaginary examples that built on the practical logic of the apparatus that would be required and its interactions with the object under study.

To talk about the position and momentum of any object, he reasoned, requires a clear, operational definition in terms of some experiment designed to measure these quantities. To illustrate this, he recalled a con-

[4] The term 'uncertainty' has since been more formally defined as 'indeterminacy', or root-mean-square deviation from the mean value of position or momentum. Also, more modern formulations of the uncertainty principle state that the product of these uncertainties cannot be smaller than $h/4\pi$.

versation he had had as a student in Göttingen. Supposing we wished to measure the path of an electron—its position and velocity (or momentum) as it passes through a cloud chamber or orbits a nucleus. The most direct way of doing this would be to follow the electron's motion using a microscope. Now the resolving power of an optical microscope increases with increasing frequency of radiation, and so a hypothetical gamma-ray microscope would be required to 'see' an electron in this way. The gamma-ray photons[5] bounce off the electron, and some are collected by a lens system and used to produce a magnified image.

But now, Heisenberg reasoned, we have a problem. Gamma-rays consist of high-energy photons. Each time a gamma-ray photon bounces off an electron, the operation of the Compton effect means that the electron is given a severe jolt. This jolt means that the direction of motion and the momentum of the electron are changed in ways that are governed by quantum mechanics. Only the probability for scattering in certain directions with certain momenta can be calculated, as Born had argued. Although we might be able to obtain a fix on the electron's instantaneous position, the sizeable interaction of the electron with the device we are using to measure its position means that we can say nothing at all about the electron's momentum. If we look with Pauli's 'q-eye' we cannot measure the electron's p.

We could use much lower-energy photons in an attempt to avoid this problem and so measure the electron's momentum, but the use of lower-energy (lower frequency or longer wavelength) photons would mean that we must then give up hope of determining the electron's position. If we look with Pauli's 'p-eye', we cannot measure the electron's q.

'Thus,' Heisenberg wrote, 'the more precisely the position is determined, the less precisely the momentum is known, and conversely. In this circumstance we see a direct physical interpretation of the equation $\mathbf{pq}-\mathbf{qp} = -i\hbar$.' When seeking to determine the path of an electron orbiting a nucleus: '...we have to illuminate the atom with light whose wavelength is considerably shorter than [the dimensions of the orbit]. However, a single photon of such light is enough to eject the electron

[5] In a speculative paper published in 1926, American chemist Gilbert N. Lewis had coined the name photon to describe a unit of radiant energy. The name was first applied to Einstein's light-quantum in 1927.

completely from its "path" (so that only a single point of such a path can be defined). Therefore here the word "path" has no definable meaning.'

Heisenberg extended similar arguments to the measurement of energy and time, and so deduced an equivalent energy–time uncertainty relation. This is usually interpreted in a practical sense to signify that the lifetime of an emission (the amount of time taken for the intensity of the light emitted by an atom in a higher-energy state to decay to some specific proportion of its initial intensity) will be uncertain by an amount related to the uncertainty in its energy. The uncertainty in the lifetime can be translated into an uncertainty in the exact moment of emission of a photon from a higher-energy state. In other words, the more sharply we can measure (in time) the lifetime and hence the moment of creation of a photon, or follow its passage through an apparatus, the more uncertain will be its energy and the energy of the state from which it was emitted, and *vice versa*.

Heisenberg's basic premise is that when making measurements on quantum scales, we run up against a fundamental limit. These are the same scales of distance and energy at which the primary measurement process itself operates. It is therefore not possible to make a measurement without disturbing the object under study in an essential, unpredictable, way. The discontinuity characteristic of quantum jumping dominates the process. At the quantum level, our techniques of measurement are simply too 'clumsy'. In this interpretation, quantum mechanics places fundamental limits on what is *measureable* and it is impossible to do anything other than speculate on what is not measureable.

The uncertainty principle sounded the death knell for causality, Heisenberg believed. The essential discontinuity at the heart of all quantum interactions means that it is impossible to predict with certainty the outcomes of such interactions. But, even in classical physics, the law of causality is itself flawed. Heisenberg wrote:

> 'When we know the present precisely, we can predict the future,' is not the conclusion but the assumption. Even in principle we cannot know the present in all detail. For that reason everything observed is a selection from a plenitude of possibilities and a limitation on what is possible in the future. As the statistical character of quantum theory is so closely linked

> to the inexactness of all perceptions, one might be led to the presumption that behind the perceived statistical world there still hides a 'real' world in which causality holds. But such speculations seem to us, to say it explicitly, fruitless and senseless. Physics ought to describe only the correlation of observations. One can express the true state of affairs better in this way: Because all experiments are subject to the laws of quantum mechanics..., it follows that quantum mechanics established the final failure of causality.

It is the theory that decides what we can observe, Einstein had said.

Nascent within Heisenberg's obituary for causality was another disturbing characteristic of measurement at the quantum level: everything observed is a selection from a 'plenitude of possibilities'. The measurement interaction creates a range of possible outcomes describable as a superposition of measurement eigenstates, with each ψ_n in the superposition contributing according to its amplitude, c_n. From all these possibilities there emerges only one actuality: the electron is 'here' or 'there', is scattered by this angle or that angle. Only the relative probabilities of each possible outcome can be known in advance and it is therefore impossible to predict with certainty the outcome actually obtained. This transition from many measurement possibilities to one measurement actuality would become known as the 'collapse of the wavefunction'.

Heisenberg wrote a long letter to Pauli, essentially the draft of a paper describing the results of his work on the uncertainty principle. The letter was dated 23 February 1927. He did not write to Bohr. 'I wanted to get Pauli's reactions before Bohr was back [from his skiing holiday],' he later explained, 'Because I felt again that when Bohr comes back he will be angry about my interpretation. So I first wanted to have some support, and see whether somebody else liked it.'

Pauli's reply was encouraging, and Heisenberg now wrote out his paper in full. But he was right to be worried about what Bohr would make of it all.

11

The 'Kopenhagener Geist'

Copenhagen, June 1927

The debate with Schrödinger had also left a deep impression on Bohr. It wasn't Schrödinger's stubbornness that had perplexed him. It was the realization that much of their difficulty had arisen simply because they had become mired in the concepts and the language of classical physics.

Schrödinger's words told of a realistic, visualizable, continuously variable world composed of classical waves. Heisenberg's words were of the world of the positivist, rejecting realism and visualizability in favour of particles and quantum discontinuity embodied within a predictive mathematical formalism. Bohr hovered between these extremes, perceiving the validity of both descriptions yet puzzled by the fact that he could find no words of his own to describe the quantum domain.

If the language of classical physics, of waves and particles, of causality, space–time, and continuity seemed inappropriate for describing the quantum world, then Bohr had to acknowledge that it remained the only language we have.

Bohr wrestled with the inherent contradictions in the classical wave and particle concepts. Schrödinger's successful approach spoke of the electron as a wave. Yet, as Born had observed of Franck's experiments in Göttingen, the particulate nature of the electron appeared undeniable. These behaviours were surely irreconcilable. In this experiment the electron is a wave—a non-localized disturbance extended over a region of space, which is neither exclusively 'here' nor 'there'. In this experiment the electron is a particle, a small, concentrated bit of electrically charged matter which at any one time can be in one location only—it is 'here' and nowhere else.

The mathematical apparatus of matrix or wave mechanics could not help to resolve this kind of conundrum. Whilst all this behaviour appeared contradictory, nature itself experienced no paradox. Bohr was determined to understand how this was possible.

He, too, used the break from his stressful dialogue with Heisenberg to do some independent thinking. Whilst on holiday in Norway he received a letter from Heisenberg dated 10 March 1927 which outlined the breakthrough he had made with the uncertainty principle.

Bohr returned from his skiing holiday about two weeks later. His thinking, though not yet complete, had reached an important stage of maturity. Heisenberg brought him up to speed with his discovery and shared the paper he had drafted. Bohr was initially thrilled with Heisenberg's new result but not with the methods he had used to arrive at it. The two physicists now became locked in a further fierce debate.

Sparks flew.

The tranquillity afforded by hours of skiing through the Norwegian countryside had given Bohr an opportunity for some uninterrupted, quality thinking. This thinking represented the culmination of at least two years' reflection on the interpretation of quantum phenomena, and he now reached an important conclusion.

The contradiction implied by the electron's wave-like and particle-like behaviours was more apparent than real, he decided. We reach for classical wave and particle concepts to describe the results of experiments because these are the only kinds of concepts with which we are familiar from our experiences as human beings living in a classical world. Bohr now realized that this, essentially classical, language was the only language available.

Whatever the 'true' nature of the electron, the behaviour it exhibits is conditioned by the kinds of experiments we choose to perform. These, by definition, are experiments requiring apparatus of 'classical' dimensions, resulting in effects substantial enough to be observed and recorded in the laboratory, perhaps in the form of an exposed photographic plate, or the deflection of a pointer in a voltmeter, or the observation of a track in a cloud chamber.

So, a certain kind of experiment will yield effects which we interpret, using the language of classical physics, as electron diffraction and interference. We conclude that in this experiment the electron is a wave. Another kind of experiment will yield effects which we interpret in terms of momentum transfer in a collision or a trajectory involving a localized electron. We conclude that in this experiment the electron is a particle. These experiments are mutually exclusive. We cannot conceive an experiment to demonstrate both types of behaviour simultaneously, not because we lack the ingenuity, but because such an experiment is simply inconceivable.

Constrained by our inability to construct experimental apparatus in anything other than classical dimensions, we are denied an insight into the 'true' nature of the quantum world. What we get instead is the quantum world as reflected in the mirrors of our classical apparatus. And, as the experimental questions we can ask will always be thus constrained, it is therefore pointless to enquire about the 'true' nature of quantum reality, as this is something we can never know.

This means that we can ask questions concerning the electron's wave-like properties and we can ask mutually exclusive questions concerning the electron's particle-like properties, but we cannot ask what the electron *really is*. What we are left to deal with is a fundamental wave–particle duality, a quantum world that is consistently different when we choose to reflect it in different classical mirrors.

Bohr resolved the dilemma of interpretation by declaring that these very different, mutually exclusive behaviours are not contradictory, they are *complementary*.

At first glance, Heisenberg's uncertainty principle appears entirely consistent with Bohr's reasoning. Indeed, Bohr may have immediately grasped the significance of the uncertainty relations in terms of complementary wave and particle concepts. But as he read through Heisenberg's paper, which Heisenberg had submitted to the *Zeitschrift* just a day before his return to Copenhagen on 23 March 1927, Bohr grew increasingly dismayed. Although the end result was compatible, the logic that Heisenberg had set forth in his paper betrayed a startlingly different philosophy.

'Then Niels Bohr returned from his skiing holiday,' Heisenberg wrote, 'and we had a fresh round of difficult discussions.'

There were several aspects of Heisenberg's treatment that Bohr found objectionable. For one thing, Bohr believed that the reasoning Heisenberg had used in his gamma-ray microscope *gedankenexperiment* was fatally flawed.

In grasping for a purely particulate interpretation, Heisenberg had traced the origin of uncertainty to the Compton effect, to an essential 'clumsiness' resulting from the substantial, discontinuous interaction between the electron and the gamma-ray photon being used to detect it. But, Bohr now pointed out, in principle the Compton effect gives rise to a precisely calculable recoil and is, in any case, applicable only to 'free' electrons (i.e. electrons that are not bound in an orbit around an atomic nucleus).

The origin of the uncertainty, Bohr now argued, should rather be traced to the wave nature of gamma-rays used to probe the properties of the electron. The resolution, or resolving power, of any microscope is limited by the effects of diffraction in the lens aperture. This diffraction results in a blurring of the image; an inability to distinguish objects that are closer than the minimum resolvable distance. Although the resolution increases as shorter and shorter wavelengths are used (thus requiring a microscope based on gamma-rays to resolve distances approaching the dimensions of an electron, as Heisenberg had assumed), the simple fact that the aperture must be of finite dimensions means that there remains a fundamental limit on the resolving power of the device. This loss of precision represents a fundamental uncertainty.

Bohr may have gone on to explain to Heisenberg that the *classical* relationships between the uncertainties derived from the theory of the resolving power of optical instruments allow both the position–momentum and energy–time uncertainty relations to be derived in a quite straightforward manner. The uncertainties in spatial extension and in reciprocal wavelength of a wave packet are such that their product cannot be smaller than unity. The spatial extension of the wave packet can be narrowed to precise dimensions by adding to the packet more and more waves of different frequency, or wavelength, concentrating the amplitude more and more sharply at a single point. In doing this, however, we lose precision in the wavelength (and, hence, reciprocal wavelength) of the wave packet. Alternatively, we can restrict the wave packet to a single, precisely known wavelength, but because this wave is extended over a region of space its amplitude is no longer focused on a single point. We therefore lose precision in its 'position'.

This relationship can be converted into the position–momentum uncertainty relation simply by replacing wavelength with momentum, using the relation $\lambda = h/p$ deduced by de Broglie. The result is that the product of the uncertainties in position and momentum cannot be smaller than h, as Heisenberg had deduced.

A second, equivalent, classical relation connects the uncertainties in frequency and time, such that their product cannot be smaller than unity. This can be simply understood in terms of the time interval within which the frequency of a classical wave form is 'sampled'. If the time interval is too short, not enough of the wave is sampled for a precise determination of its frequency. A precise determination of the frequency requires a time interval long enough to accommodate at least one up-and-down cycle, leading to a loss of precision in 'time'. The energy–time uncertainty relation can be derived from this classical relationship by substituting energy for frequency using Planck's relation $\varepsilon = h\nu$. The result is that the product of the uncertainties in energy and time cannot be smaller than h.

Heisenberg had almost failed to secure his doctorate at the University of Munich because he had been unable to derive expressions for the resolving power of a microscope, incurring the wrath of his examiner Wien, who had covered all the required background in his lectures. Heisenberg had been mortified by this experience, as had Sommerfeld, his thesis advisor. He now appeared to be struggling with the theory all over again.[1]

But Heisenberg was in any case extremely reluctant to admit any kind of wave interpretation into his discussion. Waves were identified with his rival Schrödinger, and his paper on uncertainty was a crusade against Schrödinger's continuous wave physics. To accept the legitimacy of a wave description, even within the context of Bohr's complementarity, was for him a step too far. In his paper he had rather wanted to stress both the particulate and essentially discontinuous nature of quantum phenomena.

[1] See Chapter 5, p. 45.

Bohr believed that Heisenberg was missing the point. He had by now realized that the position–momentum and energy–time uncertainty relations actually manifest the complementarity between the classical wave and particle concepts. Wave behaviour and particle behaviour are inherent in all quantum systems exposed to experiment and by choosing an experiment—choosing the wave mirror or the particle mirror—we introduce an inevitable uncertainty in the properties to be measured. This is not an uncertainty introduced through the 'clumsiness' of our measurements, as Heisenberg had argued, but arises because our choice of apparatus forces the quantum system to reveal one kind of behaviour rather than another.

This can be extended to a second level of complementarity, between an essentially causal description of phenomena in terms of momentum and energy, and a space–time description in terms of position and time. For as long as we do not interfere with it, an electron in a stationary state or stable orbit around a nucleus will behave according to causal laws, with predictable momentum and energy. But a space–time description, in which we fix the electron's position and time, requires some form of interaction which implies a discontinuity, precludes causality, and leaves us to deal with quantum probabilities.

According to Bohr, the uncertainty relations place a fundamental limit not on what is measureable, but on what is in principle *knowable*.

Bohr was insistent. Heisenberg's paper contained fundamental errors, it was premature, and considered only a special case for which a general rule in terms of complementarity could now be formulated. Heisenberg's paper—by now in press—should be withdrawn.

Heisenberg stubbornly refused to yield. As far as he was concerned, the error in the gamma-ray microscope *gedankenexperiment* did not undermine the basic argument of his uncertainty paper and he saw no reason to withdraw it. Young Swedish physicist Oskar Klein, recently arrived in Copenhagen on a visiting fellowship, was drawn into the argument. He took Bohr's side and, as can often happen in such super-heated circumstances, the argument became bitter and personal.

Their conflict was 'very disagreeable', Heisenberg later admitted. 'I remember that it ended with my breaking out in tears because I just couldn't stand this pressure from Bohr.' Heisenberg in his turn uttered regrettable statements, meant only to wound. Bohr sent a plea for help to Pauli and even offered to pay his travel expenses to Copenhagen. But Pauli was unavailable.

The debate wore on, relentlessly, through the spring of 1927. Bohr argued in favour of making the complementarity of waves and particles the centrepiece of uncertainty. Heisenberg saw no value in sticking to classical concepts with no demonstrable validity in the quantum domain. 'Well,' he argued, 'we have a consistent mathematical scheme and this consistent mathematical scheme tells us everything which can be observed. Nothing is in nature which cannot be described by this mathematical scheme.'

Bohr rejected the notion that nature follows a mathematical scheme. Heisenberg countered:

> Well, waves and [particles] are, certainly, a way in which we talk and we do come to these concepts from classical physics. Classical physics has taught us to talk about particles and waves, but since classical physics is not true there [at the quantum level], why should we stick so much to these concepts? Why should we not simply say that we cannot use these concepts with a very high precision, therefore the uncertainty relations, and therefore we have to abandon these concepts to a certain extent. When we get beyond this range of the classical theory, we must realise that our words don't fit. They don't really get a hold in the physical reality and therefore a new mathematical scheme is just as good as anything because the new mathematical scheme then tells what may be there and what may not be there. Nature just in some way follows the scheme.

But this wouldn't do. As far as Bohr was concerned our words have to fit, because understanding comes from the words we use in our descriptions, not from mathematical schemes. And our words are all we have.

'Naturally,' Heisenberg complained to Pauli in a letter dated 16 May, 'if one starts with [wave–particle duality], one can do everything without fear of contradiction.' But by now Heisenberg's resolve had been eroded by the force of Bohr's personality and conviction. He was now at least ready to accept that Bohr was right about the origin of uncertainty in the gamma-ray microscope experiment. In the same letter, he conceded to Pauli that: '...the [position–momentum uncertainty relation] indeed comes out naturally, but not entirely as I had thought.'

At Klein's suggestion, Heisenberg agreed to add a note in the proof of his soon-to-be published paper on uncertainty:

> In this connection Bohr has brought to my attention that I have overlooked essential points in the course of several discussions in this paper. Above all, the uncertainty in our observation does not arise exclusively from the occurrence of discontinuities, but is tied directly to the demand that we ascribe equal validity to the quite different experiments which show up in the [particulate] theory on one hand, and in the wave theory on the other hand.

The note corrected the misinterpretation of the gamma-ray microscope *gedankenexperiment*, and acknowledged the debt of gratitude to Bohr: 'I owe great thanks to Professor Bohr for sharing with me at an early stage the results of these more recent investigations of his—to appear soon in a paper on the conceptual structure of quantum theory—and for discussing them with me.'

Pauli finally found time to make the journey to Copenhagen in early June 1927. Wounds were healed, though the scars they left behind would be visible to the protagonists for some time to come. With Pauli's support, Bohr and Heisenberg achieved something of a rapprochement. This was not so much a united front, more a rather uneasy alliance between the very different interpretational views of the three physicists.

The pillars of this common-ground interpretation were the complementarity of waves and particles, the uncertainty principle, the interpretation of the wavefunction in terms of quantum probabilities, the correspondence between the eigenvalues of the wave theory and the measured values of observable quantities, such as momentum and energy, and the correspondence principle—the transition from quantum to classical behaviour in the limit of very large quantum numbers.

What was remarkable was the zeal with which the disciples of this new quantum orthodoxy embraced and spread the new gospel. Heisenberg spoke and wrote of the 'Kopenhagener Geist der Quantentheorie', the 'Copenhagen spirit' of quantum theory.

It became known as the *Copenhagen interpretation*.

12

There is no Quantum World

Lake Como, September 1927

By June 1927 Heisenberg had received word that he was once again to be called to a professorship in Leipzig. There followed a number of further job opportunities, in Halle, Munich, America, and Zurich. He would have greatly preferred Munich but this was an offer of an associate professorship under his former thesis advisor Sommerfeld, on the understanding that he would succeed Sommerfeld on the latter's retirement in about seven years' time. The opportunities in Leipzig and Halle were full professorships, however, and in German universities. As he had already turned down an associate professorship, protocol demanded that he now accept one of these offers. He agreed to take the Leipzig chair.

It was Debye, recently arrived in Leipzig from the ETH in Zurich, who had called him back to Germany. Pauli succeeded Debye at the ETH. Jordan succeeded Pauli in Hamburg. Klein took the position vacated by Heisenberg in Copenhagen. Within a matter of a few months, the disciples of the 'Kopenhagener Geist' had risen to tenured positions in prestigious universities, from where they could preach the Copenhagen doctrine to a new generation of quantum physicists.

There were still some voices of dissent, however. Schrödinger succeeded Planck at the University of Berlin. He arrived in Berlin, where he now joined fellow dissenter Einstein, shortly before his fortieth birthday in August 1927.

Bohr had won the argument with Heisenberg which put complementarity at the heart of the new interpretation of quantum theory. But his principle had yet to be articulated outside of the small group of physicists that had been involved in the debate.

An opportunity to present his ideas arrived in the form of an invitation to an international congress convened to commemorate the centenary of the death of Alessandro Volta, to be held in September 1927 at the Istituto Carducci on the shore of Lake Como in Italy.

He had begun work on his lecture in April, assisted by Klein. The process was tortuous. Bohr was driven by an insatiable desire for clarity in what was, he believed, to be a fundamentally important contribution to the philosophy of physics. He would establish the very basis on which we would attempt to make sense of physical reality at the quantum level. 'Bohr dictated and the next day all he had dictated was discarded and we began anew,' Klein explained. There were endless drafts.

As the Como meeting approached, Niels' brother Harald insisted that he get something properly written down.

It was an illustrious international group of physicists that assembled that September in the city of Como. Among the group arriving in Como were Born, de Broglie, Compton, Debye, the Italian physicist Enrico Fermi, Franck, Heisenberg, Max von Laue, Lorentz, Millikan, Paschen, Pauli, Planck, Rutherford, Sommerfeld, Stern, and Zeeman. There were some notable absences, however. Einstein had been invited, but could not attend. Schrödinger, too, was absent. As was Dirac.

Bohr stood to deliver his lecture on 16 September. The text of this lecture has not survived, although many earlier drafts have been preserved. It was subsequently published in a set of conference proceedings and reprinted in the British scientific journal *Nature* in 1928. He opened thus:

> ...I shall try, by making use only of simple considerations and without going into any details of technical mathematical character, to describe to you a certain general point of view which I believe is suited to give an impression of the general trend of the development of the theory from its very beginning and which I hope will be helpful in order to harmonise the apparently conflicting views taken by different scientists.

Although the arguments between Bohr and Heisenberg had hinged on the inclusion of a complementary wave description, in his lecture Bohr concentrated his attention on the complementarity of the causal and space–time perspectives. He said:

> On one hand, the definition of the state of a physical system, as ordinarily understood, claims the elimination of all external disturbances. But in that case, according to the quantum postulate [the essential discontinuity of quantum change], any observation will be impossible, and, above all, the concepts of space and time lose their immediate sense. On the other hand, if in order to make observation possible we permit certain interactions with suitable agencies of measurement, not belonging to the system, an unambiguous definition of the state of the system is naturally no longer possible, and there can be no question of causality in the ordinary sense of the word. The very nature of quantum theory thus forces us to regard the space–time co-ordination and the claim of causality, the union of which characterises the classical theories, as complementary but exclusive features of the description, symbolising the idealisation of observation and definition respectively.

Despite Bohr's efforts to achieve clarity in the endless redrafts he had worked on with Klein, his soft-spoken words remained vague and confusing. A highlight was his elegant derivation of the uncertainty relations from equations of classical wave optics, but the often obscure terminology he employed in his long and wordy lecture left his audience broadly unimpressed.

Hungarian physicist Eugene Wigner remarked to Belgian Léon Rosenfeld, one of Bohr's close associates: 'This lecture will not induce any one of us to change his own [opinion] about quantum mechanics.'

But the confusion of the moment could not mask the simple fact that complementarity—however Bohr chose to define it—represented a marked break with the past.

For the first time in its history, science was confronting what Bohr believed to be a profound limitation on our ability to acquire scientific knowledge. Classical physics had known no such limitation. The means by which we determine the magnitudes of physical quantities in our classical world has no bearing on the nature or the magnitudes of the quantities themselves. We quite happily assume that a classical object possesses these properties, with whatever magnitudes, whether we perform a measurement or not.

But now Bohr was claiming that it is not meaningful to regard a quantum wave–particle as having any intrinsic properties independent of some measuring device or means of observation. Although we may speak of the position of an electron, or its velocity, momentum, etc., as though these are independently existing properties, they are nevertheless properties that can only become 'real' when the electron interacts with an instrument specifically designed to reveal them. These concepts help us to correlate and describe our observations, but they have no meaning beyond their use as a means of connecting the object of our study with the instrument we use to study it.

'The quantum theory is characterised by the acknowledgement of a fundamental limitation in the classical physical ideas when applied to atomic phenomena,' Bohr said. 'The situation thus created is of a peculiar nature, since our interpretation of the experimental material rests essentially upon the classical concepts.'

Summarizing the situation many years later, Heisenberg put it this way:

> The Copenhagen interpretation of quantum theory starts from a paradox. Any experiment in physics, whether it refers to the phenomena of daily life or to atomic events, is to be described in the terms of classical physics. The concepts of classical physics form the language by which we describe the arrangement of our experiments and state the results. We cannot and should not replace these concepts by any others. Still the application of these concepts is limited by the relations of uncertainty. We must keep in mind this limited range of applicability of the classical concepts while using them, but we cannot and should not try to improve them.

The mathematical formalism of quantum theory then becomes an attempt to repackage complementary wave and particle descriptions in a single structure. This, Bohr insisted, does not imply that the theory is wrong or somehow incomplete.

It is, perhaps, difficult to resist the temptation to conjure up a mental picture of an individual electron existing in some kind of predetermined state independently of our measurements. But according to the Copenhagen interpretation, such a mental picture would be at best unhelpful and at worst positively misleading.

Those with some understanding of contemporary philosophy could detect a strong empiricist or even positivist flavour in the Copenhagen interpretation.

The empiricist tradition can be traced loosely from Scottish philosopher David Hume, through French philosopher Auguste Comte to the physicist Ernst Mach. Mach was professor of physics at the Universities of Prague and Vienna from 1867 to 1901. He argued that scientific activity involves the study of facts about nature revealed to us through our sensory perceptions and the attempt to understand their interrelationship through observation and experiment. According to Mach, this attempt should be made in the most economical way possible.

Mach rejected as non-scientific any statements that could be made about the world that cannot be verified by experiment or observation. He argued that there is no purpose to be served by seeking to describe a reality beyond our immediate senses. Instead, our judgement should be guided by the criteria of verifiability (does the theory agree with experiment or observation?) and simplicity (is it the simplest theory that will agree with experiment or observation?).

In constructing a physical theory we should seek the most economical way of organizing facts and making connections between them. We should not attach a deeper significance to the concepts used in a theory or the conceptual entities they describe if these are not in themselves observable. According to Mach, only those elements that we can perceive actually exist, and there is no point in searching for a physical reality that we cannot perceive: we can only know what we experience. Seeing is believing.

Mach's criterion of what constituted a verifiable statement was particularly stringent. It led him to reject the concepts of absolute space and absolute time, and to side with Ludwig Boltzmann's opponents in rejecting the reality of atoms and molecules.

Speculations that are intrinsically not verifiable, that involve some kind of appeal to the emotions or to faith, are not scientific. However, these speculations, which belong in a branch of philosophy called metaphysics (literally, 'beyond physics'), are not rejected outright. They are recognized as a legitimate part of the process of developing an attitude towards life, but they are perceived to have no place in science. This emphasis on verifiability and an unmerciful attitude to the elimination of metaphysics is a philosophical position generally known as *positivism*, a name first coined by Comte.

Mach's views were enormously influential in the development of a new school of philosophical thought that emerged in Vienna in the early 1920s. Centred around Moritz Schlick, professor of philosophy at the University of Vienna, German-born philosopher Rudolf Carnap, Austrian philosopher Otto Neurath, and others, the 'Vienna Circle' extended the positivist outlook through the use of formal logical analysis. They drew their inspiration from a wide variety of sources, particularly the work of the physicists Mach,[1] Boltzmann, and Einstein. Philosophically, their particular brand of positivism was foreshadowed in the work of Hume and Comte, and they were greatly influenced by the analytical approaches of their contemporaries Bertrand Russell in Cambridge and Ludwig Wittgenstein (a former student of Russell's) in Vienna.

The Vienna Circle began with the contention that the only true knowledge is scientific knowledge and that, in order to be meaningful, a scientific statement has to be both formally logical and verifiable. The foundations of their philosophy, which is sometimes known as *logical positivism,* was logical analysis, the criterion of verifiability and a strict demarcation between science and metaphysics.

Most importantly, the use of logical analysis leads to the elimination of all metaphysical statements as meaningless. With one stroke, the logical positivists eliminated from philosophy centuries of 'pseudo-statements' about mind, being, reality, and God. The views of the Vienna Circle came to dominate the philosophy of science in the middle of the twentieth century.

The implications for the development of physical theories are clear. Theories are merely instruments for making connections between observations or the results of experiments in the most economical way possible. If they describe the behaviour of entities that we cannot directly perceive, then the entities themselves are merely convenient theoretical devices and should not be misinterpreted as elements of an independent reality.

This does not necessarily mean that there is no such thing as reality. Scientific theories describe elements of an *empirical reality*—the reality manifested as effects that we can directly perceive and hence verify—but

[1] They called themselves the 'Ernst Mach Society'.

don't expect to be able to go beyond this empirical level. To do so is to engage in meaningless speculation.

Of the physicists of the Copenhagen school, Heisenberg was the positivist. It had been Heisenberg who in his paper on matrix mechanics had insisted on a new language, insisted on a rejection of classical concepts and their replacement with a consistent mathematical scheme. Many years later he wrote:

> Our actual situation in research work in atomic physics is usually this: we wish to understand a certain phenomenon, we wish to recognise how this phenomenon follows from the general laws of nature. Therefore, that part of matter or radiation which takes part in the phenomenon is the natural 'object' in the theoretical treatment and should be separated in this respect from the tools used to study the phenomenon. This again emphasises a subjective element in the description of atomic events, since the measuring device has been constructed by the observer, and we have to remember that what we observe is not nature in itself but nature exposed to our method of questioning.

Although they were not particularly concerned to devote much time to the elaboration of their philosophy, the physicists of the Copenhagen school were nevertheless aware that their interpretation created considerable problems for the understanding of what constitutes knowledge at the quantum level and the methods of its acquisition. Heisenberg in particular made it his business to raise awareness of these issues in his many public addresses on the subject.[2]

The Copenhagen interpretation essentially states that in quantum theory we have reached the limit of what we can know. To try to go beyond this limit is pointless: how can we ever hope to know something that is unknowable? The argument is that any attempt to introduce a new concept to describe an underlying independent reality inevitably involves a reworking of familiar classical concepts and a descent into metaphysics.

[2] Eventually, the philosophers began to take note and, in correspondence spanning the years 1930–32, Schlick of the Vienna school sought advice and direction from Heisenberg on quantum theory's implications for causality and the philosophy of knowledge. Clearly, the positivists of the Vienna Circle found much resonance with what the Copenhagen physicists were now saying.

We always return to the idealized concepts that summarize the fullest extent of our knowledge—waves and particles.

This interpretation requires that we accept that we can never 'know' quantum concepts. They are simply beyond human experience and are therefore metaphysical. A quantum entity is neither a wave nor a particle. Instead we substitute the appropriate classical concept—wave or particle—as and when necessary.

Compare the statement credited to Bohr:

> There is no quantum world. There is only an abstract quantum physical description. It is wrong to think that the task of physics is to find out how nature is. Physics concerns what we can say about nature.

with the following comment on logical positivism by British philosopher A.J. Ayer:

> The originality of the logical positivists lay in their making the impossibility of metaphysics depend not upon the nature of what could be known but upon the nature of what could be said.

A careful analysis of Bohr's philosophical influences and his writings on the Copenhagen interpretation and complementarity suggest that philosophically he was closer to the tradition known as *pragmatism* than to positivism.[3] Pragmatism, founded by the American philosopher Charles Sanders Pierce, traces its lineage to German philosopher Georg Hegel and has many of the characteristics of positivism in that they both roundly reject metaphysics. There are differences, however.

The positivist doctrine of 'seeing is believing' means that what we can know is limited by what we can observe. The pragmatist doctrine admits a more practical (or, indeed, pragmatic) approach to the reality of entities—such as electrons—whose properties and behaviour are described by theories and which produce secondary observable effects but which themselves cannot be seen. According to the pragmatist, what we can know is limited not by what we can see, but by what we can *do*.

[3] See, for example, Murdoch, pp. 231–232.

It seems logical that the father of modern atomic theory would want to accept the reality of atoms. But in his Como lecture Bohr placed limits on what a theory of the internal structure of the atom could say. He argued that we live in a classical world and our experiments are classical experiments. Go beyond these concepts and you cross the threshold between what you can know and what you cannot.

Positivist or pragmatist, the most important feature of Bohr's philosophy is that it was principally *anti-realist*. It denied that quantum theory has anything meaningful to say about an underlying physical reality that exists independently of our measuring devices. It denied the possibility that further development of the theory could take us closer to some as yet unrevealed truth. 'Accordingly, an independent reality in the ordinary physical sense can neither be ascribed to the phenomena nor to the agencies of observation,' Bohr said.

Einstein and Schrödinger, both realists, were not going to be pleased. But neither was present in Como. There was not long to wait, however. Much of the illustrious group gathered in Como convened again a month later, at the fifth Solvay Congress in Brussels, where Bohr was to repeat his lecture on complementarity.

This time, both Einstein and Schrödinger would be present.

PART III

Quantum *Debate*

13

The Debate Commences

Brussels, October 1927

The first Solvay conference on physics was held in 1911. It was sponsored by Planck and Nernst, chaired by Lorentz, and addressed problems relating to the theories of radiation and of specific heat capacities. It was supported by wealthy Belgian industrialist and philanthropist Ernest Solvay. It was the first invitation-only international conference on physics. Such was its success that a year later Solvay founded the International Institutes for Physics and Chemistry in Brussels.

The fifth Solvay conference, on 'Electrons and Photons', was also chaired by Lorentz. It was to be his last public appearance. He died a few months later, in February 1928. Despite the title of the conference, the invitation made it clear that the principal subject of discussion was to be the new quantum mechanics.

Lorentz had begun to work on the agenda for the conference with other members of the scientific planning committee as early as April 1926. He had had to revise his opinions (and extend his invitation list) in response to the rapid developments in the field that had occurred in the intervening months. So it was that the founders of quantum theory—Planck, Einstein, Bohr, de Broglie—and the new generation of quantum 'mechanics'—Born, Heisenberg, Pauli, Schrödinger, Dirac—were all to be in attendance.

Finally, on Monday 24 October 1927, Einstein, Bohr, and many other leading physicists gathered at the Institute of Physiology in Brussels to hear Lorentz's opening remarks and the first lecture of the conference, by British physicist William L. Bragg. Bohr later remarked: '...several of us came to the conference with great anticipations to learn his [Einstein's] reaction to the latest stage of the development which, to our view, went far in clarifying the problems which he had himself from the outset elicited so ingeniously.'

Bohr was about to discover just how ingenious Einstein could be.

Bragg's lecture on X-ray reflection was followed on the Monday afternoon by a report from Compton, recently awarded the Nobel Prize for the discovery of the effect which carries his name.

The following afternoon de Broglie stood to present a radical alternative approach to the interpretation of quantum phenomena based on the idea of a 'double solution'. In effect, de Broglie was interpreting wave–particle duality not in terms of waves *or* particles, but in terms of waves *and* particles. In this approach, the motions of real, individual quantum point-particles are guided along their paths by a real wave field. The second field of the double solution has the same statistical significance as the wavefunction in Schrödinger's wave mechanics and is open to the same probabilistic interpretation:

> The wave ψ then appears as both a *pilot wave* (Führungsfeld of Mr Born) and a *probability wave*. Since the motion of the [particle] seems to us to be strictly determined..., it does not seem to us that there is any reason to renounce believing in the determinism of individual physical phenomena, and it is in this that our conceptions, which are very similar in other respects to those of Mr Born, appear nevertheless to differ from them quite markedly.

It was a bold solution, one that potentially undermined complementarity and rendered pointless all the energy that had been expended on formulating and debating it. Not surprisingly, perhaps, it was not greeted with universal enthusiasm. If de Broglie had been looking for support from Einstein, he was disappointed. Einstein remained silent.

It seems that Einstein had explored a similar approach himself some months earlier. On 5 May 1927 he had read a paper to the Prussian Academy of Sciences in Berlin entitled 'Does Schrödinger's Wave Mechanics Determine the Motion of a System Completely or Only in the Sense of Statistics?' His modification of the theory was essentially a synthesis of classical wave and particle descriptions, with the wavefunction of Schrödinger's mechanics taking the role of a guiding field, guiding physically real point-particles.

Whilst Einstein was excited by his result at the beginning of May, his enthusiasm for it had evaporated just a few weeks later. On 21 May he requested that his paper, by then in print, should be withdrawn. He had begun to have doubts about attaching physical significance to a wave-function defined in a multi-dimensional configuration space. Perhaps most importantly, he further noted that whilst in this modification the particles themselves remained localized, they could still experience non-local influences from the guiding field. Such influences had implications for causality of just the kind he had been seeking to avoid.

Having committed to presenting a paper at the Solvay conference, on 17 June he had written to Lorentz begging forgiveness: 'After much reflection back and forth,' he wrote, 'I come to the conclusion that I am not competent [to give] such a report in the way that really corresponds to the state of things.'

The next morning Born and Heisenberg set out the Copenhagen–Göttingen 'stall'. In a double-act, they described matrix mechanics, appropriately extended to make the connection with Schrödinger's wave mechanics, the interpretation in terms of quantum probabilities, Heisenberg's uncertainty relations, and numerous applications of the theory. They argued provocatively that quantum theory was complete, with no further revision of its basic physical and mathematical hypotheses required:

> ...we consider *quantum mechanics* to be a closed theory, whose fundamental physical and mathematical assumptions are no longer susceptible of any modification.

In the afternoon, Schrödinger delivered the last of the formal lectures, on wave mechanics. Wary of a showdown with the 'matricians', he kept to points of principle, elaborating those aspects of wave mechanics that had caused concern, such as the meaning of waves in configuration space, and suggesting different approaches to interpretation. In the subsequent discussion, Heisenberg rejected Schrödinger's suggestion that future developments might result in a more conventional theory in four space–time dimensions: 'I see nothing in Mr Schrödinger's calculations that would justify this hope,' he said.

There followed an interruption to the proceedings in Brussels, due to an unfortunate clash with another scientific conference organized by the Académie des Sciences in Paris to commemorate the centenary of the death of French physicist Augustin Fresnel. By the time Lorentz had discovered the clash, it was too late to reschedule. He proposed a compromise. The Solvay conference was suspended for a day. Those conference participants who wished to attend the Fresnel celebrations could do so, returning to Brussels on Friday to resume their discussions.

Consequently, Lorentz opened the meeting for general discussion on Friday morning, expressing some of his own views and making an old physicists' plea for causality and determinism before calling on Bohr to address the conference. Bohr described his concept of complementarity, much as he had presented it at Como, but now directing his statements to Einstein who was hearing this argument for the first time.[1] Einstein did not respond immediately.

Eventually, Einstein stood to make some comments. 'Despite being conscious of the fact that I have not entered deeply enough into the essence of quantum mechanics, nevertheless I want to present here some general remarks,' he said.

He referred to a general experiment involving the diffraction of a beam of electrons or photons through a narrow slit.[2] The diffraction pattern appears on a second screen and is recorded (for example using photographic film). If quantum theory is assumed to be a complete theory of individual processes, the behaviour of each individual quantum particle is described by an appropriate wavefunction and it is the properties of the wavefunction that give rise to the diffraction pattern. However, at the moment the wavefunction impinges on the second screen, it 'collapses' instantaneously, producing a localized spot on the screen which indicates 'a particle struck here'.[3]

[1] Bohr did not deliver a formal lecture to the fifth Solvay conference. At his request, a translation of his *Nature* paper on complementarity was appended to the formal conference proceedings.

[2] I have elaborated Einstein's example somewhat to make it more readily understandable.

[3] Photographic emulsion is made up of millions of tiny crystals of a silver salt (called silver halides). The photon interacts with a silver salt crystal, and the crystal (and some of its neighbours) break down to give a black silver deposit. Developing chemicals are then used to break down more crystals and so amplify the initial deposit to create a visible image. The film is treated with further chemicals to convert any remaining silver halides into colourless salts, so the film is no longer sensitive to light. This is the negative. Light is then passed through the negative onto light-sensitive paper to create the positive.

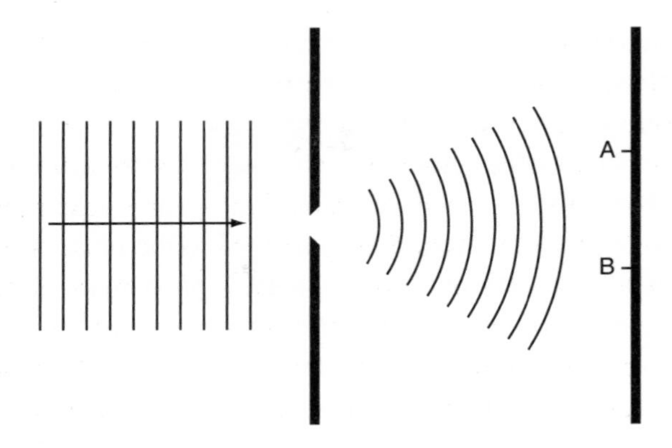

FIG 4 The simple electron or photon diffraction experiment cited by Einstein in his debate with Bohr. When we detect the particle at position A on the second screen, we learn instantaneously that it did not arrive at a second location, (B). Einstein argued that this 'collapse of the wavefunction' violates the postulates of special relativity.

Einstein objected to this way of looking at the process. Suppose, he said, that the particle is observed to arrive at some position A on the second screen. In making this observation, we learn not only that the particle arrived at A but also that it definitely did not arrive at a second location on the screen, B. What's more, we learn of the particle's non-arrival at B instantaneously with the observation of its arrival at A. Before observation, the probability of finding the particle is, supposedly, 'smeared out' over the whole screen.

He sensed that the collapse of the wavefunction implies a peculiar 'action at a distance'. The particle, which is somehow distributed over a large region of space, becomes localized instantaneously, the act of measurement appearing to change the physical state of the system far from the point where the measurement is actually recorded. He felt that this kind of action at a distance violated the postulates of special relativity.

With a nod to de Broglie's double solution approach, he said: 'In my opinion, one can remove this objection only in the following way, that one does not describe the process solely by the Schrödinger wave, but that at the same time one localizes the particle during the propagation. I think that Mr de Broglie is right to search in this direction. If one works solely with the Schrödinger waves [assuming quantum theory describes individual processes], $|\psi|^2$ implies to my mind a contradiction with the postulate of relativity.'

There is an alternative description, however. What if the wavefunction represents a probability amplitude not for a single quantum particle, but for a large collection of identically prepared particles, called an *ensemble*? According to this view, each individual particle passes through the slit along a defined, localized path, to arrive at the second screen. There are many such paths possible, and the diffraction pattern thus reflects the statistical distribution of large numbers of particles each following different but defined paths. This distribution is related to the modulus-square of the wavefunction, which expresses the probability for many particles rather than a probability for each individual particle.

We cannot discriminate between these possibilities by observing what happens to an individual quantum particle. Both descriptions say that one particle passes through the slit to arrive at one specific location on the second screen. In the first description, the point of arrival is determined at the moment the particle interacts with the photographic film, with a probability given by the modulus-square of the wavefunction. In the second description, the point of arrival is determined by the actual path that the particle follows, which is in turn obtained from a statistical probability given by the modulus-square of the wavefunction. In both cases we see the diffraction pattern only when we have detected a large number of particles.

Bohr wasn't sure what to make of Einstein's remarks. 'I feel myself in a very difficult position because I don't understand what precisely is the point which Einstein wants to [make],' he said. 'No doubt it is my fault.'

He went on to explain that: 'I do not know what quantum mechanics is. I think we are dealing with some mathematical methods which are adequate for description of our experiments.' In his view, Einstein was trying to hang on to an essentially classical, space–time description, one that had to be recognized as no longer tenable. 'The whole foundation for causal spacetime description,' he continued, 'is taken away by quantum theory, for it is based on assumption of observations without interference.'

Bohr had missed the point, but Einstein was not about to give up. The discussion continued in the dining room of the Hôtel Britannique, where the conference participants were staying. This was the scene of the opening skirmishes in one of the most important scientific debates ever witnessed, as Einstein directly challenged Bohr over the meaning of quantum theory. At stake was the way we attempt to understand the very nature of physical reality.

Otto Stern described what happened:

> Einstein came down to breakfast and expressed his misgivings about the new quantum theory, every time [he] had invented some beautiful experiment from which one saw that [the theory] did not work…Pauli and Heisenberg, who were there, did not pay much attention, 'ach was, das stimmt schon, das stimmt schon' [ah well, it will be all right, it will be all right]. Bohr, on the other hand, reflected on it with care and in the evening, at dinner, we were all together and he cleared up the matter in detail.

Einstein now chose to attack the Copenhagen interpretation by attempting to show that as a result of its incompleteness, one of the theory's principal foundations—the physical meaning of the uncertainty relations—is inconsistent. The debate took the form of a series of puzzles, developed by Einstein as *gedankenexperiments*, not to be taken too literally as practical experiments that could be carried out in the laboratory. It was enough for Einstein that the experiments could be conceived and carried out in principle.

Einstein asked the assembled audience what might happen if a quantum particle passed through a slit in a screen under conditions where the transfer of momentum between the particle and the screen is carefully controlled and observed. A particle hitting the screen as it passes through the slit would be deflected, its path beyond being determined by the conservation of momentum. We can tell in which direction the particle has been deflected by watching the deflection of the screen itself.

Now, he said, imagine that we insert a second screen—one with two slits—between the first screen and a piece of photographic film. If we are able to discover the direction in which the particle is deflected by the first screen, we can further learn which of the two slits the particle subsequently passed through. From the position on the film in which it

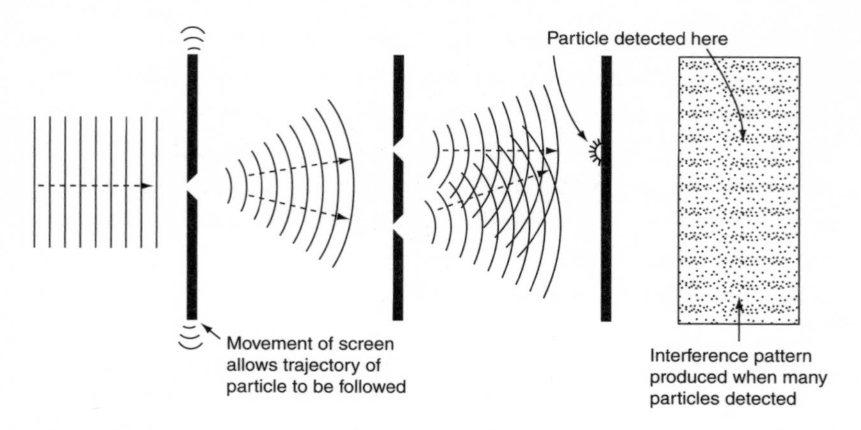

FIG 5 Einstein's variation on the classic double-slit experiment. By passing a particle through a single, narrow slit in the first screen and observing which way the screen moves, we can tell in which direction (and therefore towards which slit in the second screen) the particle is deflected. By observing where the particle subsequently strikes the third screen we can then trace its trajectory through the entire apparatus. If we now allow many particles to pass through the apparatus, one after the other, the double-slit interference pattern should in principle become visible . This thought experiment seems to allow us to observe particle-like properties (such as a trajectory) and wave-like properties (interference) simultaneously, in contradiction to Heisenberg's uncertainty principle.

is ultimately detected, we can trace the particle's trajectory through the entire apparatus.

We now leave the apparatus to detect a large number of particles—one after the other—from which we expect to see a double-slit interference pattern. Thus, Einstein concluded, we can demonstrate the particle-like (defined trajectory) and wave-like (interference) properties of quantum particles simultaneously, in contradiction to Bohr's notion of complementarity and the Copenhagen interpretation of the uncertainty relations. This proves that the Copenhagen interpretation is internally inconsistent.

Bohr reflected very carefully on Einstein's challenge. He took the thought experiment a stage further, sketching out in a pseudo-realistic style the kind of apparatus that would be needed to make the measurements to which Einstein referred. His purpose was not to try to imagine how the experiments could be done in practice, but primarily to focus on what he saw to be flaws in Einstein's arguments.

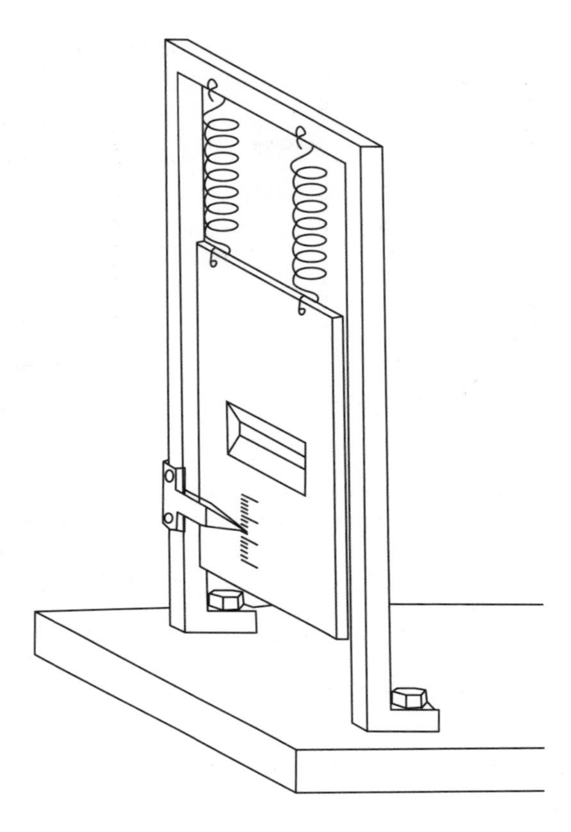

FIG 6 The hypothetical apparatus sketched by Bohr to demonstrate how measurement of the deflection of the first screen might be made. Reprinted by permission of Open Court Publishing Company, a division of Carus Publishing Company, Chicago, IL, from *Albert Einstein: Philosopher-scientist*, Vol. I, p. 220, edited by Paul Arthur Schilpp, copyright © 1949 by the Library of Living Philosophers, Inc.

Controlling and observing the transfer of momentum from the quantum particle to the first screen requires that the screen be capable of movement in the vertical plane. Observing the recoil of the screen in one direction or the other as the particle passes through the slit would then allow the experimenter to draw conclusions about the direction in which the particle had been deflected, as Einstein claimed.

Bohr envisaged a screen suspended by two weak springs. A pointer and scale inscribed on the screen allows the measurement of the amount of movement of the screen, and hence the momentum imparted to it by the particle. The fact that Bohr had in mind a macroscopic apparatus presents no problem, provided we assume that the apparatus is sufficiently sensitive to allow observation of individual quantum events. This sensitivity is important, as we will see.

Bohr had to demonstrate the consistency of the uncertainty principle, and hence complementarity, when applied to the analysis of this kind of thought experiment. Controlling the transfer of momentum to the screen in the way Einstein suggested *must* imply a concomitant uncertainty in the screen's position in accordance with the uncertainty principle. If we measure the screen's momentum in the vertical plane with a certain precision, then there must result an uncertainty in the position of the screen such that the product of the uncertainties is not less than Planck's constant, h.

Why should this be? Bohr's answer was that, in order to read the scale inscribed on the first screen sufficiently accurately, it has to be illuminated. This illumination involves the scattering of photons from the screen and hence an uncontrollable transfer of momentum, or an uncontrollable 'jolt'. 'Since, however, any reading of the scale, in whatever way performed, will involve an uncontrollable change in the momentum of the [screen], there will always be, in conformity with the [uncertainty] principle, a reciprocal relationship between our knowledge of the position of the slit and the accuracy of the momentum control.'

Bohr was able to show that the resulting uncertainty in the position of the slit in the first screen destroys the *phase coherence* of the particle's associated wave as it spreads out beyond the double slits in the second screen. The interference pattern is therefore 'washed out'. Controlling the transfer of momentum from the particle to the first screen allows us to follow the trajectory of the particle through the apparatus, but prevents us from observing interference effects, in accordance with the complementary nature of the wave and particle descriptions.

He concluded:

> ...we are presented with a choice of *either* tracing the path of a particle *or* observing interference effects, which allows us to escape from the paradoxical necessity of concluding that the behaviour of an electron or

> a photon should depend on the presence of a slit in the [second screen] through which it could be proved not to pass. We have here to do with a typical example of how the complementary phenomena appear under mutually exclusive experimental arrangements and are just faced with the impossibility, in the analysis of quantum effects, of drawing any sharp separation between an independent behaviour of atomic objects and their interaction with the measuring instruments which serve to define the conditions under which the phenomena occur.

Einstein presented further *gedankenexperiments.* He could not shake his deeply felt misgivings about the Copenhagen interpretation and forced Bohr to defend it. But Bohr rebuffed all his challenges. 'On his side, Einstein mockingly asked us whether we could really believe that the providential authorities took recourse to dice-playing ("... *ob der liebe Gott würfelt*"),' Bohr wrote, 'to which I replied by pointing at the great caution, already called for by ancient thinkers, in ascribing attributes to Providence in every-day language.'

Bohr may have been satisfied with his robust defence of the Copenhagen position, but under pressure from Einstein's insistent probing, the basis of his arguments had undergone a subtle shift. Bohr had fallen back on the notion of a 'clumsy disturbance', an uncontrollable transfer of momentum, of precisely the kind that he had earlier criticized Heisenberg for.

Although Bohr's counterarguments won the day in the eyes of the majority of physicists assembled in Brussels, the seeds of a much more substantial challenge had been sown. The fifth Solvay Conference ended with Bohr having successfully argued for the logical consistency of the Copenhagen interpretation.

Einstein remained unconvinced.

14

An Absolute Wonder

Cambridge, Christmas 1927

Pauli may have been pleased with the progress that had been made on the question of interpretation. He was nevertheless frustrated by the lack of progress on the incorporation of electron spin and the exclusion principle into the main structure of quantum theory. He called it an 'aesthetic failure'.

Furthermore, the equations that had been adopted as part of mainstream quantum mechanics suffered from the limitation that they did not meet the requirements of Einstein's special theory of relativity. There was a growing realization that the problem of electron spin was in some way connected with the problem of finding a fully relativistic expression of the equations of quantum theory.

Oskar Klein had independently rediscovered a relativistic version of Schrödinger's wave equation in the spring of 1926. With some further modifications by Hamburg theorist Walter Gordon it came to be known as the Klein–Gordon equation. But this was, in essence, the relativistic equation that Schrödinger himself had discovered and subsequently abandoned in January 1926 when he found that it did not yield predictions in agreement with experiment.

Meanwhile, the endless debate about the philosophy and interpretation of quantum theory at the Solvay conference in Brussels had left English physicist Paul Dirac cold. He did not have a great deal of time for eloquent elaboration of the finer points of interpretation if this produced no new equations. He was also aware that although his major contributions to the mathematical structure of the theory were worthy of international recognition, he had on several occasions now been pipped to the post by his European rivals. He felt strongly that he needed to discover something original, something that he was first to report and could claim as his own.

Electron spin was an obvious target for his attention. Towards the end of 1926 Dirac had agreed a bet with Heisenberg on how soon spin could be understood within the framework of quantum theory. Heisenberg had bet on three years at the least. Dirac had rashly bet three months. Three months went by, but the problem remained unresolved. But Dirac, who had become completely absorbed by the theory of relativity when he had first heard about it as an engineering student in 1919, now turned his attention to finding a fully relativistic version of quantum theory.

What he would find would simultaneously solve the riddle of electron spin. But it would also yield much more than anyone had bargained for.

Dirac was an archetype, if not a caricature, of the dry, introverted English academic. His personality had been shaped in childhood by his imposing Swiss father, Charles, who had run away from home at the age of twenty. Charles had settled in Bristol, in south-west England, in 1890. He became a schoolteacher. His second child Paul was born in 1902.

Charles, it seemed, had had an unhappy childhood. Sadly, this experience had not encouraged in him a greater generosity towards his own children. He was a tyrant. Seeing little value in social contact himself, he forbade his children from what he saw to be unnecessary socializing outside the immediate family. They became social prisoners and, within this prison, silence reigned. Charles, a native French speaker and now French teacher, insisted that Paul should speak to him only in French:

> He thought it would be good for me to learn French in that way. Since I found I couldn't express myself in French, it was better for me to stay silent than to [talk in] English. So I became very silent at that time—that started very early...

Paul compensated for his stunted social and emotional development by pouring his energies into mathematics. However, unable to see how he could forge a career in the subject he loved, and under the stern influence of his father, in 1918 he began an engineering degree at Bristol University.

A year later he was caught up in the general public excitement over Einstein's theory of relativity. The press reported Arthur Eddington's confirmation of the bending of starlight, predicted by Einstein's general theory. It was a revelation for the young student, and he went on to study deeply both the special and general theories. He became a master of the theory's mathematical structure.

Depression-era Britain was not a good time for a new engineering graduate to join the job market and on graduating in 1921 he failed to find a suitable position. When he was offered the opportunity of some free tuition in mathematics at Bristol University, he gladly accepted. He continued to study mathematics for the next two years.

Aided by a grant from the Department of Scientific and Industrial Research, in 1923 he finally escaped the suffocating emotional vice of his parental home and moved to Cambridge to study for a PhD.[1] He was initially disappointed to be assigned a research project under Ralph Fowler, as he had hoped to study relativity. Fowler's research interests were atomic and quantum physics. However, as Dirac soon discovered, there was much of interest in this emerging field, and much to do. 'The atoms were always considered as very hypothetical things by me,' he wrote years later, 'and here were people actually dealing with equations concerned with the structure of the atom.'

Fowler had good contacts with Bohr and the schools in Copenhagen and Göttingen. When he received a proof copy of Heisenberg's first paper on matrix mechanics in August 1925, he passed it to Dirac.

Dirac's fascination with relativity and his already burgeoning reputation as a quantum physicist made the search for a fully relativistic form of the new quantum theory irresistible. He had made several abortive attempts over the previous two years. At the Solvay conference in October 1927 he raised the problem with Bohr, only to be told by Bohr that Klein had already solved it. Dirac tried to explain why Klein's equation was unsatisfactory but was cut off when the lecture they had gathered to hear commenced.

Though short, the conversation with Bohr convinced Dirac that a proper solution had to be found as a matter of urgency. He once again set to work on the problem as soon as he got back from Brussels. He worked

[1] 'Escape' should be interpreted literally. Paul's older brother Reginald suffered both from his father's emotionally retarding influence and a feeling of inferiority when measured against his younger brother's intellectual achievements. Reginald committed suicide in 1924.

alone, in relative isolation, consulting nobody. It was an approach that suited his personality.

Einstein's special theory of relativity is in many ways all about the correct treatment of time as a kind of fourth dimension, on an equal footing with the three dimensions of ordinary space. The time-dependent version of Schrödinger's wave equation is 'unbalanced' in this regard, as it is a second-order differential equation in the three spatial coordinates but only a first-order differential equation in time.[2] In this sense, it actually looks more like an equation describing diffusion phenomena. This lack of balance between space and time means that Schrödinger's wave equation is non-relativistic.

Schrödinger and, subsequently, Klein and Gordon had tried to even things up by developing an equation that contains a second-order differential in time. This achieved the required balance but suffered from the problems that had led Schrödinger to abandon it. Dirac didn't like it. Neither did Heisenberg. 'Herr Pauli,' Hungarian physicist Johann Kudar reported to Dirac in a letter dated 21 December 1926, 'regards the relativistic wave equation of *second* order with much suspicion.' There remained no sign of electron spin in the Klein–Gordon equation.

It was understood that the relativistic equation had to remain a *first-order* differential equation in time. There was by now too much at stake, not least Born's probabilistic interpretation of the wavefunction, which could only continue to apply if the equation remained first-order in time. The challenge, then, was to restructure the equation in such a way as to make both its spatial and time components first-order differentials. This could easily be done, but led to some rather ugly-looking square-root operators which left the theorists scratching their heads. Dirac needed to find a way to handle these.

When in May 1927 Pauli had been faced with a similar problem dealing with square-root operators for momentum in three spatial coordinates, he had had much success using coefficients for each operator that were actually square *matrices* consisting of two rows and two columns. Pauli was trying to account for the property of electron spin, and the

[2] In a first-order differential equation, some function (*f*, say), is differentiated once with respect to some variable (*x*, say). This is written df/dx. In a second-order equation, the function is differentiated with respect to the variable a second time, written d^2f/dx^2.

coefficients, which became known as *Pauli spin matrices*, appeared to work well. This wasn't yet a full accounting for electron spin, but it was clearly a step in the right direction.

Dirac's problem was that he now had to find a way to handle *four* square-root operators; three momentum operators, and a fourth arising from a term which accounted for the variation of the mass of the electron with velocity, as demanded by the special theory of relativity. He initially figured that he just needed to find another two-by-two matrix coefficient, but this just didn't work. No such coefficient exists. Two-by-two matrices were not the answer.

Dirac was a mathematician first, a physicist second. What confronted him was a problem in mathematics. His principal concern was to solve the mathematical problem, then worry about the physical interpretation of the solution. As he played around with the equations, he was struck by an insight: 'I suddenly realised that there was no need to stick to quantities, which can be represented by matrices with only two rows and columns,' he later explained. 'Why not go to four rows and columns?'

This did the trick. Using coefficients that were structured as four-by-four matrices, he could linearize the square-root operators and go on to solve the equation.

It became known as the Dirac equation. Although it predicted no results that were not already given by earlier theories, it was a conceptual triumph. He had made his discovery; something he could call his own.

The use of matrices with the same form as Pauli's spin matrices meant that the property of electron spin automatically emerged from Dirac's relativistic equation. The introduction of a four-dimensional space–time into the equations of quantum theory had resulted in a fourth 'degree of freedom' for the electron which, in turn, *demanded* a fourth quantum number, as Goudsmit and Uhlenbeck had reasoned in November 1925.

But whatever it is, the property of electron spin does not correspond in any way to the notion of an electron spinning on its axis. It is a purely relativistic quantum property with no counterpart in classical physics. In this sense it is very different from other properties of the electron. For example, the orbital angular momentum of an electron in an atom is related to the quantum number k multiplied by h and there is in principle no restriction or upper bound on the value that k can take.[3] So, as we push to the classical limit of orbital angular momentum by reducing h to zero we can compensate by increasing k to infinity, with the result that the orbital angular momentum tends towards its classical value.

However, the same is not true for electron spin. The spin quantum number is constrained to one value for the electron and cannot be increased to infinity as h tends to zero. There is no classical counterpart for electron spin.

Although its interpretation is obscure, we do know that electron spin produces effects that give rise to a small magnetic moment. This moment can become aligned in the direction of the lines of force of an applied magnetic field or against that direction. We have learned to think of these as possibilities as 'spin-up' and 'spin-down'. In a magnetic field, the two possible orientations of the electron's magnetic moment give rise to two energy levels that are characterized by a magnetic spin quantum number, giving rise to the 'two-valuedness' corresponding to the spin-up and spin-down states. The two levels give rise to two lines in an atomic spectrum.

Reworking Dirac's equation to describe an electron moving in an electromagnetic field produced results that reproduced the Zeeman effect, in which the spin magnetic moment is twice the magnitude that would be anticipated on the basis of classical mechanics. Dirac's theory predicted that the spin magnetic moment is related to the spin quantum number of the electron multiplied by the Landé 'g-factor' for the electron. As in Kronig's original proposal, so in Dirac's theory the g-factor is exactly 2.

Charles Darwin visited Cambridge just before Christmas 1927 and was astonished to learn of Dirac's result. He immediately wrote to Bohr: '[Dirac] has now got a completely new system of equations which does the spin right in all cases and seems to be "the thing". His equations are first order, not second, differential equations!'

News spread quickly. Dirac wrote a letter to Max Born in Göttingen ahead of publication of a paper setting out his approach. Léon Rosenfeld described its reception: 'It was immediately seen as *the* solution. It was

[3] The quantum number k has been replaced in modern quantum theory by the orbital angular momentum (or azimuthal) quantum number l, which takes values $l = 0, 1, 2, \ldots, n$.

regarded really as an absolute wonder.' Fowler communicated Dirac's paper on the relativistic quantum theory of the electron to the British journal *Proceedings of the Royal Society*. The paper was received on 2 January 1928.

Dirac had been obliged to pay a price for his solution, however. His use of four-by-four matrices meant that he had twice as many solutions on his hands than he thought he had needed. The two possible orientations of the electron spin angular momentum account for half of the solutions available from Dirac's equation. The remainder correspond to electron states of negative energy and are an inevitable consequence of using the correct relativistic expression for the total energy of a freely moving particle.

The negative-energy solutions had some bizarre properties. Where the 'ordinary', positive-energy electrons would accelerate under an applied force, the particles described by the negative-energy solutions actually slowed down the harder they were pushed. The temptation of most physicists when faced with such unreasonable results is to dismiss them as 'unphysical' and continue only with the positive-energy solutions. This might have been acceptable when considering problems in classical physics, in which energy changes are assumed to be continuous and systems that start with positive energy cannot suddenly jump to negative-energy states.

However, quantum mechanics admits exactly this kind of sudden, discontinuous jump and it was perfectly feasible for a positive-energy electron to jump to a negative-energy state. This would appear in the laboratory as a jump from a 'conventional' state characterized by the familiar negative electron charge, $-e$, to a state characterized by a very unfamiliar positive electron charge, $+e$. No such transition had ever been observed.

This unusual behaviour left nagging doubts. Heisenberg wrote to Bohr: 'I am much more unhappy about the question of the relativistic formulation and about the inconsistency of the Dirac theory. Dirac was here [in Leipzig, June 1928] and gave a very fine lecture about his ingenious theory. But he has no more idea than we do about how to get rid of the difficulty $e \rightarrow -e$.'

The extra solutions had to be taken seriously, and they were a worry. The question was: what could they represent?

15

The Photon Box

Brussels, October 1930

Dirac wrestled with what became known as the '±' problem for the next two years. In December 1929 he outlined a proposed solution. Suppose, he said, that the universe is filled with a 'sea' of negative-energy states all occupied by spin-paired electrons. We would have no way of knowing of the existence of such a sea because, when filled, it wouldn't interact with anything and would merely serve as a kind of backdrop against which positive energy changes would be measured.

However, if an electron were to be promoted out of the sea—to become an observable, positive-energy electron—it would leave behind a 'hole'. The negative-energy hole would behave exactly like a positive-energy, positively charged particle.

Dirac suggested that the positively charged particle created by the hole in the negative-energy sea was, in fact, a proton. His logic was not without precedent. Rutherford had referred to the nucleus of the hydrogen atom as a 'positive electron' for six years before introducing the term 'proton' in 1920. And, making protons out of 'electron holes' had a nice symmetry to it that fed Dirac's desire to find a unitary description of the fundamental constituents of matter.

As he reported to a meeting of the British Association for the Advancement of Science in September 1930: 'It has always been the dream of philosophers to have all matter built from one fundamental kind of particle, so that it is not altogether satisfactory to have two in our theory, the electron and the proton.[1] *There are, however, reasons for believing that*

[1] The neutron had not yet been discovered.

the electron and the proton are really not independent, but are just two manifestations of one elementary kind of particle.'

But Dirac had reached for the dream too soon. His proposal was roundly criticized on all sides because, among other things, it demanded that the masses of the electron and 'hole'-derived proton should be the same. It was already well known that there is a substantial difference in the masses of these particles, the proton heavier by a factor of almost 2000. The debate continued.

As the world's most distinguished physicists gathered once again in Brussels for the sixth Solvay conference on 20 October 1930, there was much to discuss. The subject of the conference was magnetism, chaired by French physicist Paul Langevin following the death of Lorentz. But the conference would be remembered not for the formal lectures that were delivered on this subject, but for the debate on quite a different topic that took place between the formal proceedings, as Einstein and Bohr resumed their contest.

In an article published in the German journal *Die Naturwissenschaften* the year before, Bohr had expounded further on his theory of complementarity, drawing parallels with Einstein's theories of relativity.[2] The substantial (though finite) speed of light means that we can treat space and time separately for objects moving with velocities considerably smaller than that of light. Likewise, the very small (though finite) magnitude of Planck's constant of action means that for classical, macroscopic objects it is possible to apply simultaneously both space–time and causal descriptions.

However, Bohr explained, when we consider the behaviour of objects moving at speeds close to that of light, we cannot any longer ignore the effects of relativity. Similarly, when we consider objects at the quantum level, we cannot any longer ignore complementarity. For quantum objects, the space–time and causal descriptions of nature cannot any longer be applied simultaneously.

Einstein had rejected Newton's absolute space and time because in practice there is no such thing as absolute simultaneity. Why not just accept that the uncertainty relations lead us to reject the simultaneous validity of the classical concepts we are seeking to apply in the quantum domain?

[2] Bohr changed his terminology in this paper, dropping complementarity in favour of 'reciprocity', a decision he latter regretted (see Pais, *Niels Bohr's Times*, p. 426). He quickly reverted to the term complementarity. Consequently, I will stick with complementarity here.

It was a red rag to a bull.

Shortly after publication of Bohr's article, Austrian philosopher Philipp Frank pointed out that Einstein himself had invented the logic of Bohr and Heisenberg's interpretation in his seminal 1905 paper on special relativity. Einstein responded grumpily: 'A good joke should not be repeated too often.'

Einstein had been preparing his next challenge. Irrespective of what Bohr had written about the parallels between complementarity and relativity, this time he was confident that he could use his special theory of relativity to undermine the logical consistency of the energy–time uncertainty relation, and with it the consistency of the Copenhagen interpretation. Gathered once again in Brussels, Einstein described to Bohr his latest and most ingenious *gedankenexperiment*.

'At the next meeting with Einstein at the Solvay Conference in 1930,' wrote Bohr some years later, 'our discussions took quite a dramatic turn.'

Suppose, said Einstein, that we build an apparatus consisting of a box which contains a clock connected to a shutter. The shutter covers a small hole in the side of the box. We fill the box with photons and weigh it. At a predetermined and precisely known time, the clock triggers the shutter to open for a very short time interval sufficient to allow a single photon to escape from the box. The shutter closes. We reweigh the box and, from the mass difference and Einstein's theory of special relativity ($E = mc^2$) we determine the precise energy of the photon that escaped. By this means, we have measured precisely the energy and the time of release of a photon from the box, in contradiction to the energy–time uncertainty relation.

Bohr's immediate reaction was described by Léon Rosenfeld:

> It was quite a shock for Bohr... he did not see the solution at once. During the whole evening he was extremely unhappy, going from one to the other and trying to persuade them that it couldn't be true, that it would be the end of physics if Einstein were right; but he couldn't produce any refutation. I shall never forget the vision of the two antagonists leaving the club [of the Fondation Universitaire]: Einstein a tall majestic figure, walking quietly, with a somewhat ironical smile, and Bohr trotting near him, very excited...

Bohr experienced a sleepless night, searching for the flaw in Einstein's argument that he was convinced must exist. 'This argument amounted to a serious challenge and gave rise to a thorough examination of the whole problem,' Bohr wrote. By breakfast the following morning he had his answer.

On the blackboard Bohr produced another rough, pseudo-realistic sketch of the apparatus that would be required to make the measurements in the way Einstein had described them. In this sketch the whole box is imagined to be suspended by a spring and fitted with a pointer so that its position can be read on a scale affixed to the support. A small weight is added to align the pointer with the zero reading on the scale. The clock mechanism is shown inside the box, connected to the shutter.

After the release of one photon, the small weight is replaced by another, slightly heavier weight so that the pointer is returned to the zero of the scale. We suppose that the weight required to do this can be determined independently with arbitrary precision. The difference in the two weights required to balance the box gives the mass lost through the emission of one photon, and hence the energy of the photon. So far, so good.

Bohr now drew his colleagues' attention to the first weighing, before the photon escapes. Obviously, the clock is set to trigger the shutter at some predetermined time and the box is sealed. The actual reading of the clock face is, of course, not possible since this would involve an exchange of photons—and hence energy—between the box and the outside world.

To weigh the box, a weight must be selected that sets the pointer to the zero of the scale. However, to make a precise position measurement, the pointer and scale will need to be illuminated and, following Bohr's earlier arguments, this implies an uncontrollable transfer of momentum to the box and hence an uncertainty in its momentum.

How does this affect the weighing? The uncontrollable transfer of momentum to the box causes it to jump about unpredictably. Although the box's instantaneous position against the scale can be fixed, the sizeable interaction during the act of measurement means that the box will not stay in that position. Bohr argued that we can increase the precision of the measurement of the *average* position of the pointer by allowing ourselves a long time interval in which to perform the whole balancing procedure. This will give us the necessary precision in the weight of the box. Since we can anticipate the need for this, the clock can be set so that it opens the shutter after this balancing procedure has been completed.

Now comes Bohr's coup de grâce.

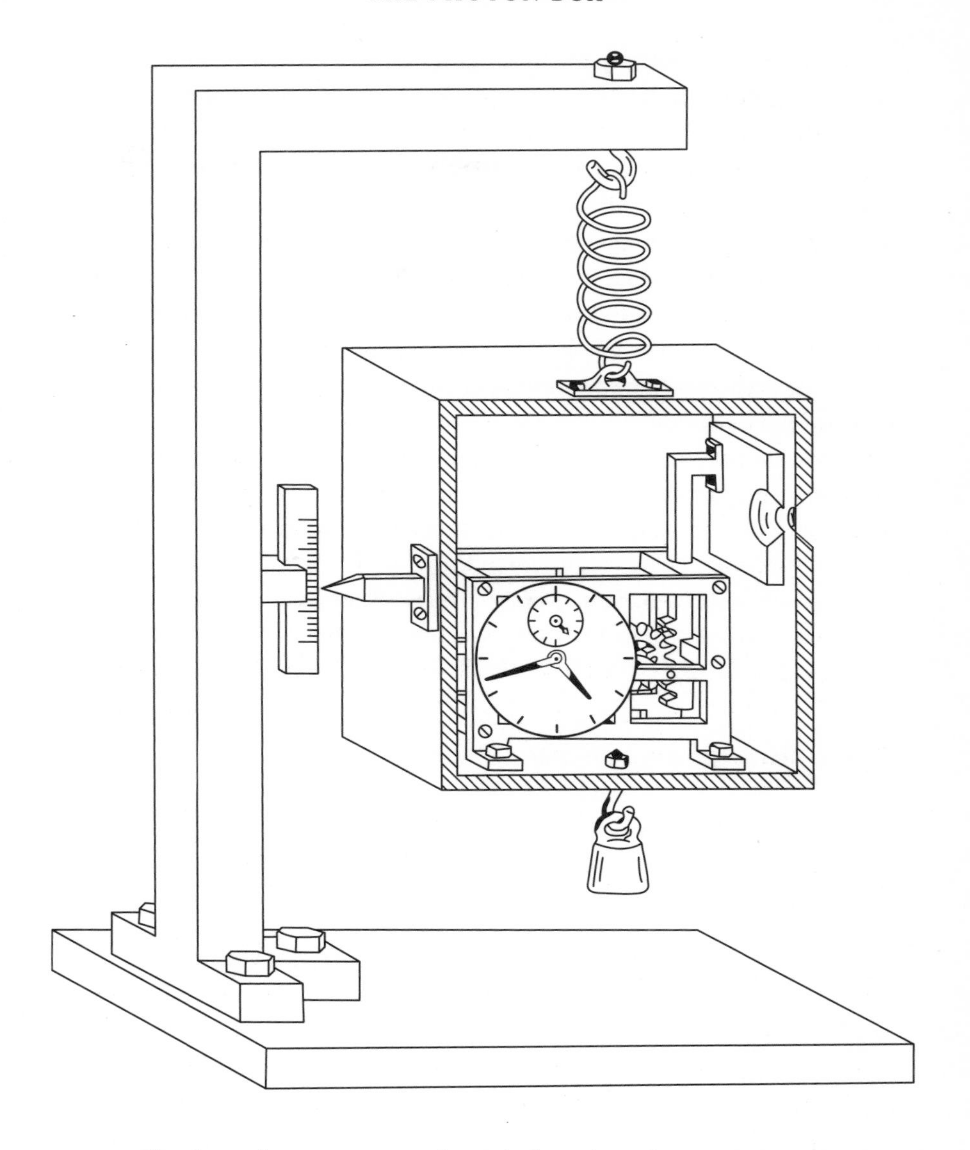

FIG 7 The photon box experiment. This is the hypothetical instrument sketched by Bohr to show how the measurements suggested by Einstein might be carried out. Reprinted by permission of Open Court Publishing Company, a division of Carus Publishing Company, Chicago, IL, from *Albert Einstein: Philosopher-scientist,* Vol. I, p. 227, edited by Paul Arthur Schilpp, copyright © 1949 by the Library of Living Philosophers, Inc.

According to Einstein's own *general* theory of relativity, a clock moving in a gravitational field is subject to time dilation effects. The very act of weighing a clock effectively changes the way it keeps time. Because the box is jumping about unpredictably in a gravitational field (owing to the act of balancing the weight of the box by measuring the position of the pointer), the rate of the clock is changed in a similarly unpredictable manner. This introduces an uncertainty in the exact timing of the opening of the shutter which depends on the length of time needed to complete the balancing procedure. The longer we make this procedure (the greater the ultimate precision in the measurement of the energy of the photon), the greater the uncertainty in its exact moment of release.

Bohr was able to show that the product of the uncertainties in energy and time for the photon box apparatus cannot be greater than Planck's constant *h*, in accordance with the uncertainty principle.

> The discussion, so illustrative of the power and consistency of relativistic arguments, thus emphasised once more the necessity of distinguishing, in the study of atomic phenomena, between the proper measuring instruments which serve to define the reference frame and those parts which are to be regarded as objects under investigation and in the account of which quantum effects cannot be neglected.

This response was hailed as a triumph for Bohr and for the Copenhagen interpretation of quantum theory. Bohr had used Einstein's own general theory of relativity against him.

Einstein remained stubbornly unconvinced. In subsequent discussions of the photon box experiment he conceded that it now appeared to be 'free of contradictions', but in his view it still contained 'a certain unreasonableness.'

Bohr had once again sought to defend the Copenhagen interpretation by arguing that the inevitable and sizeable disturbance of the observed system in any physical measurement effectively precludes the acquisition of knowledge with a precision greater than that permitted by the uncertainty principle. At first sight, there seems to be no counter-argument against this position. It seems that a measurement of any kind will always involve physical interaction on the same scale as the quantum system under investigation, and to talk about the properties of quantum systems *only* in the absence of measurement is a pointless descent into naïvety.

Einstein had somehow to find a way around this.

He believed there were still some clues in the photon box experiment. What if, he reasoned some months after the Solvay conference, the experiment is used not to challenge the *consistency* of the uncertainty relations but is used rather to derive a logical paradox arising from what he saw to be the theory's lack of *completeness*?

As before, the clock is set to trigger the release of one photon, but this time it is synchronized with a second, external clock. The box is filled with photons. Einstein now accepted that his own general theory of relativity precluded precise knowledge of the moment of release, but perhaps this was not really the point after all.

Now let the released photon travel to a fixed mirror placed a long distance from the box, say half a light-year away. Whilst the photon is on its round-trip journey of one year, we can now *choose* what measurement to make. We could open the box and compare the two clocks. They will no longer be synchronous because of the effect of the initial balancing procedure on the rate of the clock inside, but we can now correct for this by reference to the external clock. We can therefore draw a retrodictive conclusion as to the exact timing of release of the photon. We know how long the photon will take on its journey and so we can calculate its exact time of arrival back in the laboratory.

Alternatively, we could choose to keep the box sealed and rebalance it with a heavier weight. We could take as long as we need over this second balancing procedure and, as before, this would tell us the exact energy of the released photon.

We now make a further, perhaps rather obvious or even trivial assumption that a photon half a light-year distant cannot be affected by the choice we make in the laboratory back on earth. In other words, we assume that the photon is not in any way disturbed by measurements we choose to make in a laboratory over three thousand billion miles away. As we can choose to measure *either* the time of release *or* the energy of the photon, we are left to conclude that the photon must possess simultaneously exact values of both energy and time.

As there is nothing in quantum theory that tells us about the simultaneous exact values of these complementary observables, Einstein further concluded that the theory provides an incomplete description of individual quantum systems.

Bohr heard of Einstein's further adaptation of the photon box experiment in a letter from Austrian physicist Paul Ehrenfest dated 9 July 1931. Einstein was due to visit Ehrenfest in Leiden towards the end of October that year, and Ehrenfest was hopeful that Bohr could join them there for a quiet exchange of views (RUHIG, he wrote, in capital letters). He set out the details of Einstein's latest challenge.

Bohr summarized his response in his contribution to a 1949 volume dedicated to Einstein in celebration of his 70th birthday. Bohr had been unmoved:

> Since, however, according to the quantum-mechanical formalism, the specification of the state of an isolated particle cannot involve both a well-defined connection with the time scale and an accurate fixation of the energy, it might thus appear as if this formalism did not offer the means of an adequate description...In fact, we must realize that in the problem in question we are not dealing with a single specified experimental arrangement, but are referring to two different, mutually exclusive arrangements...

On the question of our ability to exercise a choice of measurement, he had this to say:

> It may also be added that it obviously can make no difference as regards observable effects obtainable by a definite experimental arrangement, whether our plans of constructing or handling the instruments are fixed beforehand or whether we prefer to postpone the completion of our planning until a later moment when the particle is already on its way from one instrument to another.

Bohr had dismissed Einstein's challenge, though he had been less than clear regarding the reasons. For the physicists of the Copenhagen–Göttingen schools, the debate had in any case already been won by Bohr and there were more pressing problems to be dealt with.

But Einstein's 1931 adaptation of the photon box experiment contained the seeds of another, more powerful challenge to the authority of the Copenhagen interpretation. Bohr would be quite unprepared for Einstein's next move.

1. In December 1900 German physicist Max Planck conjured the quantum of action from otherwise classical equations describing black body radiation.

2. In 1905 Albert Einstein formulated his special theory of relativity and introduced the light-quantum hypothesis.

3. Young Danish physicist Niels Bohr discovered a new model of the atom based on stable electron orbits each characterised by a quantum number.

4. Where had the quantum numbers come from? In 1923 French physicist Prince Louis de Broglie deduced that particles with mass might possess associated wave-like properties.

5. Werner Heisenberg, Erwin Schrödinger and Paul Dirac at the railway station in Stockholm on their way to collect the Nobel prize . Left to right: Heisenberg's mother, Schrödinger's wife, Dirac's mother, Dirac, Heisenberg, Schrödinger.

6. With the help of Wolfgang Pauli (pictured right) in 1927 Bohr and Heisenberg achieved an uneasy consensus which became known as the Copenhagen interpretation of quantum theory.

7. Einstein first heard Bohr's lecture on his principle of complementarity at the Solvay Conference in Brussels in September 1927. Back row: Paul Ehrenfest (third from left), Schrödinger (sixth from left), Pauli and Heisenberg (third and second from right). Middle row: Pieter Debye (first from left), Hendrik Kramers and Paul Dirac (fourth and fifth from left), de Broglie, Max Born, Bohr (third, second and first from right). Front row: Planck (second from left), Hendrik Lorentz (fourth from left), Einstein (centre).

8. The intellectual challenges posed by Einstein at the fifth Solvay Conference and their rebuttal by Bohr represented the opening skirmishes in a debate about the interpretation of quantum theory that would last nearly ten years and which, to a certain extent, continues to this day. This was to be one of the most profound debates in the history of science.

9. At the Shelter Island conference in 1947 young American physicist Richard Feynman heard about the problems posed by new experimental data on the hydrogen atom.

10. (*Below*) The discussion at the Shelter Island conference continued long into the night, over hastily eaten meals. Pictured here from left to right are: Willis Lamb, Abraham Pais (seated), John Wheeler (standing), Feynman, Herman Feshbach and Julian Schwinger.

11. (*Above*) Together with American Robert Mills, in 1954 Chinese physicist Chen Ning Yang (pictured here on the right, with compatriot Tsung-Dao Lee) developed an SU(2) gauge field theory of the strong force which predicted massless field particles.

12. New York physicist Murray Gell-Mann found he could accommodate the spin-0 baryons and spin-0 mesons in two octets derived from a global SU(3) symmetry group. Israeli physicist Yuval Ne'eman made a similar discovery.

13. Sheldon Glashow.

14. Steven Weinberg.

15. Abdus Salam.

Resolving the problems of the massless field particles predicted by quantum field theory took nearly twenty years and led ultimately to the Glashow-Weinberg-Salam unified SU(2) x U(1) electro-weak theory.

16

A Bolt from the Blue

Princeton, May 1935

The early 1930s saw the dramatic rise of experimental nuclear physics. In February 1932, Cambridge physicist James Chadwick obtained evidence for an electrically neutral particle—the neutron.[1] *Although it was not immediately recognized as a fundamental nuclear particle, the neutron would soon resolve many puzzles.*[2] *Later that year, Cambridge physicists John Cockcroft and Ernest Walton used a linear particle accelerator to demonstrate the first artificially induced nuclear reactions.*[3]

Other types of particle accelerators were developed. In California, American physicist Ernest Lawrence embarked on the construction of a series of ever-larger cyclotrons, designed to accelerate particles around a circular path created by an electromagnet, promising greater efficiencies and higher particle energies. In 1932 Lawrence constructed a cyclotron containing a magnet with an eleven-inch diameter pole face delivering proton energies of over a million electron volts.

In the meantime, Dirac had given up on his hole-derived proton, finally accepting in 1931 that the hole would have to have the same mass as the electron. He introduced

[1] Chadwick's letter to *Nature*, 'Possible Existence of a Neutron', *Nature*, **129**, 1932, p. 312 is reproduced, together with a handwritten letter to Bohr outlining the discovery, in Brown, pp. 365–367.

[2] Physicists were still convinced that the nucleus contained electrons as well as protons and for a time the neutron was believed to be a composite proton–electron particle. The neutron was recognized to be a fundamental particle a year after its discovery.

[3] The Cockcroft–Walton apparatus used a system of voltage multipliers to accelerate protons to energies around 700,000 electron-volts.

the positive electron, 'a new kind of particle, unknown to experimental physics, having the same mass and opposite charge to an electron.' American physicist Carl Anderson found evidence for this particle, which he named the positron, in cosmic ray experiments in 1932–33.[4]

Dirac shared the 1933 Nobel Prize for physics with Schrödinger. The Nobel Committee also announced the award of the 1932 Prize to Heisenberg. All three physicists went to Stockholm in December 1933. In his Prize lecture, Dirac speculated on the possible existence of a negative proton, and of stars composed entirely of what would later become known as anti-particles.

On 30 January 1933 Adolf Hitler became Chancellor of Germany. In April, the new National Socialist government introduced laws forbidding Jews from holding positions in the Civil Service, including academic positions in German universities. Max Planck, now the venerated president of the Kaiser Wilhelm Gesellschaft, made a direct appeal to Hitler, arguing that the forced emigration of Jewish scientists would destroy German science.

It was to no avail. There followed an exodus of near biblical proportions. A quarter of all the physicists in Germany were displaced, including many Nobel Prize winners. Among them were Max Born and James Franck. Although he was not Jewish and therefore not subject to the new laws, Schrödinger was disgusted by the policies of the Nazi regime. He left Berlin for exile in Oxford.

In October 1932 Einstein had accepted an appointment at the newly founded Institute for Advanced Study in Princeton. He had intended to spend five months of each year in Princeton and the balance of his time in Berlin. But in December 1932 he left Germany, never to return. He settled into permanent residence in Princeton in October 1933. Looking around for bright young mathematicians with whom to work, Einstein was drawn to a Russian, Boris Podolsky, and an American, Nathan Rosen.

They worked together to refine Einstein's latest and most compelling challenge on the completeness of quantum theory.

Einstein had needed to find a physical situation in which it is possible in principle to acquire knowledge of the state of a quantum particle without disturbing it in any way. Bohr would thereby be denied access to the arguments he had used to escape Einstein's earlier challenges. The modified

[4] Anderson initially thought that the particle tracks he had observed in these experiments were due to protons. In May 1933 he suggested that they were, in fact, due to positrons. These particles were subsequently confirmed to be Dirac's positive electrons by Patrick Blackett and Guiseppe Ochialini.

photon box experiment, in which the experimenter is allowed to choose which of two complementary observables to measure exactly, was a first step towards the solution.

The Einstein–Podolsky–Rosen (EPR) *gedankenexperiment* now took the approach a stage further. The experiment involves measurements made on one of two quantum particles that have interacted at some time in their recent history and have subsequently moved apart. We will denote these as particle A and particle B. The position and momentum of particle A are complementary observables and we cannot measure one without introducing an uncertainty in the other in accordance with Heisenberg's uncertainty principle. Similar arguments can be made for the position and momentum of particle B.

However, if we now consider the *difference* between the positions of particles A and B, $q_A - q_B$, and the *sum* of their momenta, $p_A + p_B$, it is possible to show that these are quantities whose operators do commute. There is therefore no restriction in principle on the precision with which we can measure these quantities simultaneously.

Einstein, Podolsky, and Rosen first established what seems to be a fairly reasonable (though philosophically loaded) definition of physical reality:

> If, without in any way disturbing a system, we can predict with certainty (i.e. with a probability equal to unity) the value of a physical quantity, then there exists an element of physical reality corresponding to this physical quantity.

The purpose of this statement is to make clear that for each particle considered individually, the measurement of one physical quantity (the position of B, say) with certainty implies an infinite uncertainty in its momentum. Therefore, according to this definition of reality, under these circumstances the position of particle B is an element of physical reality but the momentum is not. Obviously, by choosing to perform a different measurement, we can (in the language of Einstein, Podolsky, and Rosen) establish the reality of the momentum of particle B but not its position.

But we can show that the difference in the positions of particles A and B and the sum of their momenta are not subject to this constraint. Suppose we now allow the two particles to interact and move a long distance apart. We perform an experiment on particle A to measure its position with certainty. We know that the difference in position must be a physically real quantity and so we can deduce the position of particle B also with certainty. We therefore conclude that the position of B must be an element of physical reality according to the definition given above.

However, suppose instead that we choose to measure the momentum of particle A with certainty. We know that the sum of the momenta must be physically real, and so it follows that we can deduce the momentum of particle B with certainty. We conclude that it too must be an element of physical reality. Thus, although we have not performed any measurements on particle B following its separation from A, we can establish the reality of *either* its position *or* its momentum from measurements we *choose* to perform on A which, *by definition, do not disturb B.*

The Copenhagen interpretation denies that we can do this. We are forced to accept that if this interpretation of quantum theory is correct, the physical reality of either the position or momentum of particle B is determined by the nature of a measurement we choose to make on a completely different particle an arbitrarily long distance away. Einstein, Podolsky, and Rosen argued that: 'No reasonable definition of reality could be expected to permit this.'

The EPR thought experiment strikes right at the heart of the Copenhagen interpretation of quantum theory. If the uncertainty principle applies to an individual quantum particle, then it appears that we must invoke some kind of 'spooky' action at a distance if the reality of the position or momentum of particle B is to be determined by measurements we choose to perform on A.

Whether it involves a change in the physical state of the system or merely some kind of communication, the fact that this action at a distance must be exerted instantaneously on a particle an arbitrarily long distance away from our measuring device suggests that it violates the postulates of special relativity, which restrict any signal to be communicated no faster than the speed of light. Einstein, Podolsky, and Rosen did not believe that it is necessary to invoke such action at a distance. The position and momentum of particle B are defined all along and, as there is nothing in the wavefunction which tells us how these quantities are defined, quantum theory is incomplete.

They concluded:

> While we have thus shown that the wave function does not provide a complete description of physical reality, we left open the question of whether or not such a description exists. We believe, however, that such a theory is possible.

The Einstein, Podolsky, Rosen paper, entitled 'Can the Quantum-mechanical Description of Physical Reality be Considered Complete?' was published in the journal *Physical Review* in May 1935. The paper appears to have been written largely by Podolsky, and there is much in the language and nature of the argumentation employed that Einstein appears later to have regretted. In particular, the criterion of physical reality as described above unnecessarily exposed the EPR argument to a powerful counter-argument. All the more disappointing, perhaps, as the main challenge presented by Einstein, Podolsky, and Rosen does not require this (or any) criterion of physical reality, though it does rest on the assumption that, however reality is defined, it is defined to be *local*.

The EPR challenge was reported in the popular press before it was published in *Physical Review*. On Saturday, 4 May 1935 *The New York Times* carried an article entitled 'Einstein Attacks Quantum Theory' which provided a non-technical summary of the main arguments, with extensive quotations from Podolsky. He concluded that whilst quantum mechanics could be regarded as technically correct, it 'is not a complete theory.' Einstein deplored the article and the publicity surrounding it. The article was followed in the same edition by a report by American physicist Edward Condon who noted that the arguments raised a 'point of doubt' but who went on to identify the reality criterion as the argument's most significant weakness.

Bohr first heard of the EPR argument from Rosenfeld, who was at that time working with Bohr in Copenhagen. Rosenfeld later reported that:

> This onslaught came down upon us as a bolt from the blue. Its effect on Bohr was remarkable... as soon as Bohr had heard my report of Einstein's argument, everything else was abandoned: we have to clear up such a misunderstanding at once. We should reply by taking up the same example and showing the right way to speak about it. In great excitement, Bohr

> immediately started dictating to me the outline of such a reply. Very soon, however, he became hesitant. 'No, this won't do, we must try all over again...we must make it quite clear...' So it went on for a while, with growing wonder at the unexpected subtlety of the argument.

Others were devastated by the EPR challenge. Pauli was furious. Dirac exclaimed: 'Now we have to start all over again, because Einstein proved that it does not work.

Bohr and his colleagues worked on their defence day after day, week after week. They took the EPR thought experiment to pieces and reassembled it. A solution began to emerge. 'They do it "smartly",' Bohr remarked, 'but what counts is to do it right.'

Bohr's reply to the EPR argument was published in outline in *Nature* in June 1935 and subsequently in more detail in *Physical Review* in October. In the latter paper he chose to use the same title that Einstein, Podolsky, and Rosen had used in May and the abstract reads as follows:

> It is shown that a certain 'criterion of physical reality' formulated in a recent article with the above title by A. Einstein, B. Podolsky and N. Rosen contains an essential ambiguity when it is applied to quantum phenomena. In this connection a viewpoint termed 'complementarity' is explained from which quantum-mechanical description of physical phenomena would seem to fulfill, within its scope, all rational demands of completeness.

Bohr's paper is essentially a summary of complementarity and its application to quantum theory. He rejected the argument that the EPR thought experiment creates serious difficulties for the Copenhagen interpretation and stresses once again the importance of taking into account the necessary interactions between the objects of study and the measuring devices used to study them. He wrote:

> From our point of view we now see that the wording of the above-mentioned criterion of physical reality proposed by Einstein, Podolsky and Rosen contains an ambiguity as regards the meaning of the expression 'without in any way disturbing a system.' Of course there is in a case like that just considered no question of a mechanical disturbance of the system under investigation during the last critical stage of the measuring procedure. But even at this stage there is essentially the question of *an influence on the very*

> *conditions which define the possible types of predictions regarding the future behaviour of the system.* Since these conditions constitute an inherent element of the description of any phenomenon to which the term 'physical reality' can be properly attached, we see that the argumentation of the mentioned authors does not justify their conclusion that quantum-mechanical description is essentially incomplete.

In essence, Bohr argued that it does not matter that the position and momentum of particle B can be inferred from measurements we make on A. The simple fact is that we cannot conceive an experimental arrangement that will allow us to exercise the choice demanded by Einstein, Podolsky, and Rosen. By setting up an experiment to measure the position of A with certainty, we deny ourselves the possibility of measuring its momentum, and *vice versa*. And if we cannot actually exercise the choice to measure either the position or momentum of particle A, then the actual properties and behaviour of B are really rather moot. Even if there is no mechanical disturbance of particle B (as EPR assumed and Bohr accepted), its elements of physical reality are nevertheless defined by the nature of the measuring device we have selected for use with particle A.

The EPR argument pushed Bohr from his previous, perhaps rather ambiguous philosophy to a fixed anti-realist position. It was at this point that Bohr dropped the use of 'disturbance' as a counter-argument and focused instead on the nature of the experimental arrangement itself precluding the type of reality that could be exposed. Mechanical disturbance could no longer serve Bohr's purpose, just as Einstein had intended.

Many in the physics community seemed to accept that Bohr's paper put the record straight. But Bohr's wording is rather vague and unconvincing. In emphasizing the fundamental role of the measuring instrument in *defining* the elements of reality that can be observed, he offered no new insight as to how such a definition is physically established.

However, Bohr failed to respond to the real challenge posed by Einstein, Podolsky, and Rosen. What these physicists had created in their thought experiment was a two-particle wavefunction that allows for correlations to be established between quantum particles over potentially vast distances. Making a measurement then involves a notional collapse of this wavefunction, now formally enshrined in the theory of quantum measurement as Hungarian mathematician John von Neumann's 'projection postulate'.[5]

The collapse implies a spooky action at a distance that appears to violate the basic postulates of special relativity.

Does such a measurement necessarily imply an action at a distance? Certainly, if we could somehow delay our choice of measuring instrument (position *versus* momentum) until almost the last moment, then in principle the information potentially available to us about a particle some considerable distance away appears to change instantaneously. We are left to wonder how particle B is supposed to 'know' what physical property—position or momentum—it is supposed to reveal as a result of measurements we choose to make on A. An action at a distance will be required if the measurement performed on A changes the physical state of B or results in some kind of communication to B of particle A's changed circumstances.

Now if the two-particle wavefunction reflects only our state of knowledge of the quantum system, then its collapse would not necessarily seem to affect the system's physical properties. However, the problem remains that the collapse of the wavefunction requires that those physical properties become clearly defined and manifested in the quantum system where before they were vague and undefined. The physical properties of particle B suddenly become 'real' (or measurable), where before they were not.

There is simply no mechanism to account for this in the Copenhagen interpretation of quantum theory.

[5] In his book *Mathematical Foundations of Quantum Theory*, von Neumann distinguished between continuous, causal dynamics described by the time-dependent Schrödinger wave equation and the discontinuous, acausal measurement process. The book was first published in Berlin in 1932.

17

The Paradox of Schrödinger's Cat

Oxford, August 1935

Schrödinger was elected to a fellowship at Magdalen College, Oxford, in October 1933. With support in the form of a two-year grant from Britain's Imperial Chemical Industries (ICI), he and Anny moved to Oxford the following month. The Berlin Deutsche Zeitung *lamented his loss to German science: '[Schrödinger's departure] is all the more to be regretted, as only a short time ago Professor Hermann Weyl, Ordinarius for Mathematics at the University of Göttingen, also in the field of theoretical physics, accepted the call to the American university of Princeton.'*

It was at his formal welcome by other fellows of Magdalen College that Schrödinger learned that he had won the 1933 Nobel Prize for physics. As he later recalled, the president of the college, George Gordon, called him to his office and advised him that: ' The Times *had said I would be among that year's prize winners. And in his chevalieresque and witty manner he added, "I think you may believe it.* The Times *do not say such a thing unless they really know. As for me, I was truly astonished, for I thought you had the prize."'*

Schrödinger had arranged for the award of a second, temporary fellowship for a Berlin colleague, Arthur March. He had recently begun an affair with March's wife Hildegunde. On 30 May 1934 Hilde bore Schrödinger's first child. They had made no secret of their relationship but the baby, named Ruth Georgie Erica, was registered as the child of Hilde and her husband. Both Hilde and Anny cared for the baby and, to all intents and purposes, Schrödinger lived for a time as though with two wives.

Frederick Lindemann, who had worked behind the scenes to bring Schrödinger to Oxford, was furious, calling Schrödinger a 'bounder'. He struggled to persuade ICI to continue awarding grants to émigré scientists beyond the previously agreed two years, with one ICI director complaining that they had paid not only for the scientists in question but also in some cases for their mistresses.

Schrödinger grew increasingly weary of college life, referring to the colleges as 'academies of homosexuality'. Whilst his was hardly an enlightened attitude to women, he detested the thinly veiled misogyny typical of Oxford society. Although his lectures on elementary wave mechanics were well received he was discomfited by the feeling that he was being paid for doing very little or, worse still, that he was the recipient of charity.

There was some solace, at least, to be found in his correspondence with Einstein. When the Einstein, Podolsky, Rosen paper appeared in the May 1935 edition of Physical Review, *he wrote to Einstein to congratulate him.*

In the letter, dated 7 June 1935, Schrödinger wrote:

> I was very pleased that in the work that just appeared in *Physical Review* you have publicly called the dogmatic quantum mechanics to account over those things that we used to discuss so much in Berlin. Can I say something about it? It appears at first as objections, but they are only points that I would like to have formulated yet more clearly.

He concluded his letter with the following observation:

> ...My interpretation is that we do not have a [quantum mechanics] that is consistent with relativity theory, i.e., with a finite transmission speed of all influences. We have only the analogy of the old absolute mechanics...The separation process is not at all encompassed by the orthodox scheme.

Schrödinger's reference to the 'separation process' highlights the essential difficulty that the EPR argument creates for the Copenhagen interpretation. According to this interpretation, the wavefunction for the two-particle quantum state does not separate as the particles themselves separate in space–time. Instead of dissolving into two completely separate wavefunctions, one associated with each particle, the two-particle wavefunction becomes 'stretched' out and, when a measurement is made, collapses instantaneously despite the fact that it may be spread out over a considerably large distance.

Schrödinger saw that the connection or correlation between two particles resulting from the formation of a joint, two-particle wavefunction was not unique to the EPR thought experiment:

> If two separated bodies, each by itself known maximally, enter a situation in which they influence each other, and separate again, then there occurs regularly that which I have just called *entanglement* of our knowledge of the two bodies.

The definition of physical reality adopted by Einstein, Podolsky, and Rosen requires that the two particles are considered to be fully separated and distinct from each other. They are no longer described by a single two-particle wavefunction at the moment a measurement is made. The reality thus referred to is sometimes called 'local reality' and the ability of the particles to separate into two locally real independent physical entities is sometimes referred to as 'Einstein separability'.

Under the circumstances of the EPR thought experiment, the Copenhagen interpretation denies that the two particles are Einstein separable and therefore denies that they can be considered to be locally real (at least, before a measurement is made on one or other of the particles, at which point they both become localized).

In fact, Einstein had written to Schrödinger on 17 June and their letters crossed. In this letter Einstein, unsure of Schrödinger's reaction to his latest challenge, confessed that his relationship with quantum theory had changed: 'No doubt, however, you smile at me and think that, after all, many a young whore turns into an old praying sister, and many a young revolutionary becomes an old reactionary.'

However, Schrödinger's position on this latest challenge was clear from his letter of 7 June. Only two days after writing his first letter, Einstein enthusiastically drafted a second. In this letter of 19 June, he explains that the Einstein, Podolsky, Rosen paper had been written, after considerable discussion, largely by Podolsky and that: '...it did not come out as well as I had originally wanted: rather the essential thing was, so to speak, smothered by the formalism.'

Einstein now proceeded to make his point more clearly:

> The actual difficulty lies in the fact that physics is a kind of metaphysics; physics describes reality; we know it only through its physical description. All physics is a description of reality; but this description can be 'complete' or 'incomplete'. To begin with, the sense of this expression is even a problem itself. I will explain with the following analogy...

Einstein imagined that in front of him were two boxes. He can open the lids on both boxes and look inside. This business of looking inside either box is called 'making an observation'. In addition to the boxes there is also a ball. The ball can be found in one or the other of the two boxes when an observation is made. There is a 50:50 chance that the ball is in the first box. Einstein now asked himself: is this a complete description? He identified two possibilities.

The first is: no, this is not a complete description. The ball is either in the first box or it is not. The absence of a prescription by which we can predict with certainty in which box the ball will be found means that our description is incomplete, and in our ignorance we resort to probabilities.

The second possibility is: yes. Before the box is opened the ball is in some kind of undetermined state and is not physically in either of the two boxes. The ball is physically localized in one or other box only when the lid of one of the boxes is opened. When the observation is repeated again and again we deduce that the ball has a 50:50 chance of being found in either box. The result of such observations is the appearance of statistical behaviour. The state before the box is opened is completely described by the probability: nothing further is required (or possible).

Einstein continued:

> We face similar alternatives when we want to explain the relation of quantum mechanics to reality. With regard to the ball-system, naturally, the second 'spiritualist' or Schrödinger interpretation is absurd, and the man on the street would only take the first, 'Bornian' interpretation. The Talmudic philosopher dismisses 'reality' as a frightening creature of the naïve mind, and declares that the two conceptions differ only in their mode of expression.

Einstein's reference to the 'Talmudic' philosopher was a dig at the 'Kopenhagener Geist', a (religious) philosophy that is to be interpreted only through its qualified priests, which insists on its correctness and will countenance no rivals.

Schrödinger replied on 13 July. By this time the Einstein, Podolsky, Rosen paper had attracted some critical comment in the scientific press. 'What I have so far seen by way of published reactions is less witty,' Schrödinger wrote, 'It is as if one person said, "It is bitter cold in Chicago"; and another answered, "That is a fallacy, it is very hot in Florida."'

But Schrödinger was still pursuing the idea that the quantum wavefunction—and its statistical interpretation—reflected an underlying physical reality of waves, wave packets and their superpositions. Einstein insisted that the wavefunction was inadequate as a complete description of reality and reflected only the statistical probabilities of ensembles of systems.

In seeking to persuade Schrödinger to this point of view, in his reply of 8 August Einstein presented yet another thought experiment, one that was ultimately to lead Schrödinger to develop one of the most famous paradoxes of quantum theory. This thought experiment consists of a charge of gunpowder that can spontaneously combust at any time in the course of a year:

> In the beginning the ψ-function characterizes a reasonably well-defined macroscopic state. But, according to your equation, after the course of a year this is no longer the case at all. Rather, the ψ-function then describes a sort of blend of not-yet and of already-exploded systems. Through no art of interpretation can this ψ-function be turned into an adequate description of a real state of affairs; [for] in reality there is just no intermediary between exploded and not-exploded.

Einstein's gunpowder experiment was a direct challenge to Schrödinger's interpretation of the wavefunction. How could the wavefunction sensibly accommodate the contradictory components of exploded and not-exploded, being and not-being, 'blurring' them in some ridiculous superposition?

Schrödinger did not immediately back down, and their lively correspondence continued through the summer of 1935. However, in his letter of 19 August he acknowledged: 'I am long past the stage where I thought that one can consider the ψ-function as somehow a direct description of reality.' He then went on to describe a further thought experiment derived from Einstein's gunpowder example:

> Contained in a steel chamber is a Geigercounter prepared with a tiny amount of uranium, so small that in the next hour it is just as probable to expect one atomic decay as none. An amplified relay provides that the first atomic decay shatters a small bottle of prussic acid. This and—cruelly—a cat is also trapped in the steel chamber. According to the ψ-function for the total system, after an hour, *sit venia verbo*, the living and dead cat are smeared out in equal measure.

This is the famous paradox of Schrödinger's cat.

Through the summer months Schrödinger had been working on his own summary of the current situation in quantum theory in a series of three articles that was eventually published in the German journal *Die Naturwissenschaften*. Influenced by the arguments that had gone back and forth across the Atlantic in his correspondence with Einstein, he had decided to highlight the absurdity of the Copenhagen interpretation by bringing the 'measurement problem' into the macroscopic world using a 'quite ridiculous case'.

He described his cat paradox in a single paragraph. In essence this was the version he had described to Einstein, but with a little further elaboration. The radioactive substance (now unspecified) and the Geiger tube had to be secured against possible interference by the cat. On activation of the relay, a hammer is released which shatters the flask of prussic acid. 'The ψ-function of the entire system would express this [superposition] by having in it the living and dead cat (pardon the expression) mixed or smeared in equal parts.'

The diabolical scenario runs as follows. A cat is placed inside the steel chamber together with a Geiger tube containing a small amount of radioactive substance, a hammer mounted on a pivot and a phial of prussic acid. The chamber is closed. From the amount of radioactive substance used and its known half-life, we expect that within one hour there is a 50 per cent probability that one atom has disintegrated. If an atom does indeed disintegrate, the Geiger counter is triggered, releasing the hammer which smashes the phial. The prussic acid is released, killing the cat.

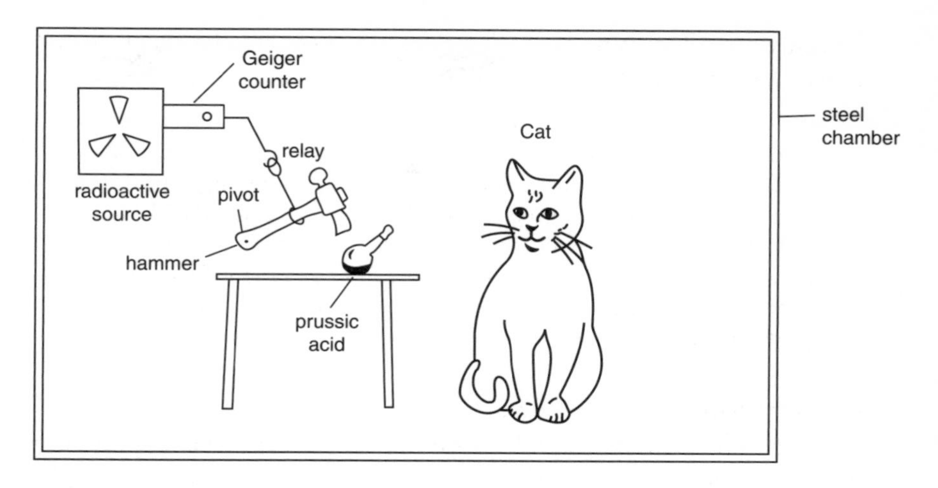

FIG 8 Schrödinger's cat. Must we suppose that, at that moment we lift the lid of the chamber, the superposition of live cat and dead cat collapses, and we observe that the cat is either completely alive or completely dead?

Prior to actually measuring the disintegration, the wavefunction of the atom of radioactive substance must be expressed as a linear combination of the possible measurement outcomes, corresponding to the physical states of the intact atom and the disintegrated atom. However, treating the measuring instrument as a quantum object and using the equations of quantum mechanics leads us to an infinite regress. We end up creating a linear combination of the two possible macroscopic outcomes of the measurement.

But what about the cat? As Schrödinger concluded, this example would seem to suggest that we should express the wavefunction of the system-plus-cat as a superposition of the products of the wavefunction describing a disintegrated atom and a dead cat and of the wavefunction describing an intact atom and a live cat. Prior to measurement, the physical state of the cat is therefore 'blurred'—it is neither alive nor dead but in some peculiar combination of both states.

We can perform a measurement on the cat by lifting the lid of the chamber and ascertaining its physical state. Do we suppose that, at that point, the wavefunction of the system-plus-cat collapses and we record the observation that the cat is alive or dead as appropriate?

Schrödinger's paradox highlighted the simple fact that in discussions of the collapse of the wavefunction, no reference had yet been made to the point in the measurement process at which the collapse is meant to occur. It might be assumed that the collapse occurs at the moment the microscopic quantum system interacts with the macroscopic measuring device. But is this assumption justified? After all, a macroscopic measuring device is composed of microscopic entities—molecules, atoms, protons, neutrons, and electrons. We could argue that the interaction takes place on a microscopic level and should, therefore, be treated using quantum mechanics.

The problem is that the collapse is itself not contained in *any* of the mathematical apparatus of quantum theory. As von Neumann had discovered, the only way to introduce such a collapse (or projection) into the theory was to *postulate* it.

Although obviously intended to be somewhat tongue-in-cheek, Schrödinger's paradox nevertheless raised an important difficulty. The Copenhagen interpretation says that elements of an empirical reality are defined by the nature of the experimental apparatus we construct to perform measurements on a quantum system. It insists that we resist the temptation to ask what physical state a particle (or a cat) is actually in prior to measurement as such a question is quite without meaning.

As far as the Copenhagen interpretation is concerned, Schrödinger's cat is indeed blurred: it is meaningless to speculate on whether it is really alive or dead until the box is opened, and we look. And, although in his paper Schrödinger asked this question in a different context, it is nonetheless legitimate to ask: What if we don't look?

This anti-realist interpretation sits uncomfortably with some scientists, particularly those with a special fondness for cats. Einstein saw the paradox as yet further evidence for the basic incompleteness of quantum theory. In his reply to Schrödinger's letter of 19 August, he wrote:

> ...your cat shows that we are in complete agreement concerning our assessment of the character of the current theory. A ψ-function that contains the living as well as the dead cat just cannot be taken as a description of a real state of affairs. To the contrary, this example shows exactly that it is reasonable to let the ψ-function correspond to a statistical ensemble that contains both systems with live cats and those with dead cats.

The cat paradox was not intended as a formal challenge to the Copenhagen view and it does not seem to have elicited any kind of formal response. Schrödinger wrote to Bohr on 13 October 1935 to tell him that he found Bohr's response to the EPR challenge, just published in *Physical Review*, to be unsatisfactory. Surely, he argued, Bohr was overlooking the possibility that future scientific developments might undermine Bohr's assertion that the measuring instrument must always be treated classically. Bohr replied briefly that, if they were to serve as measuring instruments, then these instruments simply could not belong within the range of applicability of quantum mechanics.

The infinite regress implied by the cat paradox is avoided if the measuring instruments are treated only as classical objects, as classical objects cannot form superpositions in the way that quantum objects can. This was self-evident to Bohr (as indeed it was to Schrödinger himself), but there remained no clues as to the precise origin and mechanism of the collapse of the wavefunction. It was just *supposed* to happen.

The community of physicists had in any case moved on, and probably had little appetite for endless philosophical challenges that, in the view of the majority, had already been satisfactorily addressed by Bohr. Neither Einstein nor Schrödinger offered an alternative, beyond Einstein,

Podolsky, and Rosen's instinct that a more complete theory was somehow possible.

Besides, any such theory would surely need to invoke new variables, responsible for maintaining strict causality and determinism but thus far 'hidden' from observation, much as the real motions of atoms and molecules are hidden variables in Boltzmann's statistical mechanics. In his book *Mathematical Foundations of Quantum Mechanics,* von Neumann had already provided a mathematical proof that all such hidden variable theories are impossible.

There were other things for the physicists to be worried about. In seeking to restore the conservation of energy in radioactive beta-decay, in which a high-speed electron (a beta-particle) is ejected from the nucleus, Pauli had been obliged to introduce another new particle. In the interests of energy book-keeping, this had to be a light, electrically neutral particle which interacted with virtually nothing. To distinguish it from Chadwick's 'heavy neutron', in 1934 Enrico Fermi called the new particle the *neutrino.* In an article published in *Nature* later that year, German émigré physicists Hans Bethe and Rudolf Peierls claimed: '...there is no practically possible way of observing the neutrino.'

It didn't end there. The identity of the particles responsible for some penetrating cosmic rays was debated for several years. Some physicists believed they were due to electrons. Others thought they were protons. But the mass of the particles responsible was intermediate between the mass of the electron and the proton. In 1937 Carl Anderson and Seth Neddermeyer concluded that this was another new particle, a heavy version of the electron. It was called a mesotron, shortened to meson, then later called a μ-meson or just muon. Incensed, Galician-born American physicist Isidor Rabi demanded to know: 'Who ordered that?'

In addition to the proton and electron, there were now positrons and muons. Proposed, but not yet discovered, were Dirac's anti-proton, Pauli's neutrino, and the anti-neutron. Dirac's unitary 'dream of philosophers' was in tatters.

But, as Hitler's dreams of a Third German Reich pushed Europe closer and closer to war, it would be another discovery in nuclear physics that would come to occupy the minds of many of the world's leading physicists.

Interlude

The First War of Physics[1]

Christmas 1938—August 1945

The discovery of the neutron in 1932 had not only provided the physicists with deeper insight into the structure of the nucleus, it had also given them another weapon with which to penetrate its secrets. As an electrically neutral sub-atomic particle, the neutron could be fired into a positively charged nucleus without being diverted by the force of electrostatic repulsion.

Research results on the neutron bombardment of uranium reported by Fermi in Rome in 1934 caught the attention of German chemist Otto Hahn in Berlin. Together with his assistant Fritz Strassman, Hahn now obtained results that seemed to make little sense. He described these in a series of letters to his long-time collaborator, Austrian Lise Meitner, now exiled in Sweden. Whilst on vacation in Sweden on Christmas Eve 1938, Meitner and her nephew, physicist Otto Frisch, recognized the results as evidence of nuclear fission.

Bohr heard about this discovery just as he was about to leave Copenhagen for a visit to Princeton. 'Oh what idiots we have all been! Oh but this is wonderful,' he declared, 'This is just as it must be.' He was intending to continue his debate with Einstein on the interpretation of quantum theory but the subject of discussion in Bohr's stateroom as he crossed the Atlantic was nuclear fission. At Princeton he worked with American physicist John Wheeler and deduced that nuclear fission in uranium is due to the rare isotope uranium-235, which makes up only a tiny proportion of naturally occurring uranium. Could this be used

[1] Adapted, with permission of the publisher, from Jim Baggott, *Atomic: The First War of Physics and the Secret History of the Atom Bomb, 1939–49*, Icon Books, London, 2009, pp. 86–92.

to build a 'super-bomb'? 'Yes, it would be possible to make a bomb,' Bohr told his Princeton colleagues in April 1939, 'but it would take the entire efforts of a nation to do it.'

Word spread quickly. In April, the Reich Bureau of Standards and the German War Office both established nuclear research projects. In August, Einstein wrote a letter to American President Franklin Roosevelt which warned of 'extremely powerful bombs of a new type'. A few days later Lindemann wrote to advise Winston Churchill MP that atomic super-weapons could not be expected to be available for many years.

Following the German invasion of Poland, Allied governments declared war on Germany on 3 September 1939. The two German nuclear research projects were hastily consolidated into a single Uranverein, or 'Uranium Club' under the auspices of German Army Ordnance. Heisenberg received notification of his call to serve on the Uranverein on 25 September.

The first war of physics had begun.

In the late summer of 1941 Heisenberg and his Leipzig assistant Robert Döpel set up their second experimental nuclear reactor, named L-II, in an aluminium sphere about 75 centimetres in diameter containing a little over 140 kilos of uranium oxide and about 160 kilos of 'heavy' water moderator, water in which the hydrogen atoms are replaced by heavier deuterium isotopes. The results were negative but, after correcting for the absorption of neutrons by the aluminium vessel, their calculations suggested the merest hint of neutron multiplication of the kind necessary ultimately to achieve a self-sustaining nuclear chain reaction.

The physicists sensed they were on the right track. This was little more than intuition, a 'gut feeling', but Heisenberg later stated that: 'It was from September 1941 that we saw an open road ahead of us, leading to the atomic bomb.'

Heisenberg probably did not doubt that German science was already far ahead of anything that physicists might have done in Britain or America. This left him in a unique and, possibly, quite uncomfortable position. It was his Uranverein colleague and friend Carl Friedrich von Weizsäcker who urged him to consult his former mentor, Niels Bohr. But Heisenberg's reasons for seeking such a meeting may have been quite complex.

Bohr had preferred to remain at his Institute for Theoretical Physics in Copenhagen following the German occupation of Denmark in April 1940. Of Jewish descent, Bohr was considered a 'non-Aryan' by the occupation forces but, like all eight thousand Danish Jews, was protected—at least temporarily—by the agreement reached with the Danish government designed to preserve the fiction that the Nazis were there by invitation.

That Heisenberg and Weizsäcker were concerned for Bohr's welfare is not in doubt. Growing moral qualms about the work the Uranverein was doing may also have played a part in Heisenberg's decision to visit Bohr in Copenhagen. In his memoir, written almost thirty years after the event, Heisenberg recalled Weizsäcker's proposal. 'It might be a good thing,' Weizsäcker had said, 'if you could discuss the whole subject with Niels in Copenhagen. It would mean a great deal to me if Niels were, for instance, to express the view that we are wrong and that we ought to stop working with uranium.'

According to Heisenberg, his primary purpose was to seek guidance from Bohr on the morality of working on scientific problems that could have 'grave consequences in the technique of war'. As Heisenberg's Uranverein colleague Peter Jensen later put it: Heisenberg, the high priest of German theoretical physics, sought absolution from his Pope. Or, perhaps, as German émigré physicist Rudolf Peierls later suggested: '[Heisenberg] had agreed to sup with the devil, and perhaps he found that there was not a long enough spoon.'

Heisenberg looked forward with great eagerness to the prospect of talking to his former mentor. To Heisenberg, and indeed to many other physicists of his generation, Bohr had long been something of a father-figure. As Heisenberg's wife Elisabeth later wrote:

> In Tisvilde, the beautiful vacation home of the Bohrs, he had played with their children and taken them for rides on a pony wagon: he had gone for long sailing trips on the ocean with Bohr, and Niels had visited him in his ski cottage; together they had grappled with the problems of physics, and he thought he could talk about anything with Bohr.

But there may have been other reasons for seeking a meeting. It is possible that, alarmed by reports of American efforts to build an atomic bomb that had appeared in the Swedish press, Heisenberg and Weizsäcker may have also wanted to discover what Bohr knew.

Having decided to pay Bohr a visit, Heisenberg now faced a number of practical hurdles. Although Germany dominated much of continental Western Europe, travel was restricted and the authorities were initially reluctant to let Heisenberg make the trip. Weizsäcker proposed a potential solution. He had already given several lectures in occupied Copenhagen, the most recent being a lecture on the philosophical implications of quantum theory at Bohr's institute. At Weizsäcker's prompting, an invitation was issued to both Heisenberg and Weizsäcker to participate in a symposium on astronomy, mathematics, and theoretical physics at the newly formed German Cultural Institute in Copenhagen.

The proposal was initially rejected by the Reich Ministry of Education but, after some arm-twisting by the German Foreign Office (whose officials suggested that State Secretary Ernst von Weizsäcker—Carl Friedrich's father—might intervene), approval was granted provided Heisenberg kept a low profile and stayed only a few days.

Despite this proviso, Heisenberg arrived in Copenhagen early on Monday, 15 September, four days before the conference was due to start on the following Friday. As he described it in a letter to his wife he composed at various times during his trip, it was a journey into the recent past:

> Here I am once again in the city which is so familiar to me and where a part of my heart has stayed stuck ever since that time fifteen years ago. When I heard the bells from the tower of city hall for the first time again, close to the window of my hotel room, it gripped me tight inside, and everything has stayed so much the same as if nothing out there in the world had changed. It is so strange when suddenly you encounter a piece of your own youth, just as if you were meeting yourself.

Such was his eagerness to see Bohr that he made his way to Bohr's Carlsberg residence that first night, walking through the darkened city under a clear and starry sky. He was relieved to discover that Bohr and his family were doing fine. Their conversation quickly turned to the 'human concerns and unhappy events of these times'. In his letter to his wife he expressed some dismay that 'even a great man like Bohr can not separate out thinking, feeling, and hating entirely,' but then added, 'But probably one ought not to separate these ever.'

But Heisenberg remained stubbornly insensitive to the perceptions and feelings of his former colleagues. In occupied Copenhagen in September 1941, with much of Europe conquered by Axis forces, with the German Army Group North just seven miles from Leningrad, with Kiev encircled and the assault on Moscow about to begin, it was not difficult to conceive a German victory. There was a certain startling inevitability about the impending Nazi domination of Europe, and everything this implied.

Heisenberg, the pragmatist, had made his bargain long ago. And here he now was, a representative of German culture, in Copenhagen at the behest of the German Cultural Institute, promulgating what many Danes may have perceived to be thinly veiled Nazi propaganda. Surely, Heisenberg may have reasoned, in the face of the inevitability of Nazi victory it was in the best interests of themselves and of physics for his former colleagues in occupied countries to make their bargain too?

Bohr and his colleagues boycotted the formal conference proceedings, but Heisenberg still sought them out. He visited Bohr's Institute and joined some of the physicists for lunch on a couple of occasions. Among those present were Christian Møller and Stefan Rozental. They later had bitter recollections of the discussion: '[Heisenberg] stressed how important it was that Germany should win the war... the occupation of Denmark, Norway, Belgium and Holland was a sad thing but as regards the countries in Eastern Europe it was a good development because these countries were not able to govern themselves.'

It was during Heisenberg's second meeting with Bohr on the Wednesday evening that he raised the issue of atomic weapons. Their subsequent recollections of this highly charged meeting were vague and contradictory. Heisenberg remembers that they took an after-dinner stroll, principally to avoid the risk of Gestapo surveillance. Bohr believed the conversation had taken place in his study. It would have made sense that Heisenberg would want to talk somewhere more secure, away from prying ears or listening devices, because when he raised the question of the military application of atomic energy with Bohr he was in principle committing an act of treason.

The conversation got off to a poor start. And then it got rapidly worse. Bohr had heard about Heisenberg's insensitive remarks, and became angry when Heisenberg not only defended Germany's invasion of the Soviet

Union but further argued that it would be a good thing if Germany were to win the war. When Heisenberg finally raised the question of working on an atomic bomb, Bohr was completely and utterly shocked.

As far as Bohr understood, he had already shown in 1939 that achieving an explosive nuclear chain reaction would 'take the entire efforts of a country'. Yet here was his friend and former colleague, with whom he had shared some of the most thrilling moments of scientific discovery in his life, explaining with some impatience that a bomb was possible and that he was working on it for the Nazis. In a letter that Bohr composed to Heisenberg long after the war, but which he never sent, he wrote:

> ...in vague terms you spoke in a manner that could only give me the firm impression that, under your leadership, everything was being done in Germany to develop atomic weapons and that you said that there was no need to talk about details since you were completely familiar with them and had spent the past two years working more or less exclusively on such preparations.

Heisenberg may have even drawn a sketch to explain his work, though this now seems quite doubtful. When Bohr produced this some years later it appeared to be a sketch of a reactor, but whether Heisenberg intended it or Bohr misunderstood, Bohr assumed it was a sketch of an atomic bomb. Even worse was to come. Heisenberg appeared to be probing Bohr for any information about an Allied bomb programme. Was this, after all, an intelligence mission? On whose authority was Heisenberg now acting?

After the war Heisenberg claimed he was trying—through Bohr—to establish a commitment from nuclear scientists not to develop atomic weapons. Whether this was really his intention or not, Bohr interpreted his efforts as those of a brashly confident representative of an aggressive occupying power bent on delivering the ultimate weapon to his masters. In another letter to Heisenberg that was never sent, Bohr wrote:

> It had to make a very strong impression on me that at the very outset you stated that you felt certain that the war, if it lasted sufficiently long, would be decided with atomic weapons...You added, when I perhaps looked

> doubtful, that I had to understand that in recent years you had occupied yourself almost exclusively with this question and did not doubt that it could be done.

Heisenberg later recalled Bohr's observation that it would be hopeless to try to influence the activities of physicists now working in various countries, 'and that it was, so to speak, the natural course in this world that the physicists were working in their countries on the production of weapons.'

Despite this exchange, they appeared to have parted amicably. Heisenberg met with Weizsäcker on the Langelinie, the picturesque walk near Copenhagen harbour. Heisenberg admitted: 'You know, I'm afraid it went badly wrong.'

Heisenberg and Weizsäcker attended the conference on 19 September. It was attended by just five local astronomers from the observatory in Copenhagen. In his obligatory report on the visit, Heisenberg commented that 'our relations with scientific circles in Scandinavia have become very difficult.'

After attending a lunchtime reception at the German embassy, Heisenberg joined Weizsäcker for a last visit to the Bohr household. This final visit was free of political or scientific discussions. Bohr read aloud and Heisenberg played a Mozart sonata.

If Heisenberg had really been trying to stop the development of an atomic super-weapon, he had failed. If he had been seeking absolution, he had failed. But the manner of his failure was to have profound implications. The wartime efforts of nuclear scientists in Britain and America were driven by a deep-rooted fear of what the Nazis might do with such a super-weapon. Against this background of fear, the interpretation of the *intent* of the German physicists, especially of Heisenberg, and the determination of the Nazi military authorities provided a critical moral justification for the work that Allied physicists were now doing.

In the end, it did not matter much precisely what Heisenberg had really intended to say to Bohr. As a result of the meeting one of the most respected and revered of nuclear physicists, a Danish half-Jew living under

Nazi occupation, had been left with the firm impression that Heisenberg was working earnestly to deliver an atomic bomb to Hitler's arsenal.

Heisenberg had been unable to see all possible ends. His dangerous Faustian bargain was beginning to have unexpected consequences.

American efforts to develop nuclear weapons were spurred by the discovery that just a few pounds of pure uranium-235 would be required to support an explosive fast-neutron chain reaction. For a substance as dense as uranium this was a critical mass about the size of a golf ball. Roosevelt sanctioned the American bomb project in November 1941, just days before the Japanese attack on the American fleet at Pearl Harbor. When the Army took control in September 1942, it became the Manhattan Project.

The following month, American physicist J. Robert Oppenheimer was appointed as scientific director of a new weapons research laboratory to be established at Los Alamos in New Mexico. A few months later the world's first nuclear reactor was successfully tested in a squash court at the University of Chicago.

Bohr escaped from Nazi-occupied Copenhagen in September 1942 with the aid of the Danish resistance and the British Secret Intelligence Service. He arrived at Los Alamos in January 1943. Revered by many at Los Alamos, he became a father-confessor, particularly for some of the younger physicists such as American Richard Feynman. His September 1941 conversation with Heisenberg had left him in no doubt that, under Heisenberg's leadership, everything was being done in Germany to develop atomic weapons. He provided a forceful and timely reminder of the threat of a Nazi weapon, and everything this implied.

Any Los Alamos scientist wrestling with his conscience over what he was helping to build would find the moral justification for it in Bohr's own experiences. 'He made the enterprise seem hopeful,' Oppenheimer observed after the war of Bohr's role, 'when many were not free of misgiving.'

But, in truth, Heisenberg and other leading Uranverein physicists had shied away from making a substantial bid for military funding for a large-scale weapons project when they had the opportunity in June 1942. Citing the huge technological challenges that stood in the way of a bomb, they bid for much more modest funding to support research on a nuclear reactor. Though rather put out by the German physicists' modest requests, Albert Speer, Hitler's Reichsminister for Armaments, accepted that an atomic bomb could not be produced in time to affect the course of the war.

The German physicists failed to build even a working reactor. When Heisenberg was captured by Allied forces in May 1945 and interned with other Uranverein physicists at

Farm Hall in Cambridgeshire, England, hidden microphones recorded their reaction to the news of the Allied atomic bombing of Hiroshima.

The bombing of Hiroshima and Nagasaki in August 1945 was just one more catastrophe on top of a long list of utter catastrophes; one that brought to an end a long and profoundly immoral war with such a powerful exclamation that it burned into the consciousness of all who lived then, and all who have lived since. But in ending the war the physicists had unleashed a primordial force upon the world, the threat of which would endure long after the perpetrators of evil were gone. Through their efforts they had helped to put the world in an even greater jeopardy.

'In some sort of crude sense which no vulgarity, no humor, no overstatement can quite extinguish, the physicists have known sin,' said Oppenheimer in 1947, 'and this is a knowledge which they cannot lose.'

PART IV

Quantum *Fields*

18

Shelter Island

Long Island, June 1947

The time had finally come to turn attention away from weapons of war and return to the problems of quantum theory. Theoretical physics had been in the doldrums for nearly two decades. 'The last eighteen years have been the most sterile of the century,' remarked Columbia University physicist Isidor Rabi in 1947.

The debate about interpretation was over, or so most physicists thought. But there were plenty of other things to be worrying about. Dirac's breakthrough had created a relativistic equation for the electron which predicted the property of electron spin and the existence of the anti-electron. This was indeed a triumph but it also quickly became a dead-end. Quantum mechanics could successfully predict the properties and behaviour of 'stationary' states in which the integrity of individual quantum particles such as electrons is preserved, but it could not deal with transitional situations in which particles are destroyed and new particles created.

For example, it had become apparent that when an electron collides with a positron, the two particles could be expected to 'annihilate' to form high-energy (gamma-ray) photons. Conversely, at high enough energies, gamma-ray photons could spontaneously create an electron–positron pair. If quantum mechanics couldn't handle this, what could? Some physicists had acknowledged that a quantum theory of fields was required, starting with a quantum version of Maxwell's electromagnetic field.[1]

[1] In a field theory, the strength of the field is specified at each point in space–time. Fields may be scalar (magnitude only), vector (magnitude and direction), or tensor (a generalization of a vector field in which the vector varies from one point to the next).

These physicists recognised that fields are more fundamental than particles. It was believed that a proper quantum field description should yield particles as the quanta of the fields themselves, carrying the force from one interacting particle to another. It seemed clear that the photon was the quantum of the electromagnetic field, created and destroyed when charged particles interacted. To account for the creation and annihilation of electrons, a quantum theory of an 'electron field' was needed.

Heisenberg and Pauli had developed a quantum field theory in 1929. When applied to the electron field it became a form of quantum electrodynamics (QED). It was found that the resulting field equations did not yield exact solutions, and had had to resort to an approach based on a perturbation expansion.[2] *They then ran into a major problem. Higher-order terms in the expansion should have provided smaller and smaller corrections to the result, but instead they found that some terms in the expansion yielded infinite corrections.*[3] *This was a result that made no physical sense.*

The infinities were identified to be due to the 'self-energy' of the particles. They resulted from particles interacting with their own fields, a consequence of treating them theoretically as point particles, with no volume or shape. There was no obvious solution.

And there was yet more trouble brewing. The problem with the infinities was a theoretical or mathematical problem. In 1947, new experimental data pointed to problems of a more practical kind. Closer inspection had showed that the predictions of Dirac's theory did not, after all, agree with experiment.

Quantum theory was in crisis once again.

In 1945 Duncan MacInnes, a physical chemist at the Rockefeller Institute in New York, conceived the idea of a series of small, intimate conferences at which both distinguished scientists and young, up-and-coming scholars could address the problems facing their discipline. He had grown frustrated and wary of large-scale conferences at which academics appeared to talk past each other without resolving the issues. 'The programs are

[2] In a perturbation expansion, the equation is cast in terms of a zeroth-order (or zero-interaction) expression which can be solved exactly. To this is added additional (perturbation) terms which form a power series (first-order, second-order, etc.), each in principle providing a smaller and smaller correction to the zeroth-order result. The accuracy of the final result then depends on the number of perturbation terms included in the calculation.

[3] In these early versions of QED, the first-order terms are solvable and produce physically realistic results. It is the second-order terms that mushroom to infinity.

so crowded that presentation of the papers takes most of the time and discussion is slighted if it takes place at all,' he wrote.

Much of Europe was still recovering from the aftermath of a devastating war, and European physicists were slowly picking up the pieces of their academic careers. The atom bomb was seen as a triumph of American science and ingenuity (supported, of course, by many European émigré physicists and British physicists who had contributed to the Manhattan Project). Perhaps it was time to put American physics firmly on the world stage, to show that it had come of age, by establishing an American equivalent of the Solvay conferences.

MacInnes secured financial support from the US National Academy of Sciences. The shortlist of proposed conference topics included quantum mechanics. Both Pauli, who had moved to Princeton University in 1940, and American physicist John Wheeler were asked to advise on attendance and discussion topics. After many exchanges, the first of the National Academy-sponsored conferences was held on 2–4 June 1947 at the Ram's Head Inn, a small clapboard hotel and inn on thinly populated Shelter Island, at the eastern end of New York's Long Island.

When the physicists arrived at Greenport, Long Island, they were treated like celebrities. The prestigious group was given a motorcycle escort which ran through red lights. The conference attracted the attentions of the press: 'Twenty three of the country's best-known theoretical physicists—the men who made the atomic bomb—gathered today in a rural inn to begin three days of discussion and study, during which they hope to straighten out a few of the difficulties that beset modern physics.'

Among them were Oppenheimer, Hans Bethe, who had led the Theoretical Division at Los Alamos and had returned to academia at Cornell University, Victor Weisskopf, Isidor Rabi, Edward Teller, John Van Vleck, John von Neumann, and Hendrik Kramers. The new generation was represented by Wheeler, Abraham Pais, Richard Feynman, Julian Schwinger, and former Oppenheimer students Robert Serber and David Bohm.[4] Willis Lamb, also a former Oppenheimer student and now an expert in the application of microwave techniques at Columbia University's Radiation

[4] Einstein was invited to attend but declined for reasons of ill health.

Laboratory, joined the group to report on some disturbing new results on the spectrum of the hydrogen atom.

The emphasis of the conference was on informal discussion rather than the formal presentation of prepared papers. The physicists were the only occupants of the Inn, which had opened early in the season to accommodate them. When Lamb opened the conference on the morning of 2 June, what he had to say dominated subsequent discussion.

A few months previously Lamb had announced the results of new, more detailed spectroscopic studies of the hydrogen atom. Trained as a theoretician, Lamb had realized that if he wanted to put some of his ideas to the test, then he was going to have to do the experiments himself. He focused his attention on the behaviour of two atomic states of hydrogen which share the same principal quantum number but possess different values of the orbital angular momentum quantum number.

According to Dirac's theory these states were predicted to have the same energy (such states are said to be *degenerate*). As a consequence of this degeneracy, the microwave spectrum involving transitions to these states was expected to show a single line.

Not so. Lamb had discovered with graduate student Robert Retherford that there are in fact *two* lines. One of the hydrogen-atom states is shifted in energy relative to the other, a phenomenon that quickly became known as the *Lamb shift*. The physicists gathered on Shelter Island now heard the full details of Lamb's latest experimental results. Oppenheimer and Weisskopf led the discussion that followed. Oppenheimer speculated that a purely quantum electrodynamic effect was the 'least improbable explanation' of the shift.

Rabi then stood to deliver a second experimental challenge. He reported the results of experiments conducted by his Columbia University students John Nafe and Edward Nelson and his Columbia colleagues Polykarp Kusch and H.M. Foley. They had found that the g-factor of the electron is not exactly 2, as required by Dirac's relativistic theory. It is slightly larger, more like 2.00244.

To be sure, this is a small difference; a little over 0.1 per cent. But such differences were now well outside the bounds of experimental error. For the theoreticians, these experimental findings were profoundly interesting and stimulating. It seemed that the problems with QED were nothing to do with zeros or infinities, after all. The problems were rather to do with the very finite differences between two energy states of the hydrogen atom and the very finite g-factor for the electron.

The discussion continued long into the night, over hastily eaten meals. The delegation split into groups of two and three, the corridors echoing their arguments, as the physicists regained their passion. 'It was the first time people who had all this physics pent up in them for five years could talk to each other without somebody peering over their shoulders and saying "Is this cleared?",' remarked Schwinger.

The second day of the conference was concerned with theory. Kramers opened the session with a short lecture summarizing his recent work on a classical theory of the electron. He outlined a new approach to treating the mass of an electron in an electromagnetic field. Suppose the self-energy of the electron appears as an additional contribution to the mass of the electron, he explained. The mass that we observe in experiments would then include this contribution. In other words, the observed (or 'dressed') mass of the electron is equal to an intrinsic or 'bare' mass plus an 'electromagnetic mass' arising from the electron's interaction with its own electromagnetic field.

The 'bare' mass is a purely theoretical quantity. It is the mass that the electron would possess in the absence of an electromagnetic field. Clearly, the mass that the physicists had to deal with is the observed or dressed mass. This meant that the equations had to be rewritten in terms of the observed mass. In other words, they had to be 'renormalized'.

Here was more food for thought. In subsequent discussions, Weisskopf and Schwinger suggested that a calculation of the Lamb shift based on Kramer's ideas might actually yield a finite answer.

The young New Yorkers Feynman and Schwinger were both precocious physicists and mathematical prodigies. Schwinger had spent much of his time at the City College of New York ignoring his classes, sitting in the library reading advanced mathematics and physics texts. He had written his first (unpublished) paper in 1934, at the age of sixteen. When he flabbergasted Rabi by explaining a subtle mathematical point in connection with the Einstein, Podolsky, Rosen paper, Rabi arranged for him to transfer to Columbia. He graduated in 1936 and secured his doctorate under Rabi three years later. He then worked for two years with Oppenheimer at the University of California, Berkeley, before moving to Purdue University in Indiana. He became an associate professor at Harvard University in February 1946, and was promoted to full professor a year later. He was just 29.

Where Schwinger was a quiet introvert, Feynman was a loud extravert. He was blessed with an intuitive ability to solve complex mathematical problems in his head, an ability to 'visualize' the physical interpretation of the equations and deduce the solutions. Feynman had graduated from MIT in 1938 and moved to Princeton to work with Wheeler shortly afterwards. In 1943 he was recruited to Los Alamos, where his bongo-playing and safe-cracking antics would be the subject of endless anecdotes told at the many parties hosted by the Oppenheimers, fuelled by Robert's legendary vodka martinis.

In June 1946 Feynman had borrowed Klaus Fuchs' car to drive to a hospital in Albuquerque where his wife Arline was seriously ill with tuberculosis. After suffering delays caused by three flat tyres, he arrived in time to be with her as she died.

From Los Alamos Feynman went to Cornell University, but struggled to regain his enthusiasm for physics. Eventually, his attention was captured by a seemingly trivial problem with a spinning, wobbling plate he had seen in the cafeteria. He decided to play:

> I went on to work out equations of wobbles. Then I thought about how electron orbits start to move in relativity. Then there's the Dirac equation in electrodynamics. And then quantum electrodynamics. And before I knew it (it was a very short time) I was 'playing'—working, really—with the same old problem that I loved so much, that I had stopped working on when I went to Los Alamos: my thesis-type problems; all those old-fashioned, wonderful things.

The contrasts between Feynman and Schwinger extended to the way they did physics. Schwinger was conservative. His was a thoughtful, considered, deductive approach. He was committed to doing it right, no matter how elaborate or unwieldy the mathematics. Feynman was a radical. He was more of a visionary, his approach no less considered but much more intuitive. He was happy to push through the mathematics in order to get quickly to a satisfying solution.

Schwinger had been quiet for much of the conference. On the third and final day Oppenheimer asked Feynman to present his most recent work. Feynman now stood to describe his new take on non-relativistic quantum mechanics, '...with a clear voice, great rush of words and illustrative gestures sometimes ebullient.'

The basis of Feynman's approach can be found in some of the simplest observations of classical physics. Classical light rays travel in straight lines because straight lines represent the least amount of time required for light to travel from its source to its destination. This was a principle first enunciated by Pierre de Fermat (of 'last theorem' fame) in 1657. But how is the light supposed to 'know' in advance what the path of least time is?

It is the wave nature of light that provides a solution to this particular mystery. Light does not need to know the path of least time in advance because it takes *all paths* from its source to its destination.

Imagine a wave oscillating up and down as it moves forward through space towards its destination. Clearly, a second wave with the same wavelength, starting out from the same place but heading for a place just slightly displaced from the destination of the first wave will look very similar in terms of the time and space dependences of the wave amplitude. However, as we increase the displacement of the destination of the wave further and further, the time and space dependences of the waves become increasingly 'misaligned': at specific points in time and space peak no longer lines up with peak and trough no longer lines up with trough. The result is destructive interference and a loss of *coherence* of the light.

We get constructive interference and maximum coherence for light paths that do not differ significantly in terms of distance and therefore time. The mystery is now resolved. When light travels through a single medium (such as air), the light paths that do not differ significantly in terms of distance and time are all clustered around the shortest, straight-line path from source to destination, which is also the path of least time.

The principle is one of least time and not least distance. If the light passes from one medium into another at an oblique angle (for example, from air into water), then the path of least distance remains the straight line drawn from source to destination. But light slows down in water, and this path of least distance does not minimize the amount of water that light has to travel through to get to the destination. The more water the light has to travel through, the longer it takes. We can identify a 'path of least water' which enters the water vertically and goes straight down to the destination, but this would involve a very round-about path, lengthening the distance, and hence the time travelled, through the air. The preferred path—the one of least time—represents a compromise. It is not the path of least distance (which is why a stick half-immersed in water appears bent) nor is it the path of least water.

Feynman elevated these relatively simple physical principles into an alternative formulation of non-relativistic quantum theory equivalent to wave and matrix mechanics. He represented the passage of a quantum particle from one place to another as a sum[5] over all the possible paths the particle could take or, alternatively, as the sum over all possible 'histories' of the particle's motion. The probability of finding the particle at a specific location can be determined from the amplitudes of all the various paths it can take to get there.

Here is one particularly illuminating way to think about Feynman's approach. Imagine inserting a screen between a point source of light and a photographic plate. If we drill a hole through the screen then the amplitude for the photons to propagate from the source to the photographic plate is simply the amplitude of the path through the hole. If we now drill a second hole, we need to sum the amplitudes of the paths through both holes. We can keep repeating this for three, four, five holes, each time summing the amplitudes of all the possible paths. By the time we have drilled an infinite number of holes in the screen, the screen is no longer there. We conclude that we must still sum the amplitudes of all the possible paths from the source to the photographic plate.

This is a determinedly particle representation—the wave aspect of quantum particles is captured in terms of their *phases*, whatever their origin, and the way these phases combine. A certain kind of combination results in effects which appear much like wave interference.

[5] Strictly speaking, an integral.

It was as though the Copenhagen interpretation had never been conceived.

However, apart from providing an interesting, more visualizable reinterpretation of quantum mechanics, Feynman's approach yielded no new results and, arguably, no new insights. He had tried, and so far failed, to develop a path-integral formulation for the Dirac equation. His approach was simply the result of playing with some new ideas. Abraham Pais later described the effect Feynman's lecture had on the assembled audience: 'At that time no one could follow what he was talking about.' Suffice to say, they were not overwhelmed.

Feynman was nevertheless awed by his experiences at the Shelter Island conference. 'There have been many conferences in the world since,' he explained many years later, 'but I've never felt any to be as important as this... The Shelter Island conference was my first conference with the big men... I had never gone to one like this in peacetime.'

After the conference, Bethe returned to New York and picked up a train to Schenectady, where he was engaged as a part-time consultant to General Electric. Like many of his contemporaries, he had been thinking deeply about the Lamb shift and the discussions during the conference now prompted him to attempt a calculation. As he sat on the train he played around with the equations of QED.

The existing theories of QED predicted an infinite shift, a consequence of the self-interaction of the electron with the electromagnetic field. Bethe now followed Kramers' suggestion and identified the divergent term in the perturbation expansion with an electromagnetic *mass* effect.

So, he reasoned, what if we now subtract the expression for a free electron (including the divergent self-energy term) from the expression for the electron bound in a hydrogen atom (also including the divergent self-energy term). It sounds as though subtracting infinity from infinity should yield a nonsensical answer, but Bethe now found that in a non-relativistic version of QED this subtraction produced a result that, though it still diverged, did so much more slowly. He figured that in a fully relativistic QED this normalization procedure would eliminate the divergence completely and give a physically realistic answer.

Because the procedure had greatly reduced the rate of divergence even in the non-relativistic case, he was able to make an intelligent guess and impose a relativistic limit for the divergence. He was thus able to obtain a theoretical estimate for the Lamb shift. He obtained a result just four per cent larger than the experimental value that Lamb had reported. Oppenheimer had been right. The shift was a purely quantum electrodynamic effect.

In great excitement, Bethe called his Cornell colleague Feynman from Schenectady to tell him what he had discovered. On his return to Cornell at the beginning of July, Bethe gave a lecture on the calculation and speculated on ways in which a more formal relativistic limit could be applied.

After the lecture, Feynman came up to him. 'I can do that for you,' he said, 'I'll bring it in for you tomorrow.'

19

Pictorial Semi-vision Thing

New York, January 1949

But Feynman was unable to produce an answer for Bethe the next day. There were still some things about the calculations that Feynman had to learn in order to apply his path-integral approach. He learned quickly.

Schwinger learned quickly too, as the New York physicists became rivals in the race to build a relativistic QED that could be renormalized. The Lamb shift could be almost explained using a non-relativistic approach, as Bethe had demonstrated, but the g-factor anomaly would require a fully relativistic theory. 'The magnetic moment of the electron,' Schwinger explained, 'which came from Dirac's relativistic theory, was something that no non-relativistic theory could describe correctly. It was a fundamentally relativistic phenomenon...'

Where Schwinger was logical and conventional, rigorously plodding through the terms in the perturbation expansion, Feynman was intuitive, vigorously ploughing through the mathematics using little more than inspired guesswork. He was ready to reach for bizarre methods, if needed. When faced with a particularly stubborn problem in his analysis of the situation where a field produces an electron–positron pair which then annihilate to produce a new field, he recalled a suggestion made by Wheeler. He decided to treat the problem as if the particles could travel backwards in time.

He was still wrestling with the solution when the second in the series of small National Academy-sponsored conferences was convened. This took place on 30 March 1948 at the Pocono Manor Inn in the Pocono Mountains near Scranton, Pennsylvania.

At the Pocono conference, the small gathering of physicists was looking to Schwinger for the definitive answer on relativistic QED. This time both Bohr and Dirac were present. Schwinger's presentation, on the second day of the conference, was a virtuoso, but marathon, five-hour event. Eyes glazed over and minds became numb as Schwinger derived one mathematical result after another. It was only when Schwinger tried to make connections with the underlying physics that the audience came to life and asked questions. Only Fermi and Bethe, it seemed, had followed Schwinger through to the end.

In his own lecture Feynman had intended to talk primarily about the physics. He had, after all, arrived at his mathematics largely by trial-and-error and felt that he was on shaky ground. But Bethe now advised him: 'You should better explain things mathematically and not physically, because every time Schwinger tries to talk physically he gets into trouble.'

Feynman took Bethe's advice and changed his approach to the lecture. He talked mathematics. It was a disaster. The path-integral approach was entirely foreign to his audience, and soon Dirac was asking awkward questions about the mathematics, concerned about the implications of positrons travelling backwards in time. Bohr didn't like Feynman's approach at all. The very idea of particle trajectories was anathema to the Copenhagen interpretation. 'Bohr thought that I didn't know the uncertainty principle, and was actually not doing quantum mechanics right either. He didn't understand at all what I was saying.'

When the lectures had concluded, Feynman and Schwinger got together in the hallway and compared their results. Neither understood the other's equations but, despite their very different approaches, their results were identical.

'So I knew that I wasn't crazy,' Feynman said.

Schwinger's work was seen to be definitive, but his structure was so unwieldy that it seemed that Schwinger was the only theorist able to use it. However, shortly after returning to Princeton following the Pocono conference, Oppenheimer discovered that this was not a two-horse race, after all. Japanese theoretical physicist Sin-itiro Tomonaga was also working on a relativistic QED, using methods similar to Schwinger but a lot more straightforward.

Tomonaga had learned quantum physics in Japan under the guidance of the esteemed physicist Yoshio Nishina. He graduated from Kyoto University in 1929, together with his friend and colleague Hideki Yukawa. He remained in Kyoto for the next three years, working as an unpaid assistant, and in 1931 joined Nishina's group at the Institute of Physical and Chemical Research in Tokyo. He became the Nishina group theorist, working on problems in quantum physics in close collaboration with the experimentalists.

He followed developments in quantum electrodynamics through the papers of Dirac, Heisenberg, and Pauli. In 1937 he moved to Leipzig to work with Heisenberg on nuclear physics and quantum field theory, returning to Tokyo two years later. A paper on nuclear physics written whilst in Leipzig became a major part of his doctoral thesis. He was appointed to a professorship in physics at the Tokyo College of Science and Literature in 1940.

Physics had to give way to basic survival in the aftermath of Japan's unconditional surrender on 14 August 1945, but Tomonaga managed to keep a team of young physicists together and continued to work on QED. Stimulated by reports in 1947 of Lamb's experiments and Bethe's non-relativistic calculation of the Lamb shift, the Japanese physicists worked on methods of renormalization of both the electron mass and charge.

'The method of calculation was quite new,' explained one of Tomonaga's associates. 'We were sometimes at a loss how to calculate, since there were many new things to be solved. At first we attempted to use a relativistic covariant way of calculating in various ways, but we were always unsuccessful. Then we decided to evaluate in a way which is similar to conventional perturbation theory. Prof. Tomonaga also calculated some fundamental things, and gave us many valuable suggestions. The calculation was very tedious.'

Tomonaga sent a letter to Oppenheimer outlining the group's results in April 1948. Oppenheimer immediately sent copies to all the delegates who had attended the Pocono conference. He wrote:

> When I returned from the Pocono Conference, I found a letter from Tomonaga which seemed to me of such interest to us all that I'm sending you a copy of it. Just because we were able to hear Schwinger's beautiful report, we may better be able to appreciate this independent development.

Oppenheimer sent a cable to Tomonaga urging him to write a summary of his work, for which he would arrange speedy publication in the American journal *Physical Review*. The paper appeared in the 15 July issue.

The work of Tomonaga and his group in Tokyo made a favourable impression on a young, highly creative English physicist working with Bethe at Cornell. As Freeman Dyson later explained:

> Tomonaga expressed his [version of QED] in a simple, clear language so that anybody could understand it and Schwinger did not. When you read Schwinger you had the impression it was immensely complicated from the start. Tomonaga set the framework in a very beautiful way. To me that was very important. It gave me the idea that this was after all simple.

Tomonaga's and Schwinger's approaches were similar, but Feynman's was off the wall. He had developed a very singular, appealing, and intuitive diagrammatic way of describing and keeping track of the perturbation corrections. These became known as *Feynman diagrams.*

Feynman diagrams have only two axes, a vertical one for time and a horizontal one for 'space', effectively projecting a three-dimensional view of a quantum interaction to a single dimension. Feynman used the diagrams to provide a way to visualize the interactions of quantum particles such as electrons and photons in both space and time. No wonder it didn't meet with Bohr's approval.

For example, the interaction between two electrons is represented in a Feynman diagram in terms of electron paths pictured as continuous lines, their directions marked by arrows. As the electrons move closer to each other, they 'feel' a repulsive electromagnetic force and move apart. This force is carried by a 'virtual' photon as the quantum of the electromagnetic field ('virtual' because the photon is never visible), represented as a wavy line connecting the electron paths at their point of closest approach.

Reading such a diagram logically, we might conclude that one electron emits a virtual photon which is absorbed some short time later by the second electron. However, the symmetry of the interaction is such that there is in principle nothing to prevent us from concluding that the photon is emitted by the second electron and travels backwards in time to be absorbed by the first. The wavy line of the virtual photon therefore does not show a direction, since both directions are possible.

The diagrams serve as a useful aid to the visualization of the process, but Feynman intended them primarily as a book-keeping device for all

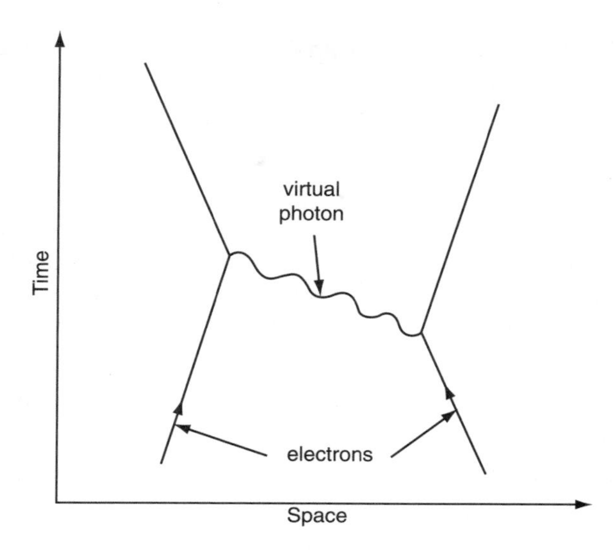

FIG 9 A Feynman diagram illustrating the interaction between two electrons. The electromagnetic force of repulsion between the electrons involves the exchange of a virtual photon at the point of closest approach.

the different ways that the quantum particles can interact in going from some initial state to some final state. For each diagram there is a term in the perturbation expansion containing an amplitude function which, when its modulus is squared, gives a probability for the process as a contribution to the total probability of conversion from initial to final state.

Feynman's path-integral approach demands that all the possible routes a quantum system can take from initial to final state be considered, no matter how seemingly improbable. The most important contribution to the interaction will be the 'direct' transition from initial to final state, but all the 'indirect' routes represent corrections to the interaction term and have to be included as well.

Feynman later tried to explain where the diagrams had come from and what they represented:

> The diagram is really, in a certain sense, the picture that comes from trying to clarify visualisation, which is a half-assed kind of vague [picture], mixed with symbols. It is very difficult to explain, because it is not clear...It is hard to believe it, but I see these things not as mathematical expressions

> but a mixture of a mathematical expression wrapped into and around, in a vague way, around the object. So I see all the time visual things associated with what I am trying to do...What I am really trying to do is to bring birth and clarity, which is really a half-assedly thought out pictorial semi-vision thing. OK?

But, no matter how vague their origin, the diagrams worked. The starting point for the evaluation of the g-factor of the electron is the interaction of an electron with a photon from a magnet. This is the simplest, most direct interaction involving an electron in some initial state at some initial time and position to some final state at a final time and position. Evaluating the electron magnetic moment from the interaction term represented by this diagram gives the Dirac result, a g-factor of exactly 2.

There are other ways this process can happen, however. Within this same interaction, a virtual photon can be emitted and absorbed by the same electron, viewed either forwards or backwards in time. This represents an electron interacting with its own electromagnetic field and, although the probability of this occurring is small, it is not zero. Including this term in the perturbation expansion makes the g-factor of the electron slightly larger than 2.

Further corrections are represented by self-interaction via two virtual photons. In a still further correction, a single virtual photon creates an electron–positron pair which then mutually annihilate to form another photon which is then subsequently absorbed.

Processes such as this last example seemed to imply that the rules of energy conservation had been abandoned once again. But it had by now become apparent that creation and destruction of virtual particles was entirely allowed within the constraints of Heisenberg's energy–time uncertainty relation. The energy required to create a photon or an electron–positron pair literally out of nothing can be 'borrowed' so long as it is 'given back' within the time frame determined by the uncertainty principle.

The probabilities of the occurrence of processes with ever-more complex and convoluted interactions may be very small, but they nonetheless provide important corrections in the perturbation expansion. A free electron doesn't simply persist as a point-particle travelling along a predetermined, classical path; it is surrounded by a swarm of virtual particles arising from self-interactions with its own electromagnetic field.

(a)

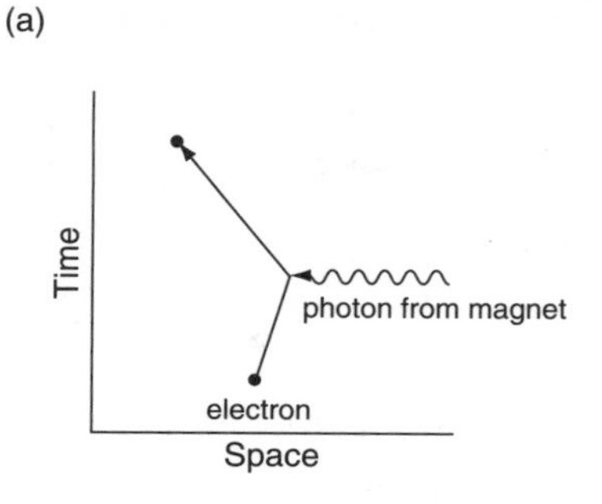

(b)

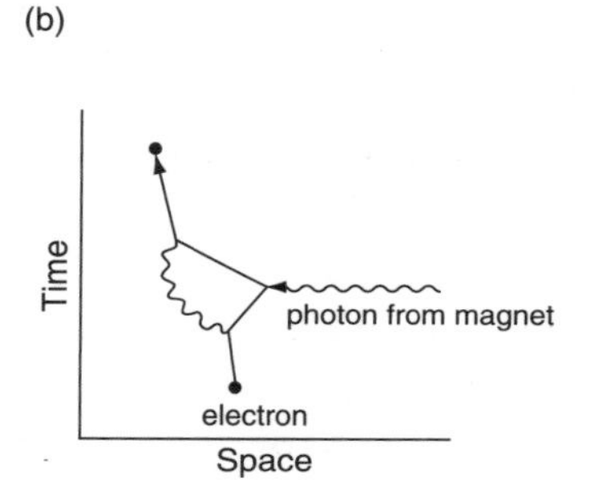

(c)

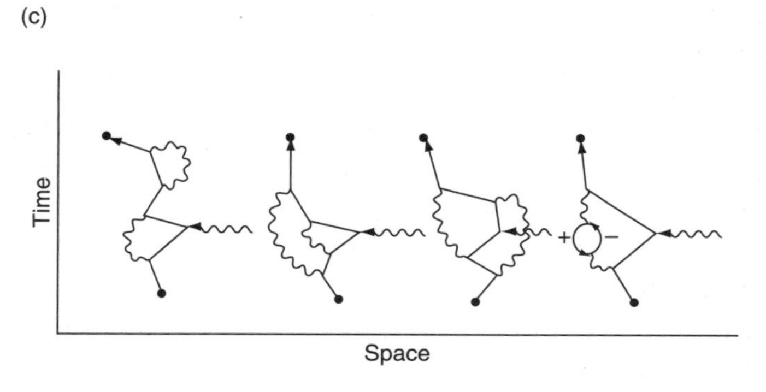

FIG 10 (a) Feynman diagram for the interaction of an electron with a photon from a magnet. Considering only this interaction would lead to the prediction of a g-factor for the electron of exactly 2, as given by Dirac's theory. (b) The same process as described in (a), but now including electron self-interaction, depicted as the emission and re-absorption of a virtual photon. Including this process leads to a slight increase in the prediction for the g-factor for the electron. (c) Further 'higher-order' processes involving emission and re-absorption of two virtual photons. The diagram on the right shows the spontaneous creation and annihilation of an electron-positron pair. Adapted from Richard P. Feynman, *QED: The Strange Theory of Light and Matter*, Penguin, London, 1985, pp. 115–117.

It was somewhat confusing. These very different approaches to relativistic QED all produced similar answers, but nobody understood why.

Enter Dyson.

Dyson's undergraduate studies at Cambridge, England, were interrupted by the war. He had already developed a reputation as a brilliant mathematician. In 1943, after just two years of undergraduate study, he joined the Royal Air Force's Bomber Command to work as a civilian scientist on the operational effectiveness of bombing raids, providing scientific advice to the commander in chief. He continued to work on problems in mathematics and physics in his spare time. After the war, he obtained a position in the mathematics department at Imperial College, London. It was his reading of the official report of the Manhattan Project, written by physicist Henry D. Smyth, that persuaded him to turn to physics.

He returned to Cambridge in 1946, this time focusing on physics. One of his friends, physicist Harish-Chandra, remarked that theoretical physics was in such a mess he had decided to switch to pure mathematics. Dyson replied: 'That's curious, I have decided to switch to theoretical physics for precisely the same reason!'

It had by now become apparent that America was the place to continue his postgraduate physics studies, and he was recommended to go to Cornell University in Ithaca to work with Bethe. With support from a Commonwealth Fellowship, he arrived at Cornell in September 1947.

Bethe had carried out his calculation of the Lamb shift using non-relativistic QED just a few months previously, and he now assigned his new English postgraduate student the task of carrying out the calculation using a fully relativistic QED but for a particle with zero spin. This was a simpler relativistic calculation than that for a spin-½ particle, like the electron, and had the advantage that the result was not likely to be greatly affected by the spin.

Dyson got to know Feynman at Cornell at this time and they became friends. He recognized Feynman as a profoundly original scientist and was struck by the contrast with Bethe:

> Hans [Bethe] was using the old cookbook quantum mechanics that Dick [Feynman] couldn't understand. Dick was using his own private quantum mechanics that nobody else could understand. They were getting the same

> answers whenever they calculated the same problems. And Dick could calculate a whole lot of things that Hans couldn't. It was obvious to me that Dick's theory must be fundamentally right. I decided that my main job, after I finished the calculation for Hans, must be to understand Dick and explain his ideas in a language that the rest of the world could understand.

As Dyson got to grips with Tomonaga's and Schwinger's theories, his mission expanded beyond providing a comprehensible translation of Feynman's approach. He realized that, if any further progress was to be made, the relationship between the Tomonaga–Schwinger and Feynman versions of QED had to be made explicit.

Bethe organized for Dyson to spend the second year of his Fellowship working with Oppenheimer at the Institute for Advanced Study in Princeton. When the summer term ended at Cornell in June 1948, Dyson had some time on his hands before he was due to attend a summer school at the University of Michigan in Ann Arbor. He therefore agreed to accompany Feynman on a road trip across the country to Albuquerque. This was hardly a direct route from Ithaca to Ann Arbor, but Dyson welcomed the opportunity to have Feynman to himself for four days.

On the journey, Feynman opened up to him about his life and his fears for the near future. Like all physicists who had witnessed an atomic explosion, he could all too easily imagine the consequences of a nuclear war. He also talked about his dead wife, Arline: 'He talked of death with an easy familiarity which can come only to one who had lived with spirit unbroken through the worst that death can do,' Dyson later wrote. They also argued about physics.

They parted in Albuquerque. Dyson boarded a Greyhound bus to head back across country, travelling mostly by night. At Ann Arbor he made new friends and attended Schwinger's lectures, a 'marvel of polished elegance.' He found Schwinger friendly and approachable, and spent the afternoons going back through every step of the lectures and every word of their conversations. He filled hundreds of pages with calculations using Schwinger's methods and by the end of five weeks he felt he had mastered the approach as well as anyone, with the exception of Schwinger himself. When the summer school was over he took a bus to Berkeley in California.

On 2 September, he boarded another bus bound for the East coast. 'On the third day of the journey a remarkable thing happened,' he wrote to his parents a few weeks later, 'going into a sort of semi-stupor as one does after 48 hours of bus-riding. I began to think very hard about physics, and particularly about the rival radiation theories of Schwinger and Feynman. Gradually my thoughts grew more coherent, and before I knew where I was, I had solved the problem that had been in the back of my mind all this year, which was to prove the equivalence of the two theories.'

In work that he characterized as neither particularly difficult nor particularly clever, Dyson worked out a way of unifying the theories, combining aspects of each that he judged most advantageous. The unifying feature of his theory was something called the scattering matrix, or S-matrix, initially proposed by Wheeler in 1937 and developed by Heisenberg in 1943. The S-matrix describes the process in which free particles approach, interact, and produce particles which then separate. Each element in the S-matrix corresponds to a Feynman diagram.

When Dyson presented his ideas to Oppenheimer back in Princeton, he was disappointed with the reception he got, and became irritated by what he perceived to be a defeatist attitude to QED on Oppenheimer's part.[1] Confident of the importance of his approach to unification, Dyson chose to make a stand. Oppenheimer agreed to have Dyson give a series of seminars on the subject. The morning after the final seminar Dyson found a note of surrender from Oppenheimer in his mailbox. It said, simply, 'Nolo contendere. R.O.'

Dyson's ultimate triumph came at the American Physical Society meeting in New York in January 1949. Oppenheimer was scheduled to deliver a presidential address, his fame filling the main hall with an audience of two thousand half an hour before the start. In his address he had nothing but praise for Dyson's work, claiming that it pointed the way

[1] Oppenheimer may have had a lot on his mind. He was still quite heavily involved in Cold War nuclear politics during what was to prove a testing period in American–Soviet relations. The Soviet Union had begun the Berlin blockade in June 1948. This was an attempt to force the Allies out of West Berlin by threatening to starve its population of two and a half million. War seemed almost inevitable.

for the immediate future. Sitting beside Dyson in the audience, Feynman loudly proclaimed: 'Well Doc, you're in.'

This statement was not strictly correct. Dyson had not yet completed his doctorate.

There was an element of madness about it. But the result was a theory of QED that predicts the results of experiments to astonishing levels of accuracy and precision. The g-factor for the electron is predicted by QED to have the value 2.002 319 304 76, with an uncertainty of plus or minus 0.000 000 000 52. The comparable experimental value is 2.002 319 304 82, with an experimental uncertainty of plus or minus 0.000 000 000 40.[2] 'To give you a feeling for the accuracy of these numbers,' Feynman wrote, 'it comes out something like this: If you were to measure the distance from Los Angeles and New York to this accuracy, it would be exact to the thickness of a human hair.'

The explanation? The photons created and destroyed in the virtual processes described by the Feynman diagrams carry away some of the mass of the electron but leave its charge unchanged, affecting the electron's magnetic moment.

And what of the Lamb shift? A very crude way of visualizing what happens to an electron in an atom in QED is to think of all the virtual processes resulting in a slight 'wobble' in the electron motion in addition to its orbital and spin motions around the nucleus. This wobble 'smears' the electron probability over a small region of space and is most noticeable when the electron occupies an orbital that keeps it close to the nucleus. The difference in the geometries of the two otherwise degenerate orbitals of the hydrogen atom is enough to account for the small difference in their energies.

[2] These numbers are subject to constant refinement, both experimental and theoretical. The values quoted here are taken from G.D. Coughlan and J.E. Dodd, *The Ideas of Particle Physics: An Introduction for Scientists*. Cambridge University Press, 1991, p. 34. The value 2.002 319 304 362 2(15), where the numbers in brackets represent the uncertainty in the last two digits, is a 2006 figure recommended by the CODATA Task Group—see http://physics.nist.gov. The value 2.002 319 304 361 46(5 6) was reported in 2008 by D. Hanneke, S. Fogwell and G. Gabrielse, *Physical Review Letters*, **100**, 2008, 120801.

Within a few years Schwinger's algebra had given way to Feynman's diagrams as the preferred approach to QED. Schwinger didn't like Feynman's approach because it was much less formal and, he believed, lacked rigour. Having stayed for a time at Schwinger's home in Cambridge, Massachusetts, theorist Murray Gell-Mann liked to report subsequently that he had searched everywhere for Feynman diagrams, but could find none.

However, one room had been locked.

20

A Beautiful Idea

Princeton, February 1954

During a visit to the University of Wisconsin in 1929, Dirac was interviewed by a journalist from the local state newspaper, the Wisconsin State Journal. *The journalist appeared to take Dirac's non-committal, monosyllabic style in his stride. Reporting on the conversation for the benefit of his readers, he wrote: '... "And now I want to ask you something more: They tell me that you and Einstein are the only two real sure-enough high-brows and the only ones who can really understand each other. I won't ask you if this is straight stuff for I know you are too modest to admit it. But I want to know this—Do you ever run across a fellow that even you can't understand?"*

"Yes," says he.

"This will make great reading for the boys down [at] the office," says I. "Do you mind releasing to me who he is?"

"Weyl," says he.'

Hermann Weyl was a German mathematician with a particular interest in symmetry and the representation of symmetry transformations as abstract 'groups'.[1] *He was also*

[1] A symmetry transformation involves the application of one or more of a variety of operations to an 'object', a physical system or a set of equations, with the aim to discover some property of the object which does not change (i.e. it is invariant). Such operations include doing nothing (called the identity operation), translations in space or time, rotations, mirror reflections, inversions, etc. Symmetry transformations can be interrelated and represented as groups, with a transformation forming each element of the group. The group describes the results of multiplying the different transformations together.

very well aware of the relationship between symmetry and physics. In 1915 his Göttingen colleague Amalie Emmy Noether had established a principle underlying all of physics. For any conserved physical quantity, such as energy or momentum, the physical laws describing the behaviour of this quantity are invariant to one or more continuous symmetry transformations. Conservation laws reflect the deep symmetries of nature.

The laws governing energy are found to be invariant to 'translations' in time, meaning that the laws are the same yesterday, today, and tomorrow. Energy is therefore conserved. For momentum, the laws are found to be invariant to translations in space—they are the same here, there, and everywhere. For angular momentum, the laws are invariant to rotational symmetry transformations: they are the same irrespective of direction.

Weyl applied group theory to quantum mechanics in a book published in 1928. It had a mixed reception. Mathematicians welcomed its rigour and beauty. However, physicists perceived it as yet another, higher level of mathematical abstraction in a quantum theory that was already hard to understand. Pauli labelled it 'die gruppenpest' (the group pestilence, or that pesty group business).

Hungarian physicist Eugene Wigner tried to make the subject more accessible in a short book on quantum mechanics and atomic spectra published in 1931. Schrödinger was dismissive of Wigner's efforts: 'This may be the first method to derive the root of spectroscopy,' he told Wigner, 'But surely no one will be doing it this way in five years.' Von Neumann was more reassuring: 'Oh, these are old fogeys. In five years, every student will learn group theory as a matter of course,' he said.

Despite the fact that physicists were uncomfortable with its abstraction, it would be symmetry considerations and group theory that would lead Chinese physicist Chen Ning Yang and American physicist Robert Mills to the next breakthrough in quantum field theory.

Though it wouldn't be recognized as a breakthrough for some time to come.

Noether's theorem led to speculation concerning the symmetry that could be identified with the conservation of another important physical property—electric charge. It had been known since the late eighteenth century that charge is conserved; it can be neither created nor destroyed in physical or chemical reactions.

Weyl had worked on the representation theory of types of symmetry groups called Lie groups, named for the eighteenth century Norwegian mathematician Sophus Lie. These are groups of *continuous* symmetry transformations, involving gradual change of one or more parameters rather than an instantaneous flipping from one form to another, as in a mirror reflection. It is continuous symmetry transformations that underpin Noether's theorem.

An example of a Lie group is the symmetry group U(1), the unitary group of transformations of one complex variable. It is fairly straightforward to picture the symmetry transformations of U(1) in the so-called complex plane, the two-dimensional plane formed by one real axis and one imaginary axis. The imaginary axis is constructed from real numbers multiplied by *i*, the square-root of minus one. We can pinpoint any complex number z in this plane using the formula, $z = re^{i\theta}$, where r is the length of the line joining the origin with the point z in the plane and θ is the angle between this line and the real axis. This expression for z can be rewritten using Euler's formula as $z = r(\cos\theta + i\sin\theta)$.

When the angle θ is zero, the point z lies on the real axis a distance $+r$ from the origin. After applying a rotation through 90°, the point z now lies on the imaginary axis a distance $+r$ from the origin. A further 90° rotation brings z back to the real axis, but this time lying $-r$ from the origin. Another 90° rotation brings z to the imaginary axis once again, now pointing a distance $-r$ from the origin. A final 90° rotation takes us back to where we started. It quickly becomes apparent that a continuous transformation through the angle θ moves the complex number z through a circle in the complex plane.

What Weyl discovered was that this symmetry property is somehow related to the conservation of charge. In Maxwell's equations of classical electromagnetism, it is the intimate relationship between the electric and magnetic fields which preserves this symmetry and ensures that charge is conserved. A static charge generates a static electric field, but a moving charge generates both an electric and a magnetic field.

But Weyl went further. Symmetries can be global or local. In a global symmetry the object is invariant to changes that are applied uniformly, at all points in space–time.[2] In a local symmetry the object is invariant to

[2] An example of a global symmetry change is a uniform shift in the lines of latitude and longitude used by cartographers to map the surface of the earth. So long as the change is uniform and applied consistently across the globe, this makes no difference to our ability to navigate from one place to another.

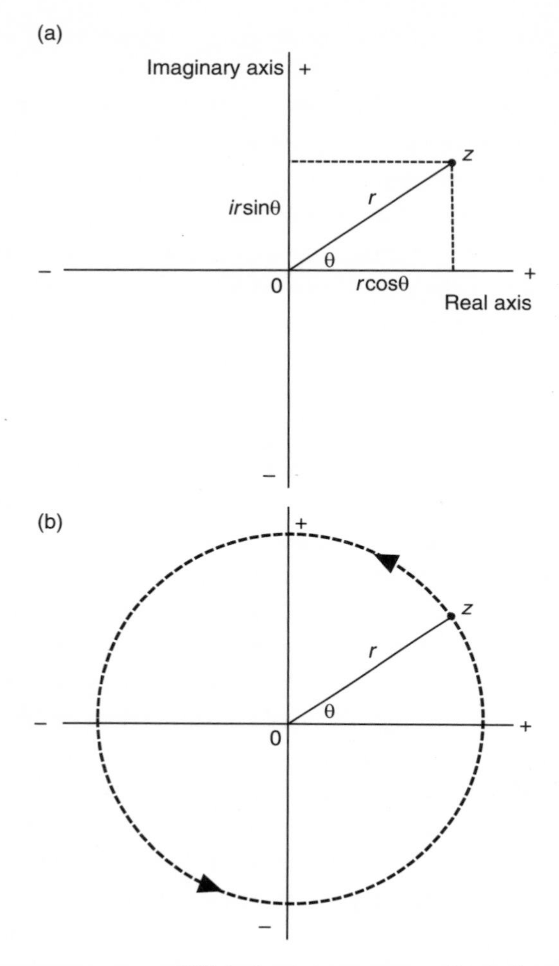

FIG 11 The symmetry group U(1) is the unitary group of transformations of one complex variable. In a complex plane formed by one real axis and one imaginary axis, we can pinpoint any complex number z using the formula $z = re^{i\theta}$, where r is the length of the line joining the origin with the point z in the plane and θ is the angle between this line and the real axis, (a). A continuous transformation through the angle θ moves the complex number z through a circle in the complex plane, (b). There is a deep connection between this continuous symmetry and simple wave motion, in which the angle θ is a phase angle, as illustrated in (c) see next page.

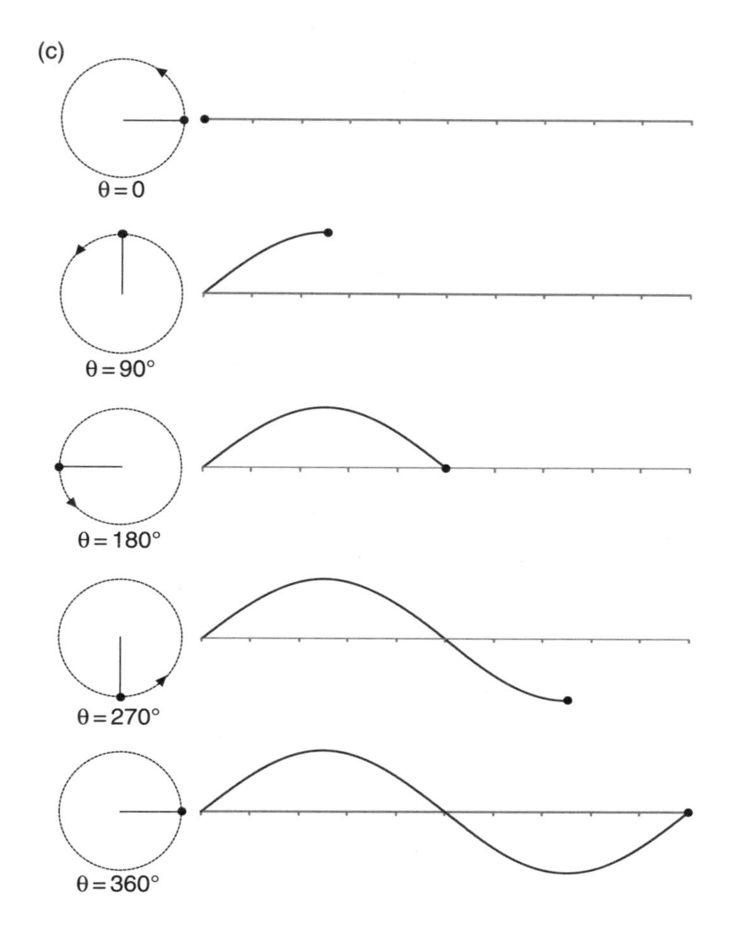

changes applied non-uniformly, with different changes at different points in space–time. It was later shown that the conservation of charge is related to invariance to *local* symmetry transformations. Requiring the symmetry to be local demands that the electric and magnetic fields be connected precisely in the way described by Maxwell's equations. Another way of looking at this is that local invariance demands a field that 'kicks back' when changes are made, restoring local symmetry and so conserving charge.

Weyl was working on these ideas before the advent of quantum mechanics, and chose to give this symmetry the generic name *gauge symmetry*. He was thinking of symmetry in relation to the distances between points in space–time, and was guided by Einstein's work on general relativity. Weyl initially ascribed the invariance to space itself, leading Einstein to point out that if this were true, a clock moved around a room would no longer keep time correctly.

In 1922 Schrödinger had identified Weyl's 'gauge factor' as a *phase* factor, pointing out that transformations of the phase through integral multiples of 360° leaves the phase unchanged. At the time it was not at all clear what this meant, other than providing a clue to the origin of the quantum numbers, the 'integral multiples'. When Schrödinger discovered wave mechanics three years later, it became clear that the phase is associated with the electron wavefunction.

Another way of representing U(1) is in terms of continuous transformations of the phase angle of a sinusoidal wave. A transformation of 180° will put the wave 'out of phase' with the original wave. A further 180° transformation will bring the wave back in phase. This is entirely analogous to the movement of the complex number z in the complex plane considered above.

But the mechanism by which symmetry is preserved is subtly different in quantum mechanics. The change in phase we are now considering affects the wavefunction of the electron as a material particle. Symmetry is preserved if changes in the phase of the material wavefunction are matched by changes in its accompanying electromagnetic field. Gauge invariance generates a force, and the force field must 'kick back'. The particle and the field are now intimately connected. What Weyl had actually discovered was the connection between the local phase symmetry of the wavefunction and the conservation of charge.

The gauge symmetry of U(1) phase transformations is a characteristic of all quantum theories of the electron, including QED. As physicists began to think about the formulation of quantum field theories to describe the strong force between protons and neutrons in the atomic nucleus, the question inevitably arose: What could be the relevant gauge symmetry for such a field?

The challenge was first to find a property that is conserved. Chinese physicist Chen Ning Yang had arrived at the University of Chicago in 1946 to study nuclear reactions under the supervision of Edward Teller. He had adopted a middle name 'Franklin' or just 'Frank', after reading the autobiography of the American inventor and politician Benjamin Franklin. He obtained his doctorate in 1948 and worked for a further year as an assistant to Enrico Fermi. In 1949 he moved to the Institute for Advanced Study in Princeton.

Yang was struck by the notion that in electrodynamics charge conservation is predicated on the gauge invariance of the phase of the electron wavefunction. It was in Princeton that he began to think about ways in which he could apply the principles of gauge invariance to the strong force that binds protons and neutrons in atomic nuclei. As far as he was concerned, the quantity that is preserved in strong nuclear interactions is *isospin*.

The concept of isospin, or isotopic spin, grew out of the observation that the masses of the proton and neutron are very similar. At the time the neutron was discovered in 1932, it was perhaps natural to imagine that the neutron was a composite particle consisting of a proton with an electron stuck on to it. After all, it was also known that beta radioactive decay involves the ejection of high-speed electrons directly from the nucleus, turning a neutron into a proton in the process. This seemed to suggest that in beta-decay, the composite neutron was somehow shedding its 'stuck-on' electron.

Heisenberg preferred to think of the neutron as a fundamental particle, but nevertheless picked up on the neutron-as-proton-plus-electron idea to develop an early theory of proton–neutron interactions in the nucleus. This was a model analogous to the chemical binding of two protons by a single electron in the ionized molecule H_2^+.[3] He hypothesized that, just as the two protons in H_2^+ are bound together by a single electron, so the proton and neutron bind together in the nucleus by exchanging an electron between them, the proton turning into a neutron and the neutron turning into a proton in the process. By the same token, the interaction between two neutrons would involve the exchange of two electrons, one in each 'direction', analogous to the chemical bond in molecular hydrogen, H_2.

[3] The neutral hydrogen atom H consists of one proton and one electron. The ionized hydrogen molecule H_2^+ therefore consists of two protons bound together by a single electron.

Eager to pursue the chemical bonding analogy, Heisenberg further identified the exchange of charge between the proton and neutron as involving a change of spin, with zero charge—the neutron—involving spin in one direction (spin-down) and positive charge—the proton—involving spin in the opposite direction (spin-up). Converting a neutron into a proton was then equivalent to 'rotating' the spin of the neutron.

It was admittedly a clumsy model, but through it Heisenberg was able to apply non-relativistic quantum mechanics to the nucleus itself. In a series of papers published in 1932 he accounted for many observations in nuclear physics, such as the stability of isotopes (atoms with the same number of protons but different numbers of neutrons) and alpha-particles (helium nuclei, consisting of two protons and two neutrons).

However, as a theory of nuclear interactions, its weaknesses were exposed in experiments performed just four years later. The electron-exchange model did not allow for any kind of proton–proton interaction, and Heisenberg had assumed that there could therefore be no strong interaction between protons. In contrast, experiments showed that the strength of the interaction between protons is comparable to that between protons and neutrons.

Despite the shortcomings of the theory, Heisenberg's electron-exchange model held at least a grain of truth. The exchange of electrons was abandoned, but the concept of isospin symmetry was retained. As far as the strong force is concerned, the proton and the neutron are two sides of the same coin, or two states of the same particle, the only difference between them being their isospin.

As Yang searched for a quantum field theory that would preserve isospin symmetry he quickly got bogged down. But, despite repeated failure, he did not give up. He became obsessed with the problem. 'Occasionally an obsession does finally turn out to be something good,' he later observed.

In the summer of 1953 he made a visit to Brookhaven National Laboratory on Long Island, New York, where he found himself sharing an office with a young American physicist called Robert Mills. Mills became caught up with Yang's obsession and together they worked on a quantum field theory for strong interactions. 'There was no other, more immediate motivation,' Mills explained some years later, 'He and I just asked ourselves "Here is something that occurs once. Why not again?"' By the end of the summer they had worked out a solution. It was a solution with some rather unphysical consequences, however.

The group U(1) is appropriate to describe the symmetry properties of the quantum field theory of the electron as this involves a single particle in a single field interacting via a single type of field quantum—the photon. There is no change of charge in the interaction and so the photon is not required to carry a charge of its own. Furthermore, the field is not limited in extent, though it weakens with distance. Consequently, the force can be carried quite happily by massless particles, able to travel long distances at high speeds.

However, a quantum field theory of the strong interaction has to account for the fact that there are now two particles whose charges and isospins change as a result of the interaction. Also, the force between the nuclear particles operates over very short distances only, within the boundary of the nucleus itself.

Yang and Mills reached for the symmetry group SU(2), the special unitary group of transformations of two complex variables. The resulting field theory satisfactorily preserved the isospin symmetry and introduced a new field analogous to the electromagnetic field in QED. They called it the B field. The theory also predicted three new field particles, responsible for carrying the strong force between the protons and neutrons in the nucleus, analogues of the photon in QED.

Three particles were needed because of the greater complexity of the interactions. Two of the three field particles were required to carry charge, accounting for the change in charge resulting from proton–neutron and neutron–proton interactions. Yang and Mills referred to these particles as B^+ and B^-. The third particle was neutral, like the photon, and was meant to account for proton–proton and neutron–neutron interactions in which there is no change in charge. This was referred to as B^0. It was found that these field particles interact not only with protons and neutrons, but also with each other.

It was here that the problems started. The renormalization methods that had been used so successfully in QED could not be applied to the Yang–Mills field theory. Worse still, the zeroth-order term in the perturbation expansion indicated that the field particles should be massless, just like the photon. But this was self-contradictory. Japanese physicist Hideki Yukawa had suggested in 1935 that the field particles of short-range forces should be heavy, arguing that virtual particles confined to short distances implied short lifetimes and short lifetimes implied large masses according to the energy–time uncertainty relation.[4] Massless field particles for the strong force made no sense.

Yang returned to Princeton, and on 23 February 1954 presented the results of the work he had done with Mills in a seminar. Oppenheimer was in the audience, as was Pauli.[5]

It turned out that Pauli had earlier explored some of the same logic and had arrived at the same puzzling conclusions concerning the masses of the field particles. He had consequently abandoned the approach. As Yang drew his equations on the blackboard, Pauli piped up:

'What is the mass of this B field,' he asked, anticipating the answer.

'I don't know,' Yang replied, somewhat feebly.

'What is the *mass* of this B field,' Pauli demanded.

'We have investigated that question,' Yang replied, 'It is a very complex question, and we cannot answer it now.'

'That is not a sufficient excuse,' Pauli grumbled.

Yang, taken aback, sat down to general embarrassment. Oppenheimer suggested that they let Yang continue. He resumed his lecture, and Pauli asked no more questions.

It was a problem that simply would not go away. Without mass, the field particles of the Yang–Mills field theory did not fit with physical expectations. If they were massless, as the theory predicted, then no such particles had ever been observed. Accepted methods of renormalization wouldn't work.

[4] Yukawa had suggested that the field particle of the strong force should possess a mass about 200 times that of the electron. When the meson (muon) was discovered in 1937 it was initially thought to be the particle that Yukawa had proposed.

[5] This was a very difficult time for Oppenheimer. His 'Q' security clearance had been withdrawn by the Eisenhower administration in December 1953. He was accused of being 'more probably than not' an agent of the Soviet Union. The Atomic Energy Commission's Personnel Security Board met in April 1954 to pass judgement on Oppenheimer's guilt or innocence.

And yet, it was still a *nice* theory.

'The idea was *beautiful* and should be published,' Yang wrote, 'But what is the mass of the gauge particle? We did not have firm conclusions, only frustrating experiences to show that [this] case is much more involved than electromagnetism. We tended to believe, on physical grounds, that the charged gauge particles cannot be massless.'

Yang and Mills published a paper describing their results in *Physical Review* in October 1954. They had made no further progress, and turned their attentions elsewhere.

21

Some Strangeness in the Proportion

Rochester, August 1960

In 1935 Japanese physicist Hideki Yukawa had suggested that the carrier of the force binding protons and neutrons together in the nucleus would be a particle about 200 times heavier than the electron. The meson, discovered just two years later, appeared to fit the bill. But this was not Yukawa's field particle. Any carrier of the nuclear force was expected to interact very strongly with matter. The meson did not.

In 1947 another particle was discovered in cosmic ray experiments performed atop the Pic du Midi in the French Pyrenees by Bristol University physicist Cecil Powell and his team. This was found to have a slightly larger mass, 273 times that of the electron, and came in positive and negative varieties. This was the particle that Yukawa had predicted.[1]

The meson was renamed the mu-meson (subsequently shortened to muon). The new particle was called the pi-meson (pion). As techniques for detecting particles produced by cosmic rays became more sophisticated, the floodgates opened. The pion was quickly followed by the positive and negative K-meson (kaon) and the neutral lambda particle. The kaons and the lambda behaved rather oddly and, for want of a more scientific descriptor, the physicists awarded the new particles the collective name 'strange'.

Flush with generous funding from a grateful post-war American administration, Ernest Lawrence redeployed to peaceful purposes his 184-inch cyclotron, which had been used during the war to separate uranium isotopes. Physicists at Berkeley's Radiation

[1] Yukawa became the first Japanese physicist to receive the Nobel Prize, in 1949. Powell was awarded the Prize in 1950.

Laboratory used the cyclotron to discover the neutral pion in 1949. As new synchrotron accelerators were constructed at Brookhaven National Laboratory in New York, Berkeley in California, Dubna in Russia, and Geneva in Switzerland there soon followed positive, negative, and neutral sigma particles, negative and neutral xi particles, the anti-proton, anti-neutron, and Pauli's long-anticipated neutrino.[2]

There was more to come. At high energies a number of nucleon 'resonances' emerged, the first associated with an increased scattering of pions from protons. Were these actually particles? They seemed hardly more than brief flirtations between pions and nucleons, surviving for no more than trillionths of trillionths of seconds. Though initially viewed as excited states of the nucleons, by the late 1950s it was generally accepted that they couldn't be ignored. They duly entered the lexicon, as delta particles.

Dirac's 'dream of philosophers' had been shattered by the brute forces unleashed in collisions between particles accelerated to greater and greater energies. Instead of revealing an underlying simplicity, the experimentalists had uncovered a bewildering complexity, a veritable 'zoo' of particles.

All cried out for an explanation.

Something new and unusual was going on with the strange particles. The physicists realized that they were witnessing the effects of two different kinds of nuclear force. The strange particles would be produced through strong-force interactions involving 'ordinary' particles, such as protons and pions. These strange particles would then travel on through the detector before disintegrating to produce characteristic 'V'-shaped tracks. Their relatively long lifetimes suggested that although they were being produced by the strong force, their decay modes were governed by a much weaker nuclear force. The same force, in fact, that governs radioactive beta-decay.

In Princeton, Dutch-born American physicist Abraham Pais became convinced that this kind of behaviour could not be explained in terms of the known quantum numbers, those for charge, spin, and isospin. He suggested that a further quantum number was needed, which he called N. Ordinary particles possessed N = 0, whereas the new strange particles possessed N = 1.

[2] The particles were not given these names initially. To avoid confusion, I will continue with their modern names and refer to their antecedents only when necessary.

He figured that in strong-force interactions, the mysterious new quantum number *N* is conserved: the total *N* for the particles that are produced must be the same as that for the reacting particles. However, once produced, the strong force cannot decompose a strange particle with an odd-numbered *N* back into ordinary particles with even *N*. It seemed that only the weak force could do this, Pais hypothesized, because the weak force somehow did not respect this even–odd conservation rule.

It all seemed a bit *ad hoc*, and American physicist Murray Gell-Mann was not convinced. Gell-Mann was another in the long and illustrious line of prodigious intellects that came to be applied to problems in quantum physics in the twentieth century. He had been born in New York in 1929, entered Yale University when he was fifteen,[3] and secured his doctorate at MIT in 1951, aged just 21. He had a short spell at the Institute for Advanced Study in Princeton before moving to Chicago to work with Fermi. It was in Chicago that he began to apply himself to the puzzle of the strange particles.

The scheme that Gell-Mann devised was no less curious, but it was more comprehensive than that of Pais. Gell-Mann invented the idea of *strangeness*, a new property of the strange particles that he later immortalized with the words of Francis Bacon: 'There is no excellent beauty that hath not some strangeness in the proportion.'

Although the origin of strangeness was not yet understood, Gell-Mann argued that, whatever it is, strangeness is conserved in strong-force interactions, much like electric charge is conserved. Pais realized that Gell-Mann had properly identified the new quantum number he had been looking for.

This was no longer about 'evenness' and 'oddness'. If a strange particle is produced in a strong-force reaction between two ordinary particles, both with zero strangeness, then conservation of strangeness meant that it could not be produced on its own. For example, collision of an accelerated negative pion with a proton produces a neutral kaon (with a strangeness arbitrarily assigned as +1) and the lambda (strangeness −1). These particles must always be produced together to conserve the total strangeness, in a phenomenon called 'associated production'. Once created, the

[3] Think about it. What were *you* doing when you were fifteen?

strange particles could not be decomposed except through weak-force interactions.

Gell-Mann called Courtney Wright, a particle physicist at Brookhaven National Laboratory, to ask about associated production. If he was right, then strange particles should always be produced in certain pairings, in preference to other possible combinations that were feasible simply on the basis of the available energy of the collision.

'What possible difference does it make?' Wright wanted to know. 'Who cares?'

'I care. Please find out,' Gell-Mann replied.

Wright found out. Gell-Mann was right.

The lack of an underlying explanation for strangeness meant that Gell-Mann's solution was not universally embraced by the physics community. Oppenheimer referred to it in 1956 as a 'temporary kind of solution'.

The behaviour of the strange particles led to renewed interest in the weak nuclear force. Fermi had elaborated a detailed theory of beta radioactivity in the early 1930s. He had recounted his new theory to his colleagues during Christmas 1933, after a full day skiing in the Italian Alps. Italian physicist Emilio Segrè described the event: '... we were all sitting on one bed in a hotelroom, and I could hardly keep still in that position, bruised as I was after several falls on icy snow. Fermi was fully aware of the importance of his accomplishment and said that he thought he would be remembered for this paper, his best so far.'

Fermi had drawn on comparisons between the weak force that governs beta radioactivity and electromagnetism. In electromagnetic theory, two electrons approaching one another experience the electromagnetic force, exchange a virtual photon, and are deflected away from each other. Likewise, Fermi surmised, a neutron experiencing the weak nuclear force transforms into a proton, exchanging 'something' (speculation on which Fermi was not prepared to be diverted), with an electron and neutrino.[4] From the resulting electromagnetic-like theory, Fermi was able to deduce the range of energies (and hence speeds) of the emitted beta-electrons.

[4] In the interests of proper book-keeping, it was subsequently agreed that the electron emitted in beta radioactive decay should be accompanied by an *anti*-neutrino.

With some small adjustments, Fermi's theory remains valid to this day. His predictions for the electron energies were shown in 1949 to be correct in experiments performed at Columbia University by Chinese-American physicist Chien-Shiung Wu.[5]

Fermi deduced that the strength of the coupling between the neutron–proton and electron–neutrino 'currents' in beta radioactivity is some ten billion times weaker than the equivalent coupling between charged particles in electromagnetism. Weak it may be, but the force has some profound consequences. Because of the weak force, neutrons, regarded as 'fundamental' nuclear particles almost from the moment of their discovery, are inherently unstable. The average lifetime of a free neutron is just eighteen minutes.

In the late 1940s it became evident that the interactions governing absorption and emission of muons were of similar strength. The force governing the decay modes of the strange particles also fell into the same range, and it seemed clear that beta radioactivity was simply one common manifestation of a more widespread phenomenon, which became known as the *Universal Fermi Interaction*.

But there was now another problem. One of the inviolate symmetries of electromagnetism relates to the behaviour of the wavefunctions of charged particles under reflection, a property known as *parity*. If reversing the signs of the particle's spatial coordinates does not change the sign of the wavefunction, the particle is said to possess even parity. If the sign of the wavefunction is reversed, the particle is said to have odd parity.[6] As far as the physicists could tell, in all electromagnetic and nuclear interactions, parity is conserved.

Aside from what was believed to be compelling experimental evidence, the assumption of parity conservation was also very much the instinct of physicists. How could it be possible for the immutable laws of nature to favour such seemingly human conventions of left *versus* right, up *versus*

[5] At Columbia University Wu had worked on the Manhattan Project, refining the gaseous diffusion technique that was used to separate U-235 from natural uranium. She is thought to be the only scientist of Chinese birth to have worked on the Manhattan Project.

[6] This symmetry operation can be thought of as a kind of mirror reflection which not only reverses left and right, but also front and back and up and down.

down, front *versus* back? Surely, no natural force could be expected to display such 'handedness'?

The problem was that two positively charged strange particles, then named the tau and the theta, looked to all intents and purposes as though they were one and the same particle. They possessed the same mass and decay rate. But the positive theta particle decayed into two pions, each with odd parity. Two particles with odd parity give a total even parity, as −1 multiplied by −1 gives +1. The tau particle, however, decayed into three pions, also each with odd parity. Three particles with odd parity give a total odd parity, as −1 multiplied by −1 multiplied by −1 gives −1.

Dark hints began to emerge that if the tau and theta really were one particle, then the weak interaction might not respect the conservation of parity. Could the weak force, after all, be 'handed'?

Yang and fellow Chinese physicist Tsung-Dao Lee decided to check the experimental record. What they found was greatly surprising. It turned out that for weak force interactions there was no experimental evidence in support of parity conservation. In June 1956 they published a speculative paper posing the question: Is parity conserved in weak interactions?

The answer came from a series of extremely careful experiments conducted towards the end of that year by Wu, Eric Ambler, and their colleagues at the US National Bureau of Standards laboratories in Washington, DC. These involved the measurement of the direction of emission of beta-electrons from atoms of radioactive cobalt-60, cooled to near absolute zero temperature, their nuclei aligned by application of a magnetic field. A symmetrical pattern of beta-electron emission would suggest that no direction was specially favoured, and that parity is conserved. An asymmetrical pattern would suggest that conservation of parity is violated.

So convinced was Pauli of parity conservation that he was ready to stake a large sum: 'I do not believe that the Lord is a weak left-hander,' he wrote to Weisskopf in January 1957, 'and I am ready to bet a very large sum that the experiments will give symmetric results.'

The experiments were unequivocal. Wu found that the Lord is indeed a 'weak left-hander'. Parity is not conserved in weak-force interactions. Within less than two weeks Pauli was eating his words and thankful that he had not placed his bet, with money he could ill-afford to lose.

The tau and theta particles became one particle, the positive kaon.[7]

There followed further skirmishes between theoreticians and experimentalists over the precise nature of the weak force. Gell-Mann had in the meantime moved to the California Institute of Technology (Caltech) in 1955, where he worked with Feynman on a theory of the weak nuclear force and became enamored of Yang–Mills field theory.

Feynman and Gell-Mann, and two other physicists, Indian-American George Sudarshan and American Robert Marshak, argued that the weak force had to possess certain so-called vector properties, in contradiction to the experimentalists' interpretation of the evidence. The theorists prevailed. Further experiments performed in late 1957 by Maurice Goldhaber and his colleagues at Brookhaven established that the neutrino is 'left-handed' and the anti-neutrino 'right-handed', confirming the universality of the Universal Fermi Interaction and establishing the weak force as a fundamental force of nature.

In his landmark paper on beta radioactivity, Fermi had drawn on the analogy between the weak force and electromagnetism. He had made his estimates of the relative strengths of the forces using the electron mass as a yardstick. In 1941, Schwinger had pondered on the implications of assuming that the weak force is carried by a much larger particle. He estimated that if this field particle was actually a couple of hundred times the proton mass, then the coupling strengths of the weak and electromagnetic forces might actually be the same.

It was the first hint that the weak and electromagnetic forces could be *unified*.

In November 1957, Schwinger published an article in the journal *Annals of Physics* in which he speculated that the weak force is mediated by three field particles. Two of these particles, the W^+ and W^-,[8] were necessary to account for the transmission of electric charge in weak interactions. The third was needed to account for instances in which no charge is transferred. This third, neutral, particle was, he believed, the photon.

[7] Yang and Lee were awarded the 1957 Nobel Prize for physics for their suggestion.

[8] Again, for the purposes of clarity I am using the currently accepted names for these particles.

Beta radioactivity would now work like this. A neutron would decay, emitting a massive W^- particle and turning into a proton. The short-lived W^- would in turn decay into a high-speed electron and an anti-neutrino.

Schwinger had manipulated his theory to be consistent with the prevailing interpretation of the nature of the weak force. When that interpretation was shown to be false, he abandoned work on weak interactions completely. However, he had in the meantime assigned one of his Harvard graduate students to work on the problem. His name was Sheldon Glashow.

An American-born son of Russian Jewish immigrants, Glashow graduated from the Bronx High School of Science in 1950 together with classmates Steven Weinberg and Gerald Feinberg. He moved to Cornell University with Weinberg, securing his bachelor's degree in 1954. From Cornell Glashow moved to Harvard, where he became one of Schwinger's graduate students.

As a doctoral project, Schwinger asked him to think about the W^+ and W^- as carriers of the weak nuclear force and for two years this is precisely what he did. He thought about it.

What he realized was that the electric charges carried by the W particles meant that it was impossible to separate the theory of the weak force from that of electromagnetism. 'We should care to suggest,' he wrote in an appendix to his PhD thesis, 'that a fully acceptable theory of these interactions may only be achieved if they are treated together...'

Glashow reached for the SU(2) field theory that had been developed by Yang and Mills, taking 'on faith' Schwinger's argument that the three field particles were the two W particles and the photon. He took care to ensure that interactions involving the W particles would violate the conservation of both parity and strangeness, but that interactions involving the photon would not. Although the result was rather ugly, by November 1958 Glashow thought he had arrived at a definitive unified theory of the weak and electromagnetic forces. What's more, he believed his theory was renormalizable.

But any euphoria he experienced was to be short-lived. He had completed a paper on the theory whilst working at Niels Bohr's Institute for Theoretical Physics in Copenhagen. In the spring of 1959 he visited

London and delivered a lecture on the subject. Sitting in the audience, Pakistan-born theoretical physicist Abdus Salam and his collaborator John Ward were stunned. Inspired by Schwinger's paper they, too, had been working on a Yang–Mills SU(2) field theory involving two charged field particles and the photon. The problem was that, try as they might, they could not make their theory renormalizable.

Glashow had made a series of errors. When these were exposed, he retreated to Copenhagen to think about it some more.

It was easy to dismiss Glashow's errors as the folly of youthful inexperience (he was still only 26, Salam was 33). And yet the result of his further reflection on the problem was nothing short of youthful bravado. As a field particle in an SU(2) theory of both weak and electromagnetic interactions, the photon had been forced to do too much. It was doubling as the carrier of both the weak and electromagnetic forces, tasks that tortured the theory in what Glashow now realized were unacceptable ways.

The solution was to enlarge the symmetry by combining the Yang–Mills SU(2) gauge field with the U(1) gauge field of electromagnetism, in a product written SU(2) × U(1). This came at the cost of full unification, but had the advantage that it freed the photon from responsibility for carrying the weak force. However, this also meant that it was necessary to introduce yet another field particle to account for neutral weak-force interactions, so-called 'weak neutral currents'.[9] In effect, Glashow now had three weak-force particles equivalent to the triplet of B particles first introduced by Yang and Mills. These were the W^+, W^-, and Z^0.[10]

Glashow lectured in Paris in March 1960, where he encountered Gell-Mann, on sabbatical leave from Caltech and working as a visiting professor at the Collège de France. Over lunch, Glashow described his SU(2) × U(1) theory and Gell-Mann offered encouragement. 'What you're doing is good,' Gell-Mann told him, 'But people will be very stupid about it.' Gell-Mann also extended an invitation for Glashow to join him at Caltech.

[9] When such weak-force interactions involve charged intermediaries they are called charged 'currents', as they involve the flow of charge from one particle to another. By analogy, the weak-force interactions involving the exchange of neutral force particles are sometimes referred to as neutral currents.

[10] Glashow originally referred to the neutral particle as B, by analogy with Yang and Mills, but it is now commonly referred to as the Z^0.

Gell-Mann decided to present Glashow's as yet unpublished[11] work at the next high-energy physics conference, to be held in Rochester, New York, from 25 August to 1 September 1960. The Rochester conferences, first established in 1950, were intended to mirror the exclusive tradition of the Shelter Island and Pocono conferences, but included experimental particle physicists as well as theoreticians. Glashow had thus far articulated the theory in Schwinger's impenetrable style, and Gell-Mann now offered a translation in a language that more physicists would understand. The reaction from conference participants was underwhelming.

In truth, the theory did not seem to have that much going for it. Nobody had yet been able to solve the problem of how the W^+, W^-, and Z^0 particles were supposed to acquire their mass. Just as Yang and Mills had discovered, the field theory predicted that the force carriers should be massless, like the photon. The carriers of the weak force needed to be massive particles, however, and adding the masses 'by hand' rendered the theory unrenormalizable.

Then there was the problem of the Z^0. The existence of a neutral force carrier could be expected to be manifested experimentally in the form of weak neutral currents. No such currents could be found in any of the strange particle decays, which by now had become the particle physicists' principal point of reference for weak-force interactions.

Glashow's explanation was never likely to find much favour with the experimentalists. He argued that the Z^0 was so much more massive than the charged W particles that interactions involving the Z^0 were simply out of reach of the largest particle accelerators.

The theoreticians had reached an impasse. High-energy physicists had run riot, triggering an explosion of particles that had left the theoreticians clutching at straws. Quantum field theory was in the doldrums and, in any case, wasn't the only theoretical game in town.

Some declared that quantum field theory was dead.

[11] Glashow's paper was published in the journal *Nuclear Physics* in November 1961.

22

Three Quarks for Muster Mark!

New York, March 1963

By the early 1960s, the physicists had notched up 30 'fundamental' particles, including the anti-particles but excluding the delta particles. If there was to be any hope of making sense of the by now bewildering array of electrons, photons, protons, neutrons, muons, pions, kaons, the lambda, sigma, xi, and delta particles, it was going to be necessary to find some rigorous way of classifying them, beyond 'ordinary' and 'strange'. Names had proliferated. Responding to a question from one young physicist, Fermi remarked: 'Young man, if I could remember the names of these particles, I would have been a botanist.'

The particles were classified according to their characteristic 'quantum numbers', their values of electric charge, spin, isospin, and strangeness. It was possible to distinguish two basic types. There are 'matter' particles that are the stuff of material substance and there are 'force' particles, the quanta of the fields responsible for transmitting forces between matter particles.

Matter particles are fermions—they possess half-integral spins.[1] These include the leptons: the electron, muon, neutrino, and their anti-particles,[2] and the baryons, including the proton, neutron, lambda (Λ), sigma (Σ), xi (Ξ) particles, and their anti-particles.[3]

[1] Fermions are named for Enrico Fermi.

[2] The leptons (from the Greek *leptos*, meaning small) are the lighter fermions which do not experience the strong nuclear force.

[3] The baryons (from the Greek *barys*, meaning heavy) are heavier fermions with masses equal to or greater than the proton mass. Baryons experience the strong nuclear force.

Force particles are bosons—they possess integral spins.[4] *They include the photon and the group of particles collectively known as 'mesons', the pions* (π) *and the kaons* (K).[5] *Of these, the mesons carry the strong force and, together with the baryons, constitute a particle category called the hadrons.*[6]

The spin-½ baryons possess positive (proton, Σ^+*), neutral (neutron,* Σ^0*,* Λ^0*, and* Ξ^0*) and negative (*Σ^- *and* Ξ^-*) electric charges and strangeness values of zero (proton, neutron), −1 (*Σ^+*,* Σ^0*,* Λ^0*, and* Σ^-*) and −2 (*Ξ^0 *and* Ξ^-*). The spin-0 mesons possess positive (*K^+*,* π^+*), neutral (*K^0*,* π^0*, and* $\bar{K}^0$*, the anti-particle of* K^0*) and negative (*π^-*,* K^-*) electric charges and strangeness values of +1 (*K^0*,* K^+*), zero (*π^+*,* π^0*,* π^-*) and −1 (*K^-*,* $\bar{K}^0$*). The spin-3/2 delta particles all possess zero strangeness and electric charges ranging from −1 to +2 (*Δ^-*,* Δ^0*,* Δ^+*, and* Δ^{2+}*).*

Baryons are classed according to their 'baryon number', B, designated as +1 for all baryons, −1 for all anti-baryons. A similar 'lepton number', l, can be defined with the value +1 for all leptons and −1 for all anti-leptons. Protons and neutrons therefore possess $B = +1$, $l = 0$. *Electrons possess* $B = 0$, $l = +1$. *All force particles (all bosons) possess* $B = 0$, $l = 0$. *These are important characterizations, as the physicists discovered that in particle interactions, baryon and lepton numbers are conserved.*

It was clear that amidst this confusion of particle categories there must be a pattern, a particle equivalent of Dmitri Mendeleev's periodic table of the elements. The question was: What is this pattern and does it have an underlying explanation? The question was answered in two stages. Gell-Mann and Israeli physicist Yuval Ne'eman independently identified the pattern in 1961.

The explanation took a little while longer.

The search began in earnest for some kind of underlying simplicity. It was the physicists' instinct that there could not be this many fundamental particles, and that somehow it should be possible to rationalize their number by finding the 'real' fundamental particles that constituted all the rest.[7] To

[4] Bosons are named for the Indian physicist Satyendra Nath Bose. Spin-zero bosons are also possible, but these are associated with matter fields rather than force fields.

[5] Note that although it was originally called the meson, the muon (the 'heavy' electron) is a lepton and does not belong in the class of mesons with pions and kaons. Confused?

[6] From the Greek *hadros*, meaning thick or heavy.

[7] This was not the only approach available. American theoretician Geoffrey Chew had offered an alternative 'bootstrap' model based on S-matrix theory, in which each particle is constructed from the others. In this 'nuclear democracy', no particle is more important or fundamental than any other.

a certain extent, physicists have adopted this kind of approach throughout the history of their science, but the beginnings of an attempt to apply this to the new particles that had begun to emerge in cosmic ray and particle accelerator experiments were first published by Fermi and Yang in 1949.

Influenced by Marxist theories of dialectical materialism, in the early 1950s Japanese physicist Shoichi Sakata and his colleagues had adopted the proton, neutron, and lambda particles as precisely this kind of fundamental 'triplet', from which they tried to build other particles. Gell-Mann tried much the same approach, but it was never really clear why these particles should be regarded as more 'fundamental' than the others. 'It was a big mess,' Gell-Mann admitted, 'I didn't like it, and I didn't publish it.'

These early attempts failed because the theorists were pushing for an underlying explanation before a proper pattern had been established. This was a bit like trying to figure out the fundamental building blocks of the elements without first appreciating the position that each element occupies in the periodic table.

Gell-Mann believed that the framework for such a pattern could be provided by a global symmetry group.[8] However, although possessed of a prodigious intellect and familiar with the work of Yang and Mills, in 1959 he was unfamiliar with group theory. He had sat through lectures on the subject, but had always thought it rather abstract and irrelevant to anything of importance in physics.

He knew he needed a larger continuous symmetry group than U(1) or SU(2) to accommodate the range and variety of particles that were then known, but was quite unsure how to proceed. By this time he was working as a visiting professor at the Collège de France in Paris. Not surprisingly, copious quantities of good French wine consumed over lunch with his French colleagues did not immediately help to point the way to a solution.[9]

[8] Note that this was a global, rather than local, symmetry group. Gell-Mann was searching for a way of classifying the particles. He was not yet seeking to develop a Yang–Mills field theory of the strong force based on a local gauge symmetry.

[9] It did not seem to aid conversation either. The colleagues with whom Gell-Mann was drinking were mathematicians who could have solved his problem almost immediately, but they never talked about it.

Glashow's visit to Paris in March 1960 therefore prompted more than noises of encouragement from Gell-Mann concerning Glashow's attempts to unify the weak and electromagnetic forces. Once he got past the relatively impenetrable Schwinger terminology that Glashow was using, Gell-Mann was intrigued by Glashow's SU(2) × U(1) theory. He began to understand how it might be possible to expand the symmetry group to higher dimensions. Not that the SU(2) × U(1) approach was the solution, as the expansion it offered (to a total of four field particles or bosons: the photon, W, and Z particles) was insufficient for his purposes.

He was nevertheless suitably impressed. Thus inspired, he now tried theories with more and more dimensions:

> I worked through the cases of three operators, four operators, five operators, six operators, and seven operators, trying to find algebras that did not correspond to what we would now call products of SU(2) factors and U(1) factors. I got all the way up to seven dimensions and found none... At that point, I said, 'That's enough!' I did not have the strength after drinking all that wine to try eight dimensions.

Glashow chose to accept Gell-Mann's offer to join him at Caltech and, shortly after his return from Paris, the two physicists searched for a solution together. But it was only after a chance discussion with Caltech mathematician Richard Block that Gell-Mann discovered that the Lie group SU(3) offered the structure he had been searching for. In Paris he had given up just as he was about to discover this for himself.

The simplest, or 'irreducible' representation of SU(3) is a fundamental triplet. Sakata and his colleagues had actually tried to construct a model based on the SU(3) symmetry group and had used the proton, neutron, and lambda particles as the fundamental representation. Gell-Mann had already been down this road, and had no wish to repeat his experiences. He simply skipped over the fundamental representation and turned his attention to the next.

In mathematical terms, the representations of the SU(3) symmetry group can be combined according to the formula: $\underset{\sim}{3} \times \underset{\sim}{3} \times \underset{\sim}{3} = \underset{\sim}{1} + \underset{\sim}{8} + \underset{\sim}{8} + \underset{\sim}{10}$. This implies a pattern consisting of a 'singlet', a couple of 'octets' of particles, and a 'decimet' of particles. Gell-Mann focused his attentions on the octets.[10]

He found that he could fit the spin-½ baryons into one such octet, essentially a hexagonal pattern with points determined by values of electric charge and strangeness and with two particles, the Σ^0 and Λ^0, positioned in the centre of the pattern.[11] He found that he needed to place the proton, neutron, and lambda particle in this pattern, and must have felt justified in his decision to resist the temptation to ascribe these to the fundamental representation.

When Gell-Mann put together an octet for the spin-0 mesons, he found that he could assign only seven particles to the pattern. One particle, the meson equivalent of the Λ^0, was 'missing'. Emboldened, he speculated that there did exist an eighth spin-0 meson with an electric charge of zero and zero strangeness. He called it the *chi*.

Gell-Mann drew back from speculating on the members of the decimet, as only the four delta particles were known at the time and to predict the existence of another six new particles must have appeared to be a step too far. In any case, he had already achieved a major breakthrough with the octets. Consequently, he chose to call his approach the 'Eightfold Way', a tongue-in-cheek reference to the teachings of Buddha on the eight steps to Nirvana.[12] He completed his work on the Eightfold Way during Christmas 1960 and it was published as a Caltech preprint in early 1961.

The particle he had predicted to complete the meson octet was found a few months later by American physicist Luis Alvarez and his team in Berkeley, as a three-pion resonance. They called the new particle the eta, η.

Yuval Ne'eman was a colonel in the Israeli army when he was granted leave by Moshe Dayan to pursue physics in London whilst working as a defence attaché at the Israeli Embassy. His was an unlikely career in

[10] The singlet representation—a one-member group—is nevertheless relevant. The phi meson, discovered in the early 1960s, is an example.

[11] In some representations, hypercharge is used instead of strangeness. The hypercharge number is simply the strangeness plus the baryon number. In the case of the spin-½ baryons, the hypercharge is then equal to the strangeness +1. This centres the pattern on zero charge and zero hypercharge.

[12] These are: right views, right intention, right speech, right action, right living, right effort, right mindfulness, and right concentration.

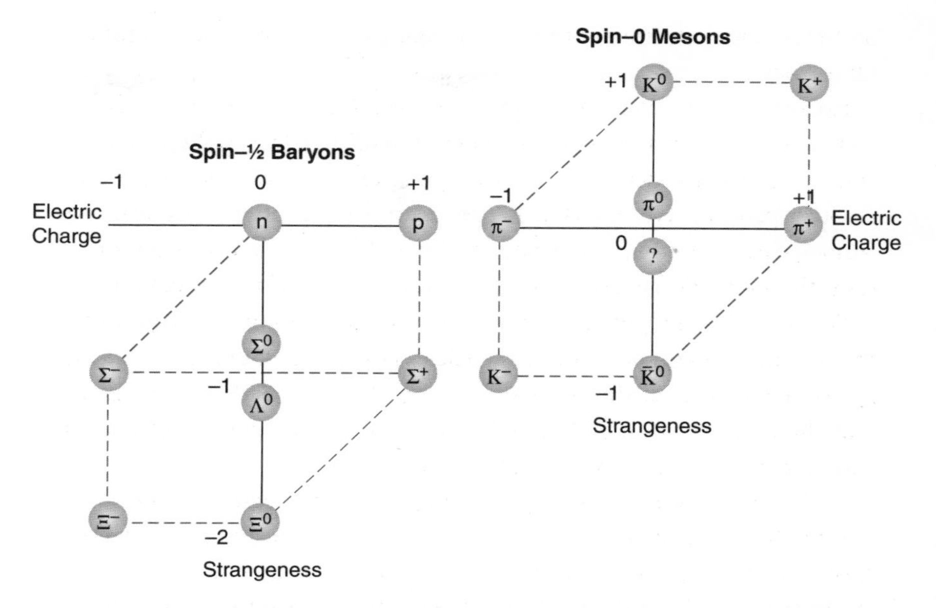

FIG 12 The Eightfold Way. Gell-Mann found that he could fit the spin-½ baryons, including the neutron (n), and proton (p) and the spin-0 mesons into two octet representations of the global symmetry group SU(3). But there were only seven particles in the representation for the spin-0 mesons. One particle, the meson equivalent of the Λ^0, was missing. This particle was found a few months later by Luis Alvarez and his team in Berkeley. They called it the eta, η.

physics. Where Gell-Mann had gone to Yale at the tender age of fifteen, Ne'eman, a native of Tel Aviv, had joined the Haganah, the Jewish underground, in what was then the British Mandate for Palestine. He had commanded an infantry battalion in the 1948 Arab-Israeli war.

Initially intending to study relativity at King's College, London, he found that the city traffic made it impossible for him to get there from the Embassy in Kensington so he switched to Imperial College and particle physics. At Imperial College he was pointed in the direction of Abdus Salam.

Ne'eman worked in the evenings and at weekends, bringing ideas to Salam which Salam would patiently point out had been worked out years previously by other, more distinguished physicists. Salam began to grow impatient, but Ne'eman's confidence with the theory was growing. A search for potential symmetry groups that might accommodate the known particles turned up five candidates, including SU(3). Initially excited by the very resonant possibilities afforded by a symmetry group that produced a Star of David pattern, Ne'eman eventually fixed on SU(3). He published his own version of the Eightfold Way in July 1961.

Both Ne'eman and Gell-Mann attended the 'Rochester' conference in June 1962, held not in Rochester that year but at the Organisation Européenne pour la Recherche Nucléaire (CERN) in Geneva. Both listened intently to reports of new particles that had been discovered, a triplet of spin-3/2 sigma-star particles (Σ^{*+},Σ^{*0}, and Σ^{*-})[13] with strangeness values of −1, and a doublet of spin-3/2 xi-star particles (Ξ^{*0} and Ξ^{*-}) with strangeness values of −2, which mirrored the spin-½ sigma and xi baryons.

Ne'eman immediately saw that these particles belonged to the decimet, to be added to the four delta particles already known. It took just a moment to realize that with nine particles, the decimet lacked just one more, yet to be discovered particle. This was a negatively charged particle with a strangeness value of −3.

He raised his hand to speak, but Gell-Mann had made precisely the same connection and was sitting closer to the front of the auditorium. It was therefore Gell-Mann who stood to predict the existence of a particle he called the omega, Ω^-. After the discussion, Nicholas Samios, a particle physicist from Brookhaven National Laboratory, asked Gell-Mann how he thought the Ω^- would decay. Gell-Mann sketched a prediction of its decay modes on a napkin. Samios tucked the napkin in his pocket.

The pattern had been found. Now came the question of the underlying explanation.

There remained the issue of the fundamental representation, the fundamental triplet at the heart of SU(3), which Gell-Mann had politely ignored. Robert Serber at Columbia University began to play around with ways of combining three fundamental 'objects' to create the two octets and the decimet of the Eightfold Way. This was equivalent in many ways to constructing the periodic table from combinations of protons, neutrons, and electrons. Serber, in effect, was asking if there might not be three fundamental particles, as yet undiscovered, underlying all the particles then known to

[13] These were originally called Y_1^* particles.

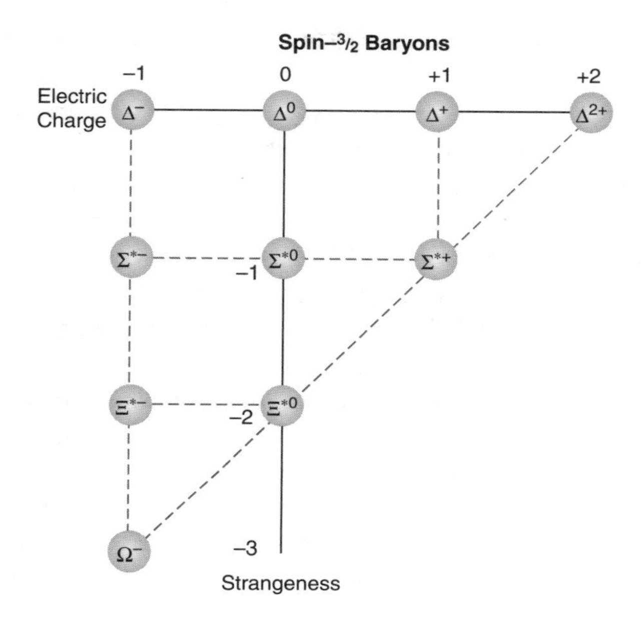

FIG 13 Both Ne'eman and Gell-Mann saw that the newly-discovered triplet of Σ^* particles and the doublet of spin-3/2 Ξ^* particles belonged to the decimet representation of SU(3). There was one further particle missing. Gell-Mann got there first. He named the missing particle the Ω^-.

physicists. When Gell-Mann arrived at Columbia in March 1963 to deliver a series of lectures, Serber asked him what he thought about this.

The conversation took place over lunch at the Columbia Faculty Club. 'I pointed out that you could take three pieces and make protons and neutrons,' Serber explained, 'Pieces and anti-pieces could make mesons. So I said "Why don't you consider that?"'

Gell-Mann had considered it, and he had already dismissed it. He asked Serber what the electric charges of this fundamental triplet of objects would be. This was something Serber hadn't considered. 'It was a crazy idea,' Gell-Mann said, 'I grabbed the back of a napkin and did the necessary calculations to show that to do this would mean that the particles would have to have fractional electric charges— $-1/3$, $+2/3$, like so—in order to add up to a proton or neutron with a charge of plus or zero.'

Serber agreed that this was an appalling result. There was absolutely no evidence for particles with fractional electric charges. In their discussion Gell-Mann referred to Serber's objects as 'quorks', deliberately chosen as a nonsense word, and mentioned them in passing during his subsequent lecture. Serber took the word as a derivative of 'quirk', as Gell-Mann had said that such particles would indeed be a strange quirk of nature.

But the discussion had set Gell-Mann thinking. The SU(3) symmetry group demanded a fundamental representation and the fact that the known particles could be fitted into the octet and decimet patterns was very suggestive of a triplet of fundamental particles. The fractional charges were problematic but, perhaps, if the 'quorks' were forever trapped or *confined* inside the larger hadrons then this might explain why fractionally charged particles had never been seen in experiments.

As Gell-Mann's ideas took shape, he happened on a passage from James Joyce's *Finnegan's Wake* which gave him a basis for the name of these mysterious new particles:

> Three quarks for Muster Mark!
> Sure he hasn't got much of a bark.
> And sure any he has it's all beside the mark.

'That's it!' he declared, 'Three quarks make a neutron and a proton!' The word didn't quite rhyme with his original 'quork' but it was close enough. 'So that was the name I chose. The whole thing is just a gag. It's a reaction against pretentious scientific language.'

Gell-Mann published a short article explaining this idea in February 1964. He referred to the three quarks as u, d, and s. Although he didn't say so in his paper, these stood for 'up' (u), with a charge of $+\frac{2}{3}$, 'down' (d) with a charge of $-\frac{1}{3}$, and 'strange' (s), also with a charge of $-\frac{1}{3}$. Baryons are formed from various permutations of these three quarks and mesons from combinations of quarks and anti-quarks.

For example, the proton consists of two up quarks and a down quark (uud), with a total charge of +1. The neutron consists of an up quark and two down quarks (udd), with a total charge of zero. The 'strange' particles, possessing strangeness quantum numbers other than zero, contain (not surprisingly) strange quarks. The octet of spin-½ baryons was explained in terms of different possible combinations of up, down, and strange quarks, the octet of spin-0 mesons in terms of combinations of quarks and anti-quarks. Isospin was now explained in terms of the numbers of up and down quarks in the composite nucleons. Beta radioactivity now involves the conversion of a down quark in a neutron into an up quark, turning the neutron into a proton, with the emission of a W^- particle.

In his paper he thanked Serber for stimulating discussions.[14]

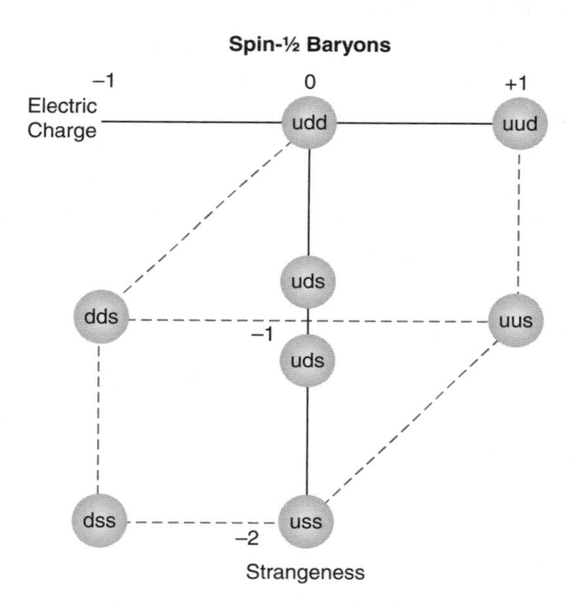

FIG 14 The Eightfold Way could be neatly explained in terms of the various possible combinations of up, down and strange quarks, illustrated here for the octet of spin-½ baryons . The Λ^0 and Σ^0 are both composed of up, down and strange quarks but differ in their isospin.

[14] An entirely equivalent scheme based on a fundamental triplet of particles was worked out around the same time by Gell-Mann's former Caltech student, George Zweig. By this time a post-doctoral student working at CERN, Zweig called his fractionally charged fundamental particles 'aces', and showed how it was possible to construct the baryons from 'treys' (triplets) of aces and the mesons from 'deuces' (doublets) of aces and anti-aces. His work was issued as a CERN preprint but he found that he could not get it published in a refereed journal. Zweig joined the Caltech faculty in late 1964, and Gell-Mann took pains to ensure he was credited with his role in the discovery of quarks.

It was a beautifully simplifying scheme, but in truth it was not much more than the result of playing with the patterns. There was simply no experimental foundation for the scheme and it was consequently poorly received. And there was another problem. As spin-½ fermions, the quarks are subject to the Pauli exclusion principle, which forbids more than one fermion from occupying a specific quantum state. Yet the proton was meant to contain two up quarks, presumably in the same state; the neutron two down quarks. How could this be squared with the exclusion principle?

Gell-Mann didn't help his cause by being rather cagey about the status of his quarks. Wishing to avoid getting tangled in philosophical debates about the reality or otherwise of particles that could in principle never be seen, he referred to the quarks as 'mathematical'. Some interpreted this to mean that Gell-Mann didn't think the quarks were made of real 'stuff', entities that existed in reality and combined to give real effects.

Samios had returned to Brookhaven National Laboratory armed with Gell-Mann's suggestions for the decay modes of the Ω^-. By mid-December 1963 the experiment to search for the Ω^- was ready. On 31 January 1964 Samios and his team of thirty particle physicists found the tell-tale tracks in a bubble chamber photograph. They fitted the predictions for the Ω^- almost exactly. Their paper was published in *Physical Review* on 11 February 1964. Gell-Mann's paper on quarks had been published in the CERN journal *Physics Letters* just ten days earlier.

The Eightfold Way was vindicated, but the quark theory was treated with derision. From Caltech Gell-Mann called Weisskopf in Geneva, mentioning almost in passing that he had had an idea which accounted for all the baryons and the mesons using particles with fractional charges. 'Oh nonsense, Murray,' Weisskopf responded, 'Don't waste time on a transatlantic call talking about stuff like that.'

It was hard to take the theoreticians seriously. Their toying with symmetries and gauge theories had turned up massless bosons that actually had to be massive, weak neutral currents that didn't exist, and now fractionally charged particles that, by the way, would never reveal themselves. Just who were these people trying to kid?

23

The 'God Particle'

Cambridge, Massachusetts, Autumn 1967

Some inklings of a possible solution to the problem of massless bosons in Yang–Mills field theories began to emerge in the early 1960s. The solution was relatively long in gestation, however, as its origin was in a different discipline, and physicists in different disciplines rarely communicate with one another. And even when they do, it tends to be in rather disparaging terms.

The quantum field theorists needed a 'trick' that would endow the massless bosons with mass. Nature itself offered an important clue. Whilst many theoretical physical descriptions depend on the identification of marvellous symmetries, related to conserved physical quantities, it is a simple fact that the natural world we live in tends to be rather more asymmetric in character, with clear preferences for this direction over that. A pencil perfectly balanced on its tip is symmetrical. But when it topples over it does so in a specific direction. The symmetry is said to be spontaneously broken.

It might seem a little curious that after expending considerable energy searching for the symmetries in nature, the physicists should now seek to find out how to break them. Considerable progress had been made in solid-state physics, a branch of physics concerned with the properties and behaviour of crystalline solids.

The archetypal example of spontaneous symmetry-breaking in a solid is afforded by a simple ferromagnet. Heat a bar magnet to a high enough temperature and the alignment of the iron atom magnetic moments becomes scrambled. The orientations of the atoms in the bar become symmetric, favouring no particular direction, and the bar loses its magnetism. As the bar cools back to room temperature, the iron atoms start to crystallize and a preferred orientation gradually emerges as their magnetic moments become aligned once more.[1] *The symmetry is spontaneously broken, and the bar becomes asymmetric as 'north' and 'south' poles are re-established.*[2]

Whilst spontaneous symmetry-breaking was familiar to solid-state physicists, it was unfamiliar to quantum field theorists or particle physicists. The latter saw themselves as purists, dealing with theories of the physical world at its most fundamental. They tended to perceive the physics of solids as horribly complex and riddled with simplifying assumptions. Gell-Mann reflected something of this attitude when he referred to the discipline as 'squalid-state' physics. In their turn, solid-state physicists tended to be rather dismissive of what they perceived as particle physicists' naïvety about how the physical world really worked.

However, a sense of desperation can help to overcome any such social barriers, especially if the unfamiliar discipline offers the hope of a solution. In 1960, Japanese-born American physicist Yoichiro Nambu drew analogies with the theory of superconductivity, an example of spontaneous symmetry-breaking in the gauge field of electromagnetism, and suggested that this might be a mechanism that could give mass to the massless Yang–Mills bosons. British-born physicist Jeffrey Goldstone went further. In 1961 he studied the effects of symmetry-breaking and concluded that one consequence was the production of yet more massless bosons.

This was not helping. In trying to find a way to give mass to the massless bosons of quantum field theories, Goldstone had found that spontaneous symmetry-breaking created even more massless particles.

It seemed like a dead end.

Abdus Salam had also identified spontaneous symmetry-breaking as a potential solution. He and Steven Weinberg attended a conference at the University of Wisconsin in Madison in the summer of 1961, where they met Goldstone.

Goldstone told them about what he had discovered. He had found that the only way to make the field theory of the symmetrical case equivalent to the field theory with broken symmetry was to introduce an additional

[1] In the absence of an externally applied magnetic field.

[2] This does not mean that all symmetry is lost. The cold bar magnet retains some intrinsic symmetries, but the symmetry group that contains these is a subset of the larger group that characterizes the hot iron bar.

massless boson, which came to be called a Nambu–Goldstone boson. Although he had derived this result for some selected symmetries, he felt instinctively that this would prove to be a general result, and elevated it to the status of a principle. It became known as the *Goldstone theorem.*

Of course, these Nambu–Goldstone particles suffered from precisely the same objections as the Yang–Mills bosons themselves. Massless particles, such as the photon, are associated with long-range forces, such as electromagnetism. The short-range nuclear forces were expected to be associated with massive particles. Any new massless particles predicted by theory could be expected to be as ubiquitous as photons. But, of course, these additional particles had never been observed. They just didn't fit the bill.

Salam and Weinberg didn't believe it. They called the new massless particle a 'snake in the grass' and worked together at Imperial College in London that autumn to prove that it didn't exist. They ended up proving precisely the opposite. Nambu–Goldstone bosons appeared whenever isospin or strangeness symmetries were broken. They published a paper jointly with Goldstone to this effect in August 1962. Weinberg was dismayed:

> I remember being so discouraged by these zero masses that when we wrote our joint paper on the subject... I added an epigraph to the paper to underscore the futility of supposing that anything could be explained in terms of a non-invariant vacuum state: it was Lear's retort to Cordelia, 'Nothing will come of nothing: speak again.'

Goldstone had assumed that empty space—the vacuum—is filled with a hypothetical field which breaks the symmetry. It was a kind of elaborate 'ether'. Weinberg's quote from *King Lear* was a sorry reflection on the futility of trying to conjure mass from the vacuum. The editors of *Physical Review,* not much given to flights of dramatic fancy, duly excised the quotation from the paper.

If any progress was to be made, some way of avoiding or beating the Goldstone theorem had to be found.

For their part, the solid-state physicists were rather nonplussed. They had been working with broken symmetries in crystalline solids for many years and had uncovered no new massless particles in systems with a local gauge symmetry. In the theory of superconductivity developed in 1957 by John Bardeen, Leon Cooper, and John Schrieffer, no new particles had to be invoked to explain why certain materials, when cooled below a critical temperature, lose all their electrical resistance.

We learn early on in our science education that like charges repel each other. However, electrons in a superconductor experience a mutual *attraction*, albeit weak. What happens is that an electron passing close to a positively charged ion in the crystal lattice exerts an attractive force which pulls the ion out of position slightly, distorting the lattice. The electron moves on, but the distorted lattice continues to vibrate. This vibration produces a region of excess positive charge, which attracts a second electron.

The upshot of this interaction is that a pair of electrons (called a 'Cooper pair'), each with opposite spin and momentum, move through the lattice cooperatively, their motion mediated or facilitated by the lattice vibrations. Electrons are fermions and, as such, they are forbidden from occupying the same quantum state by the Pauli exclusion principle. In contrast, Cooper pairs behave like bosons: there is no restriction on the number of pairs that can occupy a quantum state and at low temperatures they can experience Bose condensation, gathering in a single macroscopic quantum state.[3] The Cooper pairs in this state experience no resistance as they pass through the lattice and the result is superconductivity.

The Bardeen–Cooper–Schrieffer theory of superconductivity is not only elegant, it explains an otherwise puzzling set of phenomena entirely within the framework of quantum electrodynamics. Other than symmetry-breaking of the electromagnetic field, no new features and no new particles are required.

This kind of logic led sold-state physicist Philip Anderson to reject the Goldstone theorem. It was transparently obvious from many practical examples in solid-state physics that Nambu–Goldstone bosons are not always produced when gauge symmetries are spontaneously broken. What are produced are quantized collective excitations, vibrations of the

[3] Laser light is an example of Bose condensation involving photons.

lattice, which can be thought of as the quantum particle equivalent of wave patterns established in the solid. There was nothing particularly mysterious about these.

In 1963, Anderson suggested that the problems the quantum field theorists were wrestling with may in some way resolve themselves:

> It is likely, then, considering the superconducting analogue, that the way is now open without any difficulties involving either zero-mass Yang–Mills gauge bosons or zero-mass [Nambu–]Goldstone bosons. These two types of bosons seem capable of 'canceling each other out' and leaving finite mass bosons only.

It was to prove a prescient proposal. It was also roundly ignored. Anderson was a solid-state physicist and it was therefore assumed that he had little to say that might be of interest to quantum field theorists or particle physicists. And, as Anderson had provided no formal proof that the Goldstone theorem was invalid, those few theorists who did note what he had to say tended to disbelieve him.

Anderson's paper did provoke a minor controversy over the domain of applicability of quantum field theory. In 1964 this led, in turn, to a number of papers detailing mechanisms for spontaneous symmetry-breaking in which the various massless bosons did indeed 'cancel each other out', leaving only massive particles. These were published independently by Belgian physicists Robert Brout and François Englert working at Cornell University in New York State, English physicist Peter Higgs at Edinburgh, and Gerald Guralnik, Carl Hagen, and Tom Kibble at Imperial College in London.[4] The mechanism is commonly referred to as the *Higgs mechanism*.

In a paper published earlier in 1964, Higgs established that Anderson's instincts concerning the boson problem were essentially correct, although this happens to be the case just for Yang–Mills gauge theories. The Nambu–Goldstone bosons produced by symmetry-breaking

[4] These three papers were all published in the same volume (13) of the journal *Physical Review Letters* in 1964, on pp. 321–323, 508–509 and 585–587, respectively.

can be eliminated from the theory through their 'absorption' into the description of the Yang–Mills bosons. The Nambu–Goldstone bosons cease to exist as independent massless particles, becoming a new 'degree of freedom' of the Yang–Mills bosons.[5]

The process by which the Yang–Mills bosons are meant to acquire mass is mathematically complex, but there are several analogies that can help us to visualize what is going on. The hypothetical vacuum field required to break the symmetry (which is now called a *Higgs field*) acts selectively on the Yang–Mills field particles. To these particles, the Higgs field behaves like molasses. These particles quickly get 'bogged down', their motion impeded through interactions with the field resulting in behaviour that appears to all intents and purposes like the motion of massive particles.

CERN physicists use another analogy to explain the Higgs mechanism to politicians. Imagine a cocktail party in which a room is uniformly populated with physicists quietly drinking cocktails and chatting among themselves. This is equivalent to the vacuum containing the Higgs field. A noted celebrity physicist (no doubt a Nobel Laureate) enters the room and causes something of a stir. This is the massless Yang–Mills boson. The physicists gravitate in the direction of the celebrity in order to engage her in conversation and, before too long, a throng has gathered around her which slows down her progress as she crosses the room.[6] This resistance to accelaration of the Yang–Mills boson is equivalent to the acquisition of mass.[7]

[5] A massless particle with spin 1 (boson) has two degrees of freedom corresponding to transverse polarization states (in the photon these are associated with left-circular and right-circular polarization states). After application of the Higgs mechanism, the spin 1 boson 'absorbs' the Nambu–Goldstone boson to yield a third, longitudinal, degree of freedom. Consequently the acceleration of the particle is resisted by the field, and we recognise this as the acquisition of mass.

[6] This is a variation of the analogy devised by physicist David Miller at University College London, in which the 'field' of cocktail-drinking physicists is replaced by political party workers, and the celebrity physicist by former Prime Minister Margaret Thatcher. Miller submitted his analogy to a competition to explain the Higgs mechanism in simple terms, established in 1993 by the Science Minister, William Waldegrave. Miller's was one five winning entries. Waldegrave was convinced that the search for the Higgs boson was worth Britain's annual contribution to CERN.

[7] It's worth noting that in the symmetry-breaking of the U(1) gauge field of electromagnetism associated with superconductivity, the massless photons of the gauge field can acquire mass in precisely the same way. The result is the Meissner effect, first discovered in 1933 by Walther Meissner and R. Ochsenfeld. If a superconductor is cooled below its transition temperature in the presence of a weak-to-moderate magnetic field, the field photons become massive and the 'lines of force' of the field can no longer penetrate the interior of the superconductor. The field is forced to flow around the outside of the material.

The theory had implications not only for the massless bosons of the fledgling electro-weak theory, but for all particles. It was believed that the Higgs field was responsible for breaking the symmetry of the entire universe in the early stages of the hot 'big bang'. Just as the iron atoms start to crystallize, breaking the symmetry in a cooling ferromagnet, so the phase angle of the Higgs field 'crystallized' to a specific value as the early universe cooled. This crystallization created a 'false' vacuum, a baseline energy state higher than the zero-energy true vacuum. This symmetry-breaking mechanism affected the particles created in the early universe in different ways. The photons of the electromagnetic force were unaffected, and remained massless. But the bosons of the weak force picked up the energy trapped in the false vacuum in the form of acquired mass. Indeed, it appears that *all* massive particles acquire their mass this way.

Whilst this does seem as though something (mass) is being conjured from 'nothing' (the false vacuum, filled with the Higgs field), it is fair to say that the theory is not without its consequences. There is no quantum field without a quantum particle, and the particle of the Higgs field is called the *Higgs boson*. Stretching the cocktail party analogy a little further, the Higgs boson is like a softly spoken rumour about the celebrity, passed between the physicists as they wait for her to arrive. Now the physicists cluster together in order to hear the rumour, resulting in the creation of resistance (mass) even in the absence of the celebrity. This clustering is equivalent to a massive Higgs boson. The acquisition of mass by otherwise massless field particles is then identified as arising from interactions between these particles and the Higgs boson.[8]

The Higgs mechanism did not win converts immediately. Higgs himself had some difficulties getting his papers published. He sent the second

[8] Because of the actions of the Higgs mechanism in endowing mass to the Yang–Mills particles, and its subsequent adoption as the mechanism by which all particles acquire mass, the mechanism has acquired an almost theological status, at least outside the community of practicing physicists. In their 1993 popular book *The God Particle: If the Universe is the Answer, What is the Question?*, Nobel Laureate Leon Lederman and science writer Dick Teresi elevated the Higgs boson to the status of the 'God particle'. Higgs insists that Lederman had originally wanted to call it 'that goddamn particle' but his publisher wouldn't let him (see *The Guardian*, 30 June 2008).

of his two papers initially to the European journal *Physics Letters* in July 1964, but it was rejected by the editor as unsuitable. Years later, Higgs wrote:

> I was indignant. I believed that what I had shown could have important consequences in particle physics. Later, my colleague Squires, who spent the month of August 1964 at CERN, told me that the theorists there did not see the point of what I had done. In retrospect, this is not surprising: in 1964 the European particle theory scene was dominated by S-matrix theorists. Quantum field theory was out of fashion, and I had rashly formulated my description of the mass-generating mechanism in terms of linearised classical field theory, quantized by invoking the de Broglie relations.

Perhaps surprisingly, the mechanism had little immediate impact on those who might have benefited most from it. By his own admission, it seems that by this time Glashow had quite forgotten his earlier attempts to develop a unified theory of the weak and electromagnetic forces, a theory which predicted massless W^+, W^-, and Z^0 particles which needed somehow to be massive. 'His amnesia unfortunately persisted through 1966,' Higgs wrote. Glashow attended a lecture on the mass-generating mechanism that Higgs delivered at Harvard that year, yet failed to put two and two together.

It was Weinberg who eventually made the connection although he, too, did not immediately see the relevance of the Higgs mechanism. Weinberg spent the years 1965–67 working on the effects of spontaneous symmetry breaking in strong-force interactions described by an SU(2) × SU(2) field theory. The result of symmetry-breaking is the masses of the nucleons and Nambu–Goldstone bosons which could be approximated as the pions. At the time this all seemed to make sense and, far from trying to evade the Goldstone theorem, he now positively welcomed the predicted extra particles. This was a development '...which suddenly seemed to change the role of [Nambu–] Goldstone bosons from that of unwanted intruders to that of welcome friends.'

But after a couple of years pursuing this line of reasoning, Weinberg realized that it wasn't going to bear fruit. It was at this point that he was struck by another idea:

> At some point in the fall of 1967, I think while driving to my office at MIT, it occurred to me that I had been applying the right ideas to the wrong problem.

Weinberg had been applying symmetry-breaking to the strong force. He now realized that the mathematical structures he had been trying to apply to strong-force interactions were precisely what were needed to resolve the problems with weak-force interactions and the massive bosons these interactions implied. 'My God,' he exclaimed to himself, 'this is the answer to the weak interaction!'

Weinberg was well aware that if the masses of the W^+, W^-, and Z^0 particles were added by hand to Glashow's SU(2) × U(1) field theory of the weak and electromagnetic forces, then the result was rendered unrenormalizable. He now wondered if breaking the symmetry using the Higgs mechanism would endow the particles with mass, eliminate the unwanted Nambu–Goldstone bosons, and yield a theory that could be renormalized.

There remained the problem of the weak neutral currents, interactions involving the neutral Z^0 particle for which there was still no experimental support. He decided to avoid the problem altogether by restricting his theory to leptons—electrons, muons, and neutrinos. He had by now become wary of the hadrons, particles affected by the strong force, and especially the strange particle decays which were the principal ground for the experimental exploration of weak-force interactions.

Neutral currents would still be predicted, but in a model consisting only of leptons these would involve interactions with the neutrino. The neutrino had proved difficult enough to find experimentally in the first place, and he may have figured that finding weak neutral currents involving these particles would present sufficient experimental challenges that he could predict them with little fear of immediate contradiction.

Weinberg published a paper detailing an 'electroweak' unified theory for leptons in November 1967. This was an SU(2) × U(1) field theory reduced to the U(1) symmetry of ordinary electromagnetism by spontaneous symmetry-breaking, which gave mass to the W^+, W^-, and Z^0 particles, whilst leaving the photon massless. He estimated the mass-scales of the weak-force bosons, about 80 GeV for the W particles and 90 GeV for the Z^0 and predicted some properties of what would become known as the Higgs boson. He was not able to prove that the theory was renormalizable but felt confident that it was.

Salam, too, had finally woken to the possibilities afforded by the Higgs mechanism early in 1967 and had independently arrived at much the same formalism. He decided against publishing until he had had an opportunity properly to incorporate hadrons in the model.

Nobody took much notice. Those few who did pay attention tended to be critical. The mass problem had been fixed through some 'smoke-and-mirrors' trick involving a hypothetical field and another hypothetical boson. It seemed as though the quantum field theorists were continuing to play games with fields and particles, according to obscure rules that few understood. Physicists simply ignored them and got on with their science.

Weinberg later drew an analogy with Plato's famous allegory of prisoners in the cave, which he set out in *The Republic*. The prisoners held captive in the cave have no knowledge or experience of the outside world. All that they experience are shadows cast on one wall of the cave. Their world is a world of crude appearances of objects which they mistake for the objects themselves, 'literally nothing but the shadows of the images'.

Weinberg wrote:

> We are in such a cave, imprisoned by the limitations on the sorts of experiments we can do. In particular, we can study matter only at relatively low temperatures, where symmetries are likely to be spontaneously broken, so that nature does not appear very simple or unified. We have not been able to get out of this cave, but by looking long and hard at the shadows on the cave wall, we can at least make out the shapes of symmetries, which though broken, are exact principles governing all phenomena, expressions of the beauty of the world outside.

PART V

Quantum *Particles*

24

Deep Inelastic Scattering

Stanford, August 1968

Weinberg's use of Plato's allegory of the cave may have reflected something of the mood among quantum field theorists in the late 1960s. Theirs was a largely thankless task. They had strained to make sense of the shadows cast on the cave wall, and had spent much time blundering around in the dark. They had tried hard to make their approaches fit the shadows as best they could, patching the theory or finding 'work-arounds' when the mathematics became particularly truculent. There had been breakthroughs, but few acknowledged their success in what had by now become an unfashionable discipline. It was not at all clear where this was leading.

Experimental particle physicists were presented with a different task. Their mission was to shine more light in the cave, sharpening and clarifying the shadows so they better reflected the underlying reality that created them. In the 1960s, this meant building bigger and bigger particle accelerators.

In October 1957 the American administration had experienced a deep sense of frustrated impotence as the Soviet Union's Sputnik I satellite bleeped provocatively overhead on its 98-minute orbit of the earth.[1] This sense had been compounded by the completion that year of a 10-GeV proton synchrotron by the Joint Institute for Nuclear Research in Dubna, 120 kilometres north of Moscow. At the time this was the world's largest particle accelerator, beating the Cosmotron's 3 GeV at Brookhaven and the Bevatron's 6.2 GeV

[1] You can listen to Sputnik's telemetry on a NASA website: http://history.nasa.gov/sputnik

at Berkeley's Radiation Laboratory. CERN soon followed in 1959 with a 26-GeV proton synchrotron in Geneva.

As the race for Cold War technological supremacy reached white heat in the 1960s, American particle physicists were among the beneficiaries as funding for high-energy physics was greatly increased. The Alternating Gradient Synchrotron was constructed at Brookhaven in 1960, capable of operating at 33 GeV. These were all proton accelerators, designed to smash protons into stationary targets, including other protons. Their basic aim was the creation of exotic new particles and the study of strong- and weak-force interactions. The outcomes of these experiments involved long and complex analysis, however. As Feynman explained, proton–proton collisions were '... like smashing two pocket watches together to see how they are put together.'

This emphasis on high-energy hadron collisions came at the cost of subtlety. When construction commenced in 1962 of a new $114-million 20-GeV linear electron accelerator at Stanford University in California, it was dismissed by many particle physicists as an irrelevant machine, capable only of second-rate experiments. But American theorist James Bjorken understood that the de Broglie wavelengths of high-energy electrons would be far smaller than the diameter of the proton.[2] Such electrons could probe the interior of protons in ways that heavy-particle collisions could not. In particular, they could reveal the existence or otherwise of point-like constituents deep inside the proton.

In astonishing experiments that would mirror the fabled 1909 Geiger–Marsden experiments which revealed the interior structure of the atom,[3] the scattering of electrons from protons in the new Stanford accelerator would reveal the interior structure of the proton.

The experimentalists would finally have an opportunity to catch up with some of the more outrageous conclusions of the theorists.

The Stanford Linear Accelerator Center (SLAC) was built on 400 acres of Stanford University grounds about five kilometres west of the main campus, at the foot of the Los Altos hills about 60 kilometres south of San Francisco.[4] The three-kilometre accelerator is linear, rather than circular, because bending electron beams into a circle using intense magnetic

[2] Remember that according to the de Broglie relation, $\lambda = h/p$, the higher the speed (and hence momentum, p) of an electron, the shorter its wavelength.

[3] See Chapter 3.

[4] It is situated close to the San Andreas fault. In the event of a major earthquake it has been suggested that it might be renamed the Stanford Piecewise Linear Accelerator Center (SPLAC). See Woit, p. 23.

fields results in dramatic energy loss through emission of X-ray synchrotron radiation. The accelerator reached its 20-GeV design beam energy for the first time in 1967.

The aim was to examine electrons scattered off various targets positioned at the end of the accelerator, using large and elaborate electron spectrometers. When an electron collides with a proton, three different types of interaction may result. The electron may bounce relatively harmlessly off the proton, exchanging a virtual photon, changing the electron's velocity and direction, but leaving the particles intact. This, so-called 'elastic' scattering yields electrons with relatively high scattered energies clustered around a peak.

In a second type of interaction, the collision with the electron may exchange a virtual photon which kicks the proton into one or more excited energy states. The scattered electron comes away with less energy as a result, and a chart of scattered energy *versus* yield shows a series of peaks or 'resonances' corresponding to different excited states of the proton. Such scattering is 'inelastic', as new particles (such as delta particles and pions) may be created, although both electron and proton emerge intact from the interaction. In essence, the energy of the collision, and of the exchanged virtual photon, has gone into the creation of new particles.

The third type of interaction is called 'deep inelastic' scattering, in which much of the energy of the electron and the exchanged virtual photon goes into destroying the proton completely. A spray of different hadrons emerges and the scattered electron recoils, now with considerably less energy.

It was this last type of scattering that interested James Bjorken. Now in his early thirties, Bjorken had obtained his doctorate at Stanford University and had recently returned to California after a spell at the Niels Bohr Institute in Copenhagen. Just before SLAC was completed, he had developed an approach to predicting the outcomes of electron–proton collisions using a rather esoteric quantum field theoretic approach known as current algebra.

In this model, the electron is treated as a point-particle surrounded by a cloud of virtual particles which flicker into and out of existence. The proton is a different matter, however. At the time it was possible to conceive the structure of the proton in two distinct ways. It could be considered as a solid 'ball' of material substance, with mass and charge distributed evenly, much like snow and ice is distributed in a snowball. Or it could be thought of as a region of largely empty space containing discrete, point-like charged constituents. High-energy electrons encountering such point-like constituents could be expected to be scattered in greater numbers and at larger angles.

Bjorken found that, for these different conceptions of the proton structure, the theory predicted quite different results in the region of deep inelastic collisions. He realized that an experimental test was possible, comparable in many ways to the Geiger–Marsden experiments which supported the view that the mass of an atom is concentrated in a small positively charged nucleus. Just as the scattering of alpha-particles from their target gold atoms revealed the interior structure of the atom, so the observations of electrons scattered from target protons would reveal the interior structure of the proton.

Bjorken speculated that such point-like constituents might be quarks and performed some calculations for the spectrum of scattered electron energies that would result. But his heart wasn't yet in it. The quark model was still treated with derision by the physics community and there were alternative theories available that were better regarded. He quickly backed off from proposing that these experiments could be used as a test of the quark model. 'I brought it in mainly as a desperate attempt to interpret the rather striking phenomenon of point-like behaviour,' he explained when challenged on this view at a conference held at Stanford in September 1967.

Studies of deep inelastic scattering at relatively small angles from a liquid-hydrogen target began at SLAC around this time and continued through October 1967. They were carried out by a small experimental group including MIT physicists Jerome Friedman and Henry Kendall and Canadian-born SLAC physicist Richard Taylor. The experiments did not initially go smoothly.

> The analysis programs had to be rewritten as we continued to test the workings of the spectrometer hardware. After a couple of weeks of total chaos, things began to settle down. We took a full set of data at 6° with several initial energies and with scattered electron energies down to 2–3 GeV.

The raw data had to be corrected for all the radiative processes involving the swarm of virtual particles surrounding the electron. Although these corrections were significant, accounting for up to half of the measurements recorded in the SLAC experiments, they could be handled satisfactorily using quantum electrodynamics.

Even at the relatively low scattering angles of 6° and 10°, the physicists observed many more scattered electrons than they had anticipated. They focused their attention on the behaviour of something called the 'structure function' as a function of the difference between the initial electron energy and the scattered electron energy. This difference is related to the energy lost by the electron in the collision or the energy of the virtual photon that is exchanged. They saw that as the virtual photon energy was increased, the structure function showed marked peaks corresponding to the expected proton resonances. However, as the energy increased further, these peaks gave way to a broad, featureless plateau that fell away gradually as it extended well into the range of deep inelastic collisions.

Bjorken had anticipated something like this. It was indeed the signature of point-like proton constituents. He made a further suggestion. If the structure function declined gradually with increasing energy of the virtual photon, then the product of these (called F), plotted against the photon energy (or frequency, ν) divided by the square of something called the four-momentum transfer (q^2), might show some interesting behaviour.[5]

Kendall drew the graph by hand, using two sheets of orange graph paper that had been taped together. It showed F increasing to a maximum for ν/q^2 around 3 GeV, and remaining fairly constant thereafter. Bjorken called it 'scaling'. The curve was the same irrespective of the electron energy. The mathematics said it had to be this way, but it was not at all clear what this meant.

Irrespective of what it meant, Kendall was struck by the way the experimental data had so closely conformed to Bjorken's predictions. Years later he recalled '...wondering how Balmer may have felt when he saw, for the first time, the striking agreement of the formula that

[5] The four-momentum transfer is calculated from the initial and final electron energies and the scattering angle.

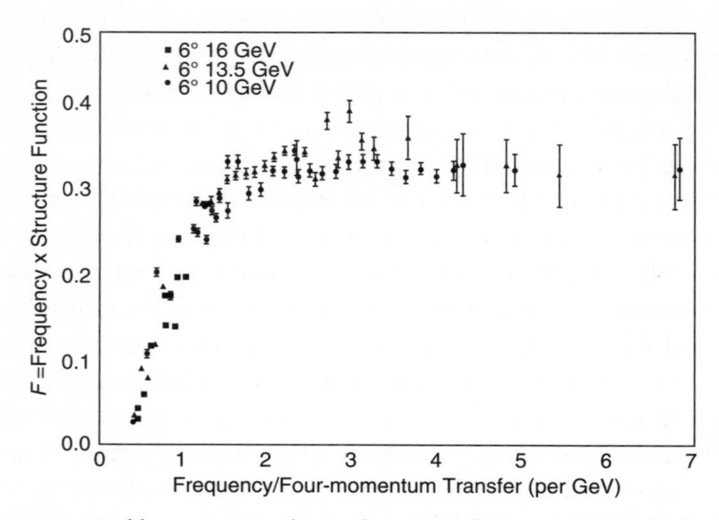

FIG 15 Prompted by suggestions by Bjorken, Kendall drew up this graph by hand. The shape of the resulting curve was the same irrespective of the electron energy (in this case, 10, 13.5 and 16 GeV, with a scattering angle of 6°). Bjorken called it 'scaling'. It was the tell-tale signature of point-like constituents inside the proton. Adapted from Michael Riordan, *The Hunting of the Quark: A True Story of Modern Physics*, Simon & Shuster, New York, 1987, p. 146.

bears his name with the measured wavelengths of the atomic spectra of hydrogen.'

There was as yet no rush to embrace these experimental results as proof of the existence of an internal structure for the proton, and certainly no rush to declare the results as evidence for the existence of quarks. There were other possible theoretical explanations for the scattering results and for the phenomenon of scaling, and arguments broke out even within the MIT-SLAC group. Nevertheless, on Kendall's insistence the graph that Bjorken had suggested was included in a paper on the group's results presented at a conference in Vienna in August 1968.

Richard Feynman had collected a Nobel Prize in December 1965 for his contribution to the development of quantum electrodynamics, alongside Julian Schwinger and Sin-Itiro Tomonaga. In the years that followed he had adapted to the celebrity status it afforded, but had at the same time lost touch with contemporary high-energy physics. He had helped resolve some of the problems of QED and, with Gell-Mann, had made important contributions to the theory of the weak force. But the strong force had eluded him.

He had resolved to play catch-up. In June 1968 he developed an outline of a theory of the hadrons in which he imagined them to consist of point-like constituents which, for want of a better name, he had dubbed *partons*, meaning simply the 'parts' of hadrons. The parton model was not grounded in a detailed quantum field theory, it was simply an idea he was toying with. What he wanted to know was that if the hadrons could be thought to be composed of partons, what were the consequences?

He had also developed an image in his mind of two hadrons colliding together. At high speeds, special relativity says that each hadron 'sees' the other not as a fully fleshed-out three-dimensional object but rather as a two-dimensional flat pancake. If these pancakes were dotted with tiny, hard partons, then a hadron–hadron collision would really be the sum of a series of individual collisions between the constituent partons.

Feynman's sister Joan lived in a house just across the road from SLAC, and during visits to her he would take the opportunity to 'snoop around' the facility. He would chat to the physicists at a picnic table in the shade of the oak trees near the SLAC cafeteria, regaling them with stories from his Los Alamos days. He would also listen carefully as they told him about the latest developments in high-energy physics.

In August 1968 he heard about the work of the MIT-SLAC group on deep inelastic scattering. A second round of experiments was about to begin, but the physicists were still puzzling over the interpretation of the data from the previous year. Bjorken was out of town, but his new postdoctoral student Emmanuel Paschos told Feynman about scaling and asked him what he thought.

When Feynman saw Kendall's graph he collapsed to his knees, his hands raised to the heavens. 'All my life,' he exclaimed, 'I've looked for an experiment like this, one that can test a field theory of the strong force!'

Feynman knew that the graph was telling him something, but he was not quite sure what it was. He figured it out that night in his motel room. The graph was related to the distribution of momentum of the point-like constituents—the partons—in the proton. In fact, simply by inverting the function that Kendall had plotted along the horizontal axis (in other words, by plotting q^2/ν), the graph became a probability distribution,

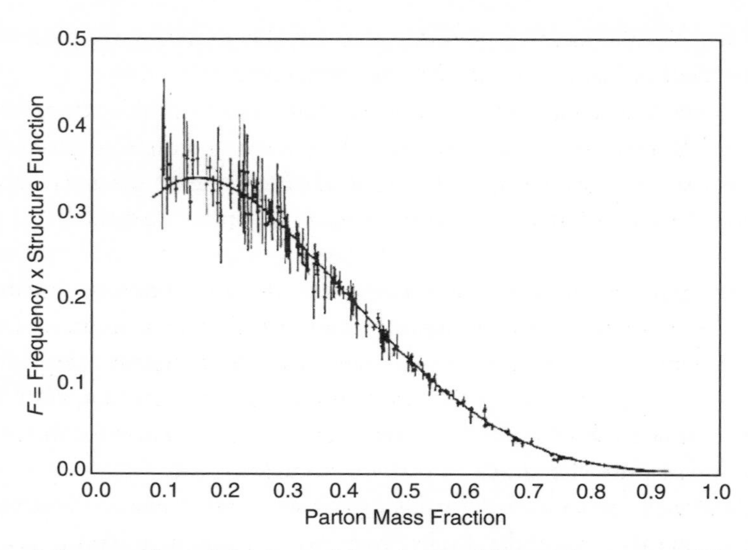

FIG 16 Feynman figured that he could re-plot Kendall's graph in terms of the fraction of the proton mass carried by the point-like constituents, which he called 'partons'. The result would be a distribution of the probability for the electron colliding with a parton carrying a specific fraction of the proton's total momentum. Adapted from Michael Riordan, *The Hunting of the Quark: A True Story of Modern Physics*, Simon & Shuster, New York, 1987, p. 152.

each point a measure of the probability of the electron colliding with a parton carrying a specific fraction of the total momentum of the proton. This could be pushed a little further. When divided by the proton mass, the quantity q^2/ν is related to the fraction of the proton mass carried by the parton.

'I've really got something to show youse guys,' Feynman declared to Friedman and Kendall the next morning, 'I figgered it all out in my motel room last night!' This was followed by much enthusiastic arm-waving.

Bjorken had arrived at most of the conclusions that Feynman had now drawn, but Bjorken had expressed them in the esoteric language of current algebra. When he had had a chance to read Bjorken's papers, Feynman acknowledged his priority. But, once again, Feynman was describing the events in a far simpler, yet richer, more visual way. When he returned to SLAC in October 1968 to deliver a lecture on the parton model, it was like setting a fire. Suddenly, all the talk among SLAC physicists was of partons. Nothing is more guaranteed to breed confidence in a bold idea than its enthusiastic reception by a Nobel Laureate.

Were partons actually quarks? Feynman didn't know and didn't care. Some of the SLAC theorists developed models in which the partons were pions and protons, stripped of their clouds of virtual particles. But Bjorken and Paschos soon had a detailed model of partons based on a triplet of quarks.

Further experiments at scattering angles of 18°, 26°, and 34° confirmed the original impressions and reproduced the scaling behaviour. By 1970 experiments had been performed using a liquid-deuterium target. By carefully separating the contributions to scattering from the proton, it was possible to determine the equivalent structure function for the neutron and so probe its interior structure in the same way.

These were difficult experiments, and it was not all plain sailing. One theoretical model describing collisions between electrons and protons and electrons and neutrons suggested that the extent of deep inelastic scattering should depend on the relative strengths of the electric charges of the parton constituents. If the partons were considered to be quarks, then the ratio of scattering events for the neutron *versus* the proton should be given as the ratio of the sum of the squares of the charges of their quarks constituents.

For the neutron, thought to consist of an up quark and two down quarks (udd), the sum of the squares of the fractional charges is $6/9$ or $2/3$. For the proton, consisting of two up quarks and a down quark (uud), the sum is $9/9$, or 1. The ratio of neutron *versus* proton scattering events (N/P), was therefore expected to fall with increasing q^2/ν to a limiting value of $2/3$.

Early results appeared to be consistent with this simple model. But as q^2/ν increased, the ratio N/P was discovered to fall substantially below $2/3$. These results ruled out a variety of competing theories, but seemed also to tell against the model of fractionally charged quarks.

Theorists had a field day trying to rationalize these results with existing models. There was still some hope for quarks, however. The model that predicted a limiting value of N/P of $2/3$ was rather naïve. It assumed that the momentum of the nucleon was evenly distributed over the three quark constituents. There were other models in which the three quarks which characterize the nucleon (called 'valence' quarks) swim in a 'sea' of quark–anti-quark pairs.

In addition, the quarks (or, more generally, partons) were now thought to be bound together through the exchange of strong force field particles called *gluons*, the 'glue' that holds the 'parts' together. These short-range force particles were required to be massive. So, the momentum of the nucleons may be shared among quark–anti-quark pairs and gluons, and is in any case not necessarily obliged to be shared equally among the valence quarks. It was recognized that N/P could indeed fall below ⅔. But it could not fall below a lower limit of ¼, in which only a down quark from the neutron and an up quark from the proton contribute to the scattering.

When experimental results suggested that the ratio actually did fall below ¼, both Bjorken and Feynman insisted that no further modifications could save the quark theory. If the results were right, then the partons simply could not be fractionally charged quarks.

The results were wrong. An error was discovered in the computer program used to correct for the effects of 'smearing' due to the internal motions of the protons and neutrons in the hydrogen and deuterium target atoms. When this error was removed, the ratio of N/P was seen to fall to a limiting value close to ¼.

Earlier experimental studies had revealed that the partons were spin-½ fermions. The N/P data now strongly supported the idea of fractional charges. Experimental results from studies of deep inelastic scattering of neutrinos from protons at CERN provided further supporting evidence. By mid-1973, quarks had officially 'arrived'. They might have been conceived partly in jest as a strange quirk of nature, but they had now taken a decisive step towards acceptance as real constituents of the hadrons.

Some questions remained unanswered. If 20-GeV electrons were striking fractionally charged quarks inside protons or neutrons, resulting in the destruction of the target nucleons, how come no free quarks had ever been seen? More worrying, perhaps, was the fact that the quarks inside the proton and neutron behaved as though they were largely independent of each other. The phenomenon of scaling suggested that the quarks could be considered to roam freely inside the nucleons.

How was this possible? Surely, the strong force must keep the quarks tightly bound inside the nucleons; so tightly bound that they are forever confined?

25

Of Charm and Weak Neutral Currents

Harvard, February 1970

The SLAC results on deep inelastic scattering of electrons from protons were still interpreted with some caution. There were many questions that remained unanswered. But, for the advocates of the quark model, here was the first evidence that these ultimate constituents of matter might actually be real.

Sheldon Glashow was an early convert to the quark model. In fact, based on little more than an appeal to an essential tidiness in nature, he had speculated with James Bjorken in 1964 that there should be a fourth quark, which they had called charm. At that time the four leptons—electron, electron neutrino, muon, and muon neutrino—seemed at odds with just three quarks—up, down, and strange. A fourth quark was surely needed to even things up.

Glashow had collaborated with Bjorken whilst visiting the Niels Bohr Institute in Copenhagen. For Bjorken, this was an unusual line of reasoning. More used to anticipating the properties and behaviour of particles likely to have implications for experiment, this was for him, at best, a highly speculative proposal. 'You have to remember the flavour of the time,' he explained, 'It was an easygoing time for models. Models came and went.' He was nevertheless rather ambivalent about the proposal. Instead of using his own name, James Daniel Bjorken, he signed the paper 'B.J. Bjørken', using his nickname, 'BJ', and a Danish interpretation of his originally Swedish surname.

Glashow's attempts to build an $SU(2) \times U(1)$ field theory of the weak and electromagnetic forces had foundered on the problem of the Z^0 boson and the weak neutral currents it implied. With Bjorken he had actually arrived at the solution to this problem, but he had *failed to make the connection. It seems that Glashow had forgotten all about his earlier efforts. 'The problem which was explicitly posed in 1961 was solved, in principle, in 1964. No one, least of all me, knew it.'*

He returned to the problems of the SU(2) × U(1) theory in 1970, in the company of two Harvard postdoctoral students, Greek physicist John Iliopoulos and Italian Luciano Maiani. Glashow had first met Iliopoulos at CERN and had been impressed with his efforts to find ways to renormalize a field theory of the weak force. Maiani arrived at Harvard with some curious ideas about the strength of weak-force interactions. All three realized that their interests converged. At this stage none was aware of Weinberg's use of spontaneous symmetry-breaking and the Higgs mechanism.

The three physicists now wrestled with the theory once again. But they kept running up against the same problems. Admitting the field particle masses by hand produced unruly divergences. Then there was the problem of the weak neutral currents. The theory predicted that a neutral kaon should in principle decay by emission of a Z^0 boson, changing the strangeness of the particle in the process and producing two muons. However, this was a decay mode for which there was absolutely no experimental evidence. For some reason, this particular reaction was being suppressed. But how?

Glashow was finally ready to put two-and-two together. In doing so he would help to sow the seeds of a major revolution in particle physics.

Even among strange particles, the neutral kaon, K^0, and its antiparticle, $\bar{K}^0$, can be regarded as 'odd'. According to the quark model, the positive kaon is a composite of an up quark and an anti-strange quark ($u\bar{s}$), with charges of $+2/3$ and $+1/3$ adding to a total of +1. Its anti-particle is the negative kaon, a composite of a strange quark and an anti-up quark ($s\bar{u}$), with charges of $-1/3$ and $-2/3$ adding to a total of −1.

For the neutral kaon, however, we have two possibilities. Both the combinations down/anti-strange ($d\bar{s}$) and strange/anti-down ($s\bar{d}$) yield zero net charge and, with no means to distinguish them, these two particles can form a quantum 'superposition', their wavefunctions overlapping and combining. The result is two new neutral kaon states. The first is formed from the sum of the wavefunctions of K^0 and $\bar{K}^0$ and is referred to as K^0_1. The second is formed from the difference of the wavefunctions and is referred to as K^0_2.

The particles suffer something of an identity crisis. Although the kaons are formed in their initial K^0 and $\bar{K}^0$ states by strong-force interactions, by the time the weak force acts on them they have transformed into K_1^0 and K_2^0.

However, this is not quite the end of the story. When parity was shown to be violated in weak-force interactions, it was assumed that a combination of charge conjugation (the symmetry between a particle and its anti-particle) and parity would nevertheless be preserved.[1] Thus, the superposition K_1^0, which is even under the charge conjugation-parity (CP) symmetry operation, can decay only into particles which are also even under this symmetry. K_1^0 was found to decay into two pions, consistent with this view. Likewise K_2^0, which has odd CP symmetry, was found to decay into three pions.

However, experiments performed at Brookhaven National Laboratory by Princeton physicists James Cronin and Val Fitch showed that K_2^0 kaons formed in a beam would very occasionally decay into two pions. These events were few in number, about 50 in a total of 23,000 observed decays, but they were much higher than expected from any background K_1^0 particles left in the beam. Cronin and Fitch concluded that CP symmetry, too, is violated by weak-force interactions.[2]

The explanation for this is that the states K_1^0 and K_2^0 'contaminate' each other slightly. What are actually seen in experiments are not pure K_1^0 and K_2^0, but K_L^0, which is mostly K_2^0 contaminated with a little bit of K_1^0, and K_S^0, which is mostly K_1^0 with a little bit of K_2^0. The subscripts L and S stand for 'long' and 'short', identifying these kaon states as long-lived and short-lived neutral kaons, respectively.

The problem confronting Glashow, Iliopoulos, and Maiani in February 1970 was principally concerned with taming the divergences in their SU(2) × U(1) field theory. They tried all kinds of different approaches. Iliopoulos later described their plight. 'Every day,' he said, 'invariably, one of us would come up with an idea. Then the other two would prove to him he was wrong. We tried all sorts of different recipes, and nothing worked.'

[1] For example, the application of the CP symmetry operation to a spin-up electron transforms it into a spin-down positron.

[2] The reason for CP violation is still the subject of active research. Cronin and Fitch were awarded the Nobel Prize for physics in 1980.

In frustration, they started to tinker with the model of four leptons and three quarks, initially by adding more leptons. Eventually they hit on the idea of adding a fourth quark. This was essentially a heavy version of the up quark with spin-½ and charge +⅔. Glashow now recalled his 1964 paper with Bjorken, and proposed to call it the charm quark. Many (though not all) of the unruly divergences disappeared.

They also realized that they had potentially solved another problem. The introduction of a charm quark had switched off the weak neutral currents. It turned out that the probability that a neutral kaon would emit a Z^0 boson and transform into two oppositely charged muons was almost equal to the probability that it would emit a Z^0 boson and become a charmed particle, containing a charm quark. A difference in signs relating to these two possible decay paths meant that they virtually cancelled each other out. Caught like a rabbit in headlights, the neutral kaon can't decide which way to jump, until it's too late.

It was a neat solution. Kaons, the romping ground for studies of weak-force interactions which should have exhibited weak neutral currents, almost never did so because of alternative decay modes involving the charm quark.

But then they realized that there were other ways that neutral kaons could decay into two muons, mimicking neutral currents, that had nothing to do with the Z^0. The down quark in a K^0_L kaon could be expected to transform into an up quark with the emission of a W^- boson. This mirrors the process by which a neutron transforms into a proton in beta radioactive decay. The up quark could then be expected to transform into a strange quark with the emission of a W^+ particle. The two W bosons would then combine through exchange of a muon neutrino to produce two oppositely charged muons. The result looks just like a weak neutral current.

Glashow, Iliopoulos, and Maiani now reasoned that the fact that this kind of decay mechanism had not been seen in experiments was also evidence for the charm quark. An equivalent mechanism involving a similar series of transformations, from down to charm to strange, produced the same particles but cancelled the mechanism involving the intermediate up quark.

Excited by their discovery, the physicists piled into a car and headed across to MIT and the office of American physicist Francis Low, who had also been working on the problem. Weinberg joined them and together they debated the merits of the Glashow–Iliopoulos–Maiani (GIM) mechanism.

What followed was a spectacular breakdown in communication. Almost all the ingredients for a unified theory of the weak and electromagnetic forces were assembled in the minds of the theorists gathered in Low's office. Weinberg had figured out how to apply spontaneous symmetry-breaking using the Higgs mechanism in an SU(2) × U(1) field theory of leptons, allowing the masses of the field particles to be deduced rather than input by hand. Glashow, Iliopoulos, and Maiani had found a potential solution to the problem of weak neutral currents in strange particle decays and held out the promise that the SU(2) × U(1) theory could be extended to weak-force interactions involving the hadrons. But they were still inputting the masses of the field particles by hand and struggling with divergences.

Glashow, Iliopoulos, and Maiani knew nothing of Weinberg's 1967 paper, and Weinberg said nothing about it. He later confessed to a 'psychological barrier' regarding the theory's renormalisability. He also did not look kindly on the charm quark proposal. What Glashow, Iliopoulos, and Maiani were invoking was not just one new particle, part of an extended family of particles of possibly dubious relevance, but an entirely new collection of 'charmed' mesons and baryons. It was an awful lot to swallow just to explain the *absence* of weak neutral currents in strange particle decays.

'Of course, not everyone believed in the predicted existence of charmed hadrons,' said Glashow.

'I do not care what and how,' Dutch theorist Martinus Veltman told his young graduate student Gerard 't Hooft early in 1971, as they walked between buildings on the campus of the University of Utrecht, 'but what we must have is at least one renormalisable theory with massive charged vector bosons, and whether that looks like nature is of no concern, [those] are details that will be fixed later by some model freak. In any case, all possible models have been published already.'

'I can do that,' 't Hooft said.

Veltman, who had spent several years trying to find ways to renormalize Yang–Mills field theories containing massive particles, almost walked into a tree. 'What do you say?' he asked.

'I can do that,' 't Hooft replied.

'Write it down and we will see,' Veltman said.

In seeking a suitable topic for his doctoral thesis at Utrecht, 't Hooft had identified Yang–Mills theories as fertile, if unfashionable, ground. Veltman had picked up the challenge of renormalizing Yang–Mills theories in 1968, and this was now a major preoccupation. 't Hooft had worked quickly. He had learned about spontaneous symmetry-breaking at a summer school in Cargèse, Corsica, in 1970 and by early 1971 had shown that massless Yang–Mills field theories could be renormalized.

't Hooft's confidence regarding the renormalizability of Yang–Mills theories with massive particles was based on his understanding of what spontaneous symmetry-breaking could do. Within a short time he had indeed written it down.

Veltman was uneasy about the symmetry-breaking mechanism, a concern he was later able to trace to its implications for cosmology.[3] They argued back and forth. In the end, Veltman decided simply to accept 't Hooft's results and ignore how he had arrived at them. He took the results with him to CERN and checked them using a computer program he had developed. He remained sceptical, but was soon calling 't Hooft on the phone, declaring: 'It nearly works. You just have some factors of two wrong.'

But 't Hooft wasn't wrong. Veltman had ignored a factor of four in 't Hooft's equations that could be traced back to the Higgs boson. 'So then he realized that even the factor of four was right,' 't Hooft explained, 'and that everything canceled in a beautiful way. By that time he was as excited about it as I was.'

It was a major breakthrough. '...the psychological effect of a complete proof of renormalisability has been immense,' wrote Veltman some years later. 't Hooft had independently re-created the broken SU(2) × U(1) field theory that Weinberg had developed in 1967, and had shown how it could be renormalized. In fact, what 't Hooft had done was demonstrate that

[3] In particular, he realized three years later that the Higgs mechanism had implications for the cosmological constant. See Martinus Veltman, Nobel Prize lecture, 8 December 1999, p. 392.

Yang–Mills gauge theories in general are renormalizable. Local gauge theories are actually the only class of field theories that can be renormalized.

't Hooft was just 25 years old. Initially, Glashow didn't understand the proof. Of 't Hooft he said: 'Either this guy's a total idiot or he's the biggest genius to hit physics in years.' Weinberg didn't believe it, but when he saw that fellow theorist Benjamin Lee was taking it seriously he decided to look more closely at 't Hooft's work. He was quickly convinced.

Now *all* the ingredients were available. A renormalizable, spontaneously broken SU(2) × U(1) field theory of the weak and electromagnetic forces was now at hand. The masses of the W and Z^0 bosons emerged 'naturally' from the operation of the Higgs mechanism. Anomalies that remained could be removed by adding another quark using the GIM mechanism.[4]

Weinberg carefully reconstructed the electro-weak theory towards the end of 1971. Weak neutral currents in the decays of neutral kaons were now supposedly ruled out by the GIM mechanism, but neutral currents should still be seen in the interactions between muon neutrinos and protons. In such collisions, exchange of a virtual W^- boson turns the muon neutrino into a negative muon and the proton into a neutron—a charged current. Exchange of a virtual Z^0 boson leaves both the muon neutrino and proton intact—a neutral current. Weinberg now estimated that the ratio of neutral to charged currents should be of the order of 0.14–0.33, the lower limit of which he believed should be observable in neutrino scattering experiments. The challenge for the experimental particle physicists was to search for events in the scattering of muon neutrinos from protons which did not involve the production of muons.

This was not as easy as it sounds. A muon neutrino does not leave a track in a bubble chamber detector, so the only signature of a 'muonless' event of the kind Weinberg was seeking would be a spray of hadrons appearing in the detector seemingly out of nowhere. However, stray neutrons

[4] 't Hooft admitted that: 'My own paper said in a footnote that the anomalies do not render the theory nonrenormalisable. Of course, this should be interpreted as saying that renormalisability can be restored by adding an appropriate amount of various kinds of fermions (quarks), but I admit that I also thought that perhaps this was not even necessary.' See Gerard 't Hooft in Hoddeson, *et al.*, p. 192.

chipped from atoms in the walls of the detector by the muon neutrinos could also produce sprays of hadrons that could be easily mistaken for muonless events. Worse, if a muon produced in a charged current event was scattered with a large recoil angle, it could easily be missed, leading to the possibility of misinterpretation as a neutral current event.

It might, perhaps, be easier to identify weak neutral currents in the collision of a muon neutrino with an electron. Again, an exchange of a virtual Z^0 boson would leave the neutrino and the electron intact, but the result would be the sudden appearance of an energetic electron, spiralling in the detector's magnetic field. Again, no charged muon would be produced. This was an event that was not so fraught with potentially misleading signals, but it was also expected to be about two thousand times rarer than equivalent events from the scattering of muon neutrinos from protons.

But Weinberg had thrown down the gauntlet at a propitious time. The world's largest bubble chamber, dubbed 'Gargamelle', had been built in France and installed at CERN in 1970, alongside the proton synchrotron.[5] It had taken six years to construct, and was designed specifically for studying collisions involving neutrinos.

The simple truth was that if weak neutral currents occurred as frequently as Weinberg suggested, then perhaps the evidence for them existed in Gargamelle photographs that had already been taken. The CERN physicists had been in the habit of dismissing low-energy events as the result of stray neutrons. But as there was no way of knowing how much collision energy was carried away by the unchanged muon neutrino, it was quite possible that some of the events were the result of neutral currents.

French physicist Paul Musset spent nearly a year examining photographs in the basement scanning room with a team of CERN colleagues. Early in 1972, the physicists believed they had found a single event: an electron sent spinning out of a Freon molecule in the detector, dislodged

[5] It was named for the mother of the giant Gargantua, from French Renaissance author François Rabelais' sixteenth-century novels *The Life of Gargantua and Pantagruel.*

by an unseen muon neutrino. No charged muon was produced. This was not an event that could be caused by a stray neutron, or a cosmic ray.

The CERN physicists were not alone. A team from universities in Harvard, Pennsylvania, and Wisconsin were also on the trail of weak neutral currents at the National Accelerator Laboratory (NAL)[6] in Chicago, home of the world's largest proton synchrotron, which achieved its design energy of 200 GeV in March 1972. A team led by Italian physicist Carlo Rubbia now sought evidence for weak neutral currents in collisions of muon neutrinos with protons.

If it was a race, it was not without its moments of comedy. Rumours bounced back-and-forth across the Atlantic. By August 1973 the CERN group had identified muonless events in the collisions of muon neutrinos with protons and had estimated that the ratio of neutral to charged current events is 0.21. For collisions involving muon anti-neutrinos, the ratio was 0.45. The NAL group had found the combined ratio for both neutrino and anti-neutrino collisions to be 0.29.

Rubbia's visa expired in July 1973 and, despite holding a professorship at Harvard, he was deported. The NAL physicists now began to back-track. Concerned to prove that what they were observing were real neutral currents and not experimental artefacts, the NAL physicists modified their detector. The neutral currents promptly disappeared, with ratios of neutral to charged current events falling as low at 0.05. When the CERN physicists heard about this, they were greatly discomfited. Had they too been mistaken?

In fact, pions creeping in from other neutrino collisions had been misidentified in the NAL detector as muons. When this was realized, the neutral currents were restored. Some in the particle physics community began to remark with amusement on the discovery of 'alternating neutral currents'.

By April 1974, the confusion had cleared. Weak neutral currents were an established experimental fact.

Now, the particle physicists wanted to know, why weren't weak neutral currents observed in strange particle decays? 'I put them out of their misery and *told* them,' said Glashow. 'It was charm.'

[6] This was renamed the Fermi National Accelerator Laboratory (Fermilab) in 1974.

26

The Magic of Colour

Princeton/Harvard, April 1973

Although the existence of quark–partons was implied by the phenomenon of scaling in the deep inelastic scattering of electrons from nucleons, in the early 1970s the quark model was still far from gaining wide acceptance. There were three obstacles.

The quarks were meant to be spin-½ fermions, and yet the model insisted that hadrons should contain two or more quarks apparently in the same quantum state, in violation of the Pauli exclusion principle. The proton, for example, was thought to consist of two up quarks and a down quark. This was a bit like saying that an atomic electron orbit should contain two spin-up electrons and one spin-down electron. It was just not possible. The symmetry properties of the electron wavefunctions forbid it. There could be only two electrons, one spin-up and one spin-down. There was no room for a third.

The phenomenon of scaling itself implied that the quarks inside protons and neutrons acted like independent point-particles, almost as if they were free. How could this be? If, as was reasonable to assume, the quarks were bound together by the strong, short-range nuclear force, surely they must be held tightly together in the nucleon? It was difficult to imagine a physical mechanism by which the quarks could be bound in the nucleon, yet free to move around independently inside.

And finally, irrespective of how free or tightly held the quarks were in the nucleon, the deep inelastic scattering of electrons should surely have liberated them. Why weren't free quarks seen in the debris from such collisions? The complete absence of the expected tell-tale tracks of fractionally charged particles was a major puzzle.

The problem with the exclusion principle had been identified shortly after publication of Gell-Mann's first quark paper. Whilst on leave from the University of Maryland, physicist Oscar Greenberg had moved to the Institute for Advanced Study in Princeton. He suggested in 1964 that quarks might be parafermions, *which was tantamount to saying that the quarks could be distinguished by other 'degrees of freedom' apart from the one for which the quantum numbers were up, down, and strange. As a result there would be different kinds of up quarks, for example. So long as they were of different kinds, two up quarks could sit happily alongside each other in a proton without occupying the same quantum state.*

Yoichiro Nambu tried a similar scheme, suggesting that there should be three different kinds of up quark (and three different kinds of down and strange quarks), each possessing a new type of 'charge' that was different from electric charge. Confusingly, he called this new type of charge charm. A young graduate student from Syracuse University in New York, Korean-born Moo-Young Han, wrote to him in 1965 outlining much the same idea. Together they wrote a paper which was published later that year. This was not a simple extension of Gell-Mann's quark theory, however, as Nambu had thought to use the opportunity to get rid of the fractional electric charges, introducing instead charges of +1, 0 and −1 alongside the charm charge.

Nobody took much notice. However, in 1969 new experimental results on the decay of neutral pions into two energetic photons were shown to be incompatible with the quark model.[1] *Specifically, the decay rate predicted by the quark model was a factor of three too small. Just as scaling promised to put quarks firmly on the map, so anomalies in the decay of neutral pions threatened to wipe them off again.*

It was clear something had to be done. But Gell-Mann, recently returned from Stockholm where he had collected the 1969 Nobel Prize for physics, did not seem all that concerned.

Harald Fritzsch was born in Zwickau, south of Leipzig in East Germany. Together with a colleague he had defected from Communist East Germany, escaping from the authorities in Bulgaria in a kayak fitted with an outboard motor. They had travelled 200 miles down the Black Sea to Turkey.

[1] Like the neutral kaon, the neutral pion evolves into a quantum superposition. It is formed from the superposition of a state containing an up quark and an anti-up quark and a state containing a down quark and an anti-down quark.

He had begun studying for a doctorate in theoretical physics at the Max Planck Institute for Physics and Astrophysics in Munich, West Germany, where one of his professors was Heisenberg. On his way through America heading for SLAC in the summer of 1970 he encountered Gell-Mann at the Aspen Center for Physics in Aspen, Colorado. Gell-Mann had retreated to Aspen with his family after, in typical style, missing the deadline for submission of his Nobel Prize lecture for publication in the Swedish Academy's *Le Prix Nobel*.[2] It was one among many missed deadlines.

The Aspen Center was a peaceful place, where Gell-Mann would hike in the nearby Rocky Mountains, relax, forget all about his work pressures, and think only about physics. But Fritzsch was about to wake Gell-Mann from his reverie. Fritzsch was a fervent believer in the quark model and excited by what the SLAC results were beginning to reveal about scaling and point-like quark–partons. He was surprised to find that Gell-Mann was curiously ambiguous about his 'mathematical' creation.

It seemed that, even now, Gell-Mann had not grasped the full significance of his original proposal. As a student in East Germany, Fritzsch had become convinced that quarks must lie at the heart of a quantum field theory of the strong nuclear force. These things were much more than mathematical devices. They were real.

Gell-Mann was impressed by the young German's enthusiasm and agreed to have Fritzsch join him at Caltech, visiting about once a month. Together they began to work on a field theory constructed from quarks. When Fritzsch completed his graduate studies in early 1971, he transferred to Caltech. His arrival date, 9 February 1971, coincided with an earthquake that struck the San Fernando Valley early that morning near Sylmar, with a magnitude of 6.6 on the Richter scale. 'In memory of that occasion,' Gell-Mann later wrote, 'I left the pictures on the wall askew, until they were further disturbed by the 1987 earthquake.'

Results from SLAC on the ratio of neutron-to-proton deep inelastic scattering events continued to point in the direction of quarks. The science journalist Walter Sullivan wrote of the discovery in the *New York*

[2] The Nobelprize.org website states flatly that: 'Professor Gell-Mann has presented his Nobel Lecture [on 11 December 1969], but did not submit a manuscript for inclusion in this volume.'

Times in April 1971. He remarked: 'Specifically these findings suggest the presence, in protons and neutrons, of points of electric charge that, in several respects, resemble the elusive and long-sought quarks.'

Gell-Mann organized grants for himself and Fritzsch and in the autumn of 1971 they both travelled to CERN. It was here that William Bardeen, son of John Bardeen of the Bardeen–Cooper–Schrieffer theory of superconductivity, told them about the anomalies in the decay of neutral pions. Here was a direct experimental challenge to the original quark model. The Han–Nambu model of integral-charged quarks actually did a better job of predicting the decay rate.

Working together, Gell-Mann, Fritzsch, and Bardeen began to explore the options. They wanted to see if it was possible to reconcile the results on neutral pion decay with a model of fractionally charged quarks. They tinkered with a parafermion model, before alighting on a variation of the Han–Nambu idea.

They acknowledged that what they needed was a new quantum number. Han and Nambu had used 'charm' but this was now associated with Glashow's hypothetical fourth quark. Instead, Gell-Mann decided to call the new quantum number 'colour'. This use of an abstract interpretation of colour was not a particularly new idea—Gell-Mann and Feynman had in the past referred to different neutrinos as 'red' and 'blue', and other physicists had used the term. In this new scheme, quarks would possess three possible colour quantum numbers: blue, red, and green.[3] Baryons are constituted from three quarks of different colour, such that their total 'colour charge' is zero and their product is 'white'. For example, a proton could be thought to consist of a blue up quark, a red up quark, and a green down quark ($u_b u_r d_g$). A neutron would consist of a blue up quark, a red down quark, and a green down quark ($u_b d_r d_g$). The mesons, such as pions and kaons, could be thought to consist of coloured quarks and their anti-quarks, such that the total colour charge is zero and the particles are also 'white'.

[3] In their original scheme Gell-Mann, Fritzsch, and Bardeen called them red, white, and blue (inspired by the French national flag). However, it soon became clear that the three primary colours would work better as, when blended, they produce the colour white. To avoid confusion I have adopted the currently accepted terminology from the outset.

It was a neat solution. The different quark colours provided the extra degree of freedom and meant that there was no violation of the Pauli exclusion principle. Tripling the number of different types of quarks meant that the decay rate of the neutral pion was now accurately predicted. And nobody could expect to see the colour charge revealed in experiments as this was a property of quarks and the quarks were 'confined' inside white-coloured hadrons. Colour cannot be seen because nature demands that all observable particles are white.

'We gradually saw that that [colour] variable was going to do everything for us!' Gell-Mann explained. 'It fixed the statistics, and it could do that without involving us in crazy new particles. Then we realized that it could also fix the dynamics, because we could build an SU(3) gauge theory, a Yang–Mills theory, on it.'

This was a different application of the SU(3) symmetry group. Gell-Mann had derived the Eightfold Way from a *global* SU(3) symmetry which transforms one kind of 'flavour'—up, down, and strange—into another, leaving the quark colour unchanged. What was now proposed was to build an SU(3) *local* gauge theory from the quark colour charge.[4]

By September 1972, Gell-Mann and Fritzsch had elaborated a model consisting of three fractionally charged quarks which could take three flavours and three colours, bound together by a system of eight coloured gluons, the carriers of the strong 'colour force'. Gell-Mann presented the model at a Rochester conference on high-energy physics held to mark the opening of the National Accelerator Laboratory in Chicago.

But he was already beginning to have second thoughts. Once more troubled particularly by the ontological status of the quarks and the mechanism by which they are permanently confined, Gell-Mann gave the theory a somewhat muted fanfare. He mentioned a variation of the model featuring a single gluon. He emphasized that the quarks and gluons were 'fictitious'. By the time he and Fritzsch came to write up the lecture, they had been overtaken by their doubts. 'In preparing the written version,' he later wrote, 'unfortunately, we were troubled by the doubts just mentioned, and we retreated into technical matters.'

[4] Although these applications of SU(3) are different, the fact both are SU(3) is no coincidence. If the masses of the quarks were zero, then the SU(3) flavour symmetry would be an exact symmetry.

Colour had removed one of the obstacles, but another two remained.

The theorists called it *asymptotic freedom*. The phenomenon of scaling appeared to show that the quarks behaved almost as free point-particles inside nucleons, such that in the asymptotic limit of zero separation they would be completely free. And yet for some reason they couldn't be pulled out. As the separation between them increased beyond the boundary of the nucleon, somehow the strong force held them in check.

This was quite counter-intuitive. Physicists were used to thinking about force fields that behaved like electromagnetism or gravity, in which the effects of the force falls off with increasing distance from its source. What was needed instead was a field that did the opposite, one that fell away to zero the closer the quarks approached and increased as they moved apart.

In early 1972, 31-year old American theorist David Gross at Princeton decided that he would kill off quantum field theory once and for all. He was going to do this first by showing that the scaling revealed in the deep inelastic scattering experiments demanded a field theory that was asymptotically free. He would then go on to show that asymptotic freedom was simply impossible in a quantum field theory.

He succeeded with the first task. He then worked his way through various classes of renormalizable field theories, ruling them out one after another by showing that they were not asymptotically free. Finally, in the spring of 1973, he came to the less tractable class of local gauge theories. 'The one hole left in this thing was gauge theories, which didn't fit into the same line of proof,' he later explained. 'So that hole I was going to close with Frank Wilczek, who had started work with me as a graduate student.'

Wilczek had arrived in Princeton in the autumn of 1972 as a mathematics student, but discovered that his real passion was for particle physics. He was Gross's first graduate student. 'He spoiled me,' Gross later said of Wilczek, 'I thought they'd all be that good.'

They set to work, but it was a long and tedious proof. For 21-year old Wilczek, this was to be the subject of his doctoral thesis and he was wary of competition from other theorists. In fact, 't Hooft had already concluded that Yang–Mills gauge theories could show this counter-intuitive behaviour. But he was still busy working on renormalization and did not have the time to follow this up.

By April 1973 Gross and Wilczek thought they had the answer. It was as Gross had expected. Local gauge theories could *not* be asymptotically free. Another approach was needed if there was to be any hope of providing a theoretical description of scaling.

Unknown to both of them, 23-year old Harvard graduate student David Politzer had been working on the same problem. Politzer had graduated from the Bronx High School of Science in 1966, securing a bachelor's degree at the University of Michigan before moving to Harvard in 1969. His thesis adviser was theorist Sidney Coleman, who had studied under Gell-Mann at Caltech.

What Politzer had found was rather different, however. He concluded that asymptotic freedom *was* possible in local gauge theories. Coleman had taken some leave from Harvard and had moved to Princeton for the spring semester. Politzer now called him and explained what he had discovered.

'Um, hum,' Coleman replied. 'That's very interesting—except for one problem, which is that David Gross and a student of his [have] worked on the same calculation, and they said it comes out the other way.'

'I checked it,' Politzer protested, 'and I think I got it right.'

'Those guys don't make mistakes,' Coleman said.

Politzer retreated for a short holiday in Maine with his wife. It rained a lot, so he used the time to check his calculations through once more. Convinced that he hadn't made a mistake, on his return he told Coleman that he had got the same results. 'Yes, I know,' Coleman now admitted. 'David and Frank found their mistake, and they've submitted a paper to *Physical Review Letters* because it's important.'

Unbidden by Coleman, Gross and Wilczek had checked their results. They had found a single sign error.

Politzer dashed off a short paper of his own. Both were published back-to-back in the June 1973 issue of *Physical Review Letters.*

It was a remarkable discovery. The strong nuclear force is not like other, more familiar, forces. It weakens as the quarks move closer together, so that inside hadrons they can barely sense each others' presence. Another obstacle had suddenly been removed.

Gell-Mann retreated to the Aspen Center in June 1973, clutching preprints of the Gross–Wilczek and Politzer papers. He was joined by Fritzsch and Heinrich Leutwyler, a Swiss theorist from the University of

Bern on study leave at Caltech. Together they crafted a Yang–Mills field theory of three spin-½ coloured quarks and eight spin-1 coloured gluons. The gluons were now *required* to carry colour charge in order to account for asymptotic freedom. The theorists also predicted that there should be small *violations* of scaling in deep inelastic scattering.

Further papers from Gross, Wilczek, and Politzer later that year confirmed that such scaling violations were a signature of asymptotic freedom. Graphs of the structure function for different electron energies tended to converge with increasing q^2/ν. But in the region of such scaling it was now expected that for some electron energies the graph would show a gentle increase, for other energies a gentle decline.

At SLAC, MIT physicist Michael Riordan was noting precisely this kind of behaviour. 'It was a tremendous eye-opener,' Riordan wrote. 'Gross, Wilczek, and Politzer had no possible way of knowing what our data looked like, yet here they were predicting essentially what we were just then noticing.'

The new theory needed a name. In 1973 Gell-Mann and Fritzsch had been referring to it as quantum hadron dynamics, but the following summer Gell-Mann thought he had come up with a better name. 'The theory had many virtues and no known vices,' he explained. 'It was during a subsequent summer at Aspen that I invented the name quantum chromodynamics, or QCD, for the theory and urged it upon Heinz Pagels and others.'

A great synthesis, combining the theories of the strong and electroweak forces in a single SU(3) × SU(2) × U(1) quantum field theory, seemed at last to be at hand.

But what of the final obstacle? Asymptotic freedom explains why quarks interact only very weakly inside hadrons, but it does not explain why the quarks are confined. Arguments were advanced to the effect that the strength of the strong colour force increases as the quarks are pulled apart, confining the quarks through 'infrared slavery', but nothing could be *proved*. The usually dependable mathematical tools that theorists reach for tend to become unusable at the limits of very strong forces.

Various picturesque models were introduced in an attempt to explain confinement. In one of these, the gluon fields surrounding the quarks are imagined to form narrow tubes or 'strings' of colour charge between the quarks as they separate. As the quarks are pulled apart, the string first tenses and then stretches, the resistance to further stretching increasing with increasing separation.

Eventually the string breaks, but at energies sufficient to conjure quark–anti-quark pairs spontaneously from the vacuum. So, pulling a quark from the interior of a hadron cannot be done without creating an anti-quark which will immediately pair with it to form a meson, and another quark which will take its place inside the hadron.

Quarks are not so much confined as never, but never, seen without a chaperone.

And where, in all this, was the charm quark?

27

The November Revolution

Long Island/Stanford, November 1974

Quantum chromodynamics was in trouble even before the theory had been so named.

Particle physicists had begun to shift their attention away from firing high-energy particles at stationary targets. In such experiments, the energy available for the creation of new particles scales with the square-root of the energy of the particles in the beam, with the remaining collision energy 'wasted'. The physicists had figured that if they fired two high-energy particle beams directly at each other from opposite directions, the net momentum of two colliding particles in the intersecting beams is zero and so all the beam energy is in principle available to do interesting new physics. Electron–electron and electron–positron colliders were among the first to be built, in Frascati in Italy, in Cambridge, Massachusetts, and at SLAC.

When an electron and a positron collide, they may annihilate to produce a high-energy virtual photon, perhaps best thought of as a tiny electromagnetic 'fireball'. The photon cannot hold on to all this energy and produces a spray of different particles. These can include any particle–anti-particle combination allowed by conservation rules: another electron–positron pair, a muon–anti-muon pair or a quark–anti-quark pair. The quark–anti-quark pairs quickly find chaperones and go on to form hadrons.

So long as total mass-energy and spin are conserved, the probabilities of each of these different decay routes of the photon depend simply on the square of the electric charges of the particles created. The ratio of hadron to muon production then depends on the ratio of the sum of the squares of the three quark charges (⅔), divided by the square of the muon charge (1). But now we have to remember that there are three types of up, down, and strange quarks, so this ratio is further multiplied by 3. The result is a QCD prediction for the ratio, R, of hadron to muon production, of 2.

For collision energies between 1 and 3 GeV, the Italian collider ADONE seemed to confirm R = 2. But for collision energies between 4 and 5 GeV, the collider at the Cambridge Electron Accelerator (CEA) yielded values of R between 4 and 6. When the Stanford collider, known as the Stanford Positron Electron Asymmetric Rings, or SPEAR, came on stream in mid-1973, Stanford physicist Burton Richter and his team used it to check these results. By December, Richter was able to show that the CEA results were correct. The ratio of hadron to muon production increased substantially above 2 with increasing collision energy, and showed no sign of leveling off.[1]

What was going on? Could it be that there were more quarks to be discovered? Or was it possible that electrons and protons could themselves transform into quarks and antiquarks, so creating an excess of hadrons, as Richter believed?

In April 1974, Glashow addressed a conference of meson spectroscopists—specialists in the study of the properties and behaviour of mesons. He urged them to look for charmed mesons, or risk the possibility that these would be found by 'outlanders', other experimental physicists who were not specialists. 'There are just three possibilities,' Glashow declared. 'One, charm is not found, and I eat my hat. Two, charm is found by spectroscopists, and we celebrate. Three, charm is found by outlanders, and you eat your hats.'

Iliopoulos was similarly bullish about the charm quark. At the seventeenth Rochester conference held in London in July 1974 he delivered a memorable oration in favour of an SU(3) × SU(2) × U(1) quantum field theory which, he declared, would be shown to be '...the broken remnant of a single unified gauge group that existed in the distant past.'

He then bet a case of fine wine that discussions at the next Rochester conference would be dominated by the discovery of charm particles.

Glashow had bent the ears of virtually every particle physicist who would listen. At a farewell party organized for Maiani in 1970, aboard an ocean liner in Boston harbour, Glashow, Iliopoulos, and Maiani had cornered MIT physicist and Brookhaven group leader Samuel Chao Chung Ting. They had rather drunkenly muttered about the charm quark and the Nobel Prize. Ting, who outwardly professed a disdain for the wilder ramblings

[1] Theories other than QCD predicted different values for R, ranging from 0.36 to infinity.

of the theorists, was inwardly a shrewd judge of new ideas suggestive of interesting experiments. Outwardly, Ting seemed unconvinced.

Glashow had better luck with Brookhaven's Samios. He had explained that indirect evidence for the charm quark might be available in the debris from collisions of neutrinos with protons. Exchange of a virtual W^+ particle would, he believed, transform a down quark in the proton into a charm quark, creating, for the merest instant, a charmed baryon. This would decay quickly into a shower of hadrons, one of which would be the neutral lambda, Λ^0, consisting of an up quark, a down quark, and a strange quark (uds), the result of the further transformation of the charm quark into a strange quark.

This was the clincher. In studies of lambda production, the particle is always produced alongside its anti-particle, $\bar{\Lambda}^0$. Yet in this kind of collision, it would be produced alone. The appearance of a lone Λ^0 particle, noticeable from the characteristic 'V'-shaped tracks produced by its decay products, would be evidence for the charm quark.

Samios and his colleague Robert Palmer now searched for this kind of event in the hundreds of thousands of photographs of particle collisions that had been gathered through the year using a new bubble chamber detector installed at Brookhaven. Just such a candidate event was identified in late May 1974 by Helen LaSauce, a former switchboard operator now employed as a scanner. In the sketch that LaSauce produced from the photograph, the invisible neutrino had struck a proton in the bubble chamber, causing a spray of pions and an invisible particle that exploded into the 'V'-shaped signature of the Λ^0.

One photograph was not enough, however, and there were other possible explanations for the appearance of the tracks that had nothing to do with the charm quark. 'You don't want to say you've found a new state of matter on the basis of one event,' Palmer explained, 'So we kept going back and redoing the calculations.' They continued their search through the summer and early autumn.

Glashow visited Brookhaven in August 1974, once more to urge the search for the charm quark and to describe the various ways in which evidence might be found. Ting was preparing to use the 30 GeV Alternating Gradient Synchrotron (AGS) to study high-energy proton–proton collisions and watch carefully for electron–positron pairs emerging amidst

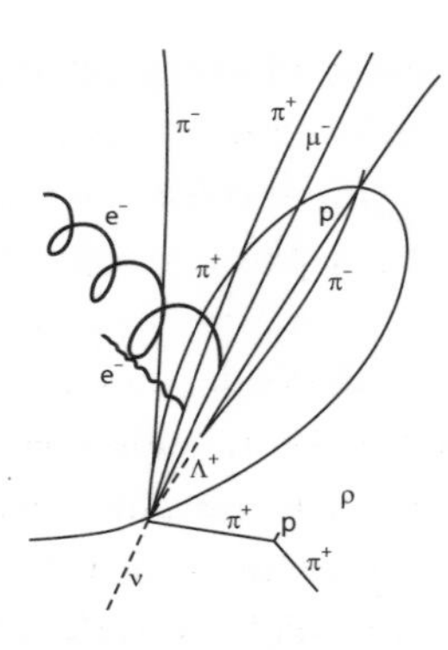

FIG 17 A possible candidate event involving the production of a charmed particle. This event, from Brookhaven's bubble chamber in May 1974, appears to show the production of a a lone Λ° particle, identified through the appearance of the tell-tale 'V'-shaped tracks produced by its decay products. This figure shows the reconstruction of the particle trajectories. Reproduced courtesy of Brookhaven National Laboratory.

the chaos of hadrons produced. These were tricky experiments, requiring that the tracks of individually detected electrons and positrons be traced back to their source to confirm that the pair originated in the same collision. No, no, Glashow urged. Evidence for the charm quark would be gained not by watching electron–positron pairs, he told Ting's research team, but by looking for strange particles—pairs of pions and kaons.

Ting's experiments began on 22 August, after a number of false starts. On 31 August, he directed that the team look for electron–positron pairs arising from 'parent' particles with energies between 2.5 and 4.0 GeV. Ting had built a reputation for thoroughness, and he had arranged for two groups within his research team to analyse the raw data from the experiment, each independently of the other and using their own computer programs. They were not to communicate to each other except in wider team meetings.

The small minicomputer connected directly to the experiment had insufficient capacity for the analysis, and the researchers were therefore obliged to feed their data into Brookhaven's mainframe. One of Ting's postdoctoral students, Terence Rhoades, was the first to notice something unusual in the results of the analysis, on 2 September.

The data suggested that electron–positron pairs were 'piling up' at an energy of around 3 GeV. Rhoades wasn't sure quite what to make of this. Fearing an error in the analysis, he did not inform Ting. The experiments were halted by a planned shutdown of the accelerator two days later.

Something had gone wrong. Rhoades and his colleague, German physicist Ulrich Becker, had worked on the analysis program together. They now argued, each convinced that the other had made mistakes. They descended on Brookhaven's computer centre, turfing everyone out while they rechecked the analysis. It made no difference. The peak remained stubbornly fixed at 3.1 GeV, and stubbornly narrow.

MIT assistant professor Min Chen worked with Glen Everhart on the second analysis during the first few weeks in September. As he surveyed the consolidated results for the first time, Chen was initially alarmed, then immensely curious. The data suggested an incredibly strong yet narrow resonance around 3.1 GeV. He suspected a problem with the computer program, but quick checks revealed no obvious errors. He then suspected that this was new physics, the resonance indicative of a previously unknown particle.[2]

Ting was intrigued, but immensely cautious. He had also acquired a reputation for showing up errors in other physicists' experiments, and he didn't want to fall victim to the same treatment. The experiments had to be repeated and various checks made to determine the authenticity of the 3.1 GeV resonance. That meant getting more time on the AGS when the routine maintenance was completed in early October. He was in competition for beam time with American physicist Melvin Schwartz, who had moved to Stanford in 1966.

Ting felt strongly that the pressure to publish the results, thereby claiming priority for the discovery, had to be resisted in the interests of accuracy. Ting swore his colleagues to absolute secrecy. Publication would wait until they had had a chance to reconfirm the data.

[2] There was a precedent. American physicist Leon Lederman and his Columbia University team had observed a 'shoulder' around 3.1 GeV in similar experiments in 1968. But they had gone on to study higher energy regimes and did not return to investigate.

It was a risky strategy, not least because all the physicists working on the Stanford electron–positron collider had to do was tune the energy of their beams to about 1.5 GeV (half the energy of the resonance) and they would undoubtedly make the same discovery. However, Ting was aware that the Stanford physicists were still exploring the total energy range 4.0–6.0 GeV, making measurements of the ratio of hadron to muon production. For as long as they stayed at these energies, he could rest easy.

But Ting could not stop the rumours. On 22 October he regained access to the AGS and was able to recommence experiments. That same day he was approached by Schwartz.

'I hear you got a bump at 3.1,' Schwartz congratulated him.

'No, absolutely not,' Ting lied. 'Not only do I not have a bump, it's absolutely flat.'

Schwartz was annoyed. He could have accepted the need for secrecy but didn't like being lied to. 'I'll make you a bet,' he said. 'Ten dollars you got a bump.'

'Absolutely. I'll bet,' Ting replied. They shook hands. Schwartz departed, quietly seething. Ting went to his office and pinned a note to his noticeboard. It said: 'I owe M. Schwartz $10.'

Stanford physicist Roy Schwitters had a problem. He had found an error in one of the computer programs used to analyse data from the SPEAR experiments. When he corrected it, data reanalysed from experiments carried out in June now showed hints of structure—small bumps at 3.1 and 4.2 GeV. Two experimental runs in particular, runs 1380 and 1383, showed unusually high numbers of hadron-producing events at 3.1 GeV. It would have been easy to dismiss these as spurious results, but Schwitters could find no good reasons to reject them. On 22 October he asked Gerson Goldhaber, head of a collaborative group from the Lawrence Berkeley Laboratory, to take a closer look.[3]

[3] The Lawrence Berkeley Laboratory is the former Berkeley Radiation Laboratory, renamed following the death of Ernest Lawrence in 1959. It is now the Lawrence Berkeley National Laboratory.

Ting was convinced they had something real by early November. His research team had made numerous changes to the experimental configuration, but the narrow peak at 3.1 GeV remained unmoved. There could now be no doubt that the Brookhaven physicists had discovered a new particle. Ting had called it the J, on the basis that really interesting particles had been accorded names based on Roman letters (such as the W and Z bosons).

The pressure to publish was now really intense, but Ting was aware that if the new particle did indeed contain the charm quark, then it was merely the first in what should prove to be a series of new particle discoveries. He now wanted to lay claim to the entire family of charmed particles. Against the better judgement of Becker and Chen in his team, and many of his colleagues who were now in on the secret, he continued to withhold publication. He figured that the SPEAR researchers wouldn't find the new particle unless they knew precisely where to look.

On 4 November, Richter returned to SLAC from Harvard, where he had been delivering a series of lectures. He found the SPEAR researchers in a state of some excitement over the interpretation of the bump at 3.1 GeV. They urged him to arrange for the collider to be reconfigured so that they could examine it in more detail. But this was no trivial task. By this time SPEAR had been upgraded to explore energies above 5 GeV, and to go back to 3.1 GeV would be time-consuming and, if this turned out to be some wild goose chase, a waste of valuable time, effort, and money. The debate continued through the week.

In the meantime Goldhaber had become convinced that run 1383 showed a remarkable increase in kaon production. This was precisely what Glashow had suggested the experimentalists should look for as the signature of the charm quark. Richter held a meeting in his office on 8 November to review the situation. Goldhaber's arguments won the day: SPEAR would be turned back to a total collision energy of 3.1 GeV.[4]

That something very unusual was happening was apparent the next morning, as the first results began to emerge from the now reconfigured collider. The researchers probed energies around 3.1 GeV and identified a

[4] Though this was to prove fortunate, as Goldhaber had in fact misinterpreted the data. There was actually no evidence for excess kaon production.

narrow peak in hadron production. Events were initially logged at about one per minute, already more than a factor of three higher than normal. By the next morning, with the collider energy tuned to 3.11 GeV, events were being logged at an even greater rate, a factor of seven above the baseline.

Goldhaber sat down to start writing a short paper about the discovery, only to be told an hour later that at 3.104 GeV the event rate had jumped by another factor of ten. A hadron-producing event was now being recorded every second. This was simply unprecedented. At 3.105 GeV they finally found the peak, a factor of one hundred above the baseline. Champagne was liberated from the refrigerator. Cookies were consumed. As word about the discovery spread around SLAC, the SPEAR control room became jammed with visitors. The phones began ringing.

The new particle had to have a name. After some investigation, Goldhaber and Richter fixed on the Greek letter ψ (psi). The physicists agreed to make a public announcement about the discovery the next day, Monday 11 November.

Ting left New York on Sunday 10 November to attend a Program Advisory Committee meeting at SLAC. Back at Brookhaven, news of the SPEAR discovery was beginning to filter through to some in Ting's research team. They left a message for him at the TWA desk at San Francisco airport, but when he called back at 1 am they had decided that this was surely some kind of practical joke.

He received a further call shortly after checking into the Flamingo Motel (where Feynman had worked out what Kendall's graph was telling him in August 1968). This time there seemed little doubt. A major discovery had been made by the SPEAR researchers. Ting made a quick call to a SLAC colleague and heard that an announcement was to be made the next day.

Ting got little sleep. He now had no choice but to make a simultaneous announcement. He organized his team back in Brookhaven to prepare copies of the experimental data for release.

When Ting and Richter met at SLAC around 8 am on 11 November, their conversation went as follows:

'Burt, I have some interesting physics to tell you about,' Ting said.

'Sam,' Richter replied, 'I have some interesting physics to tell *you* about.'

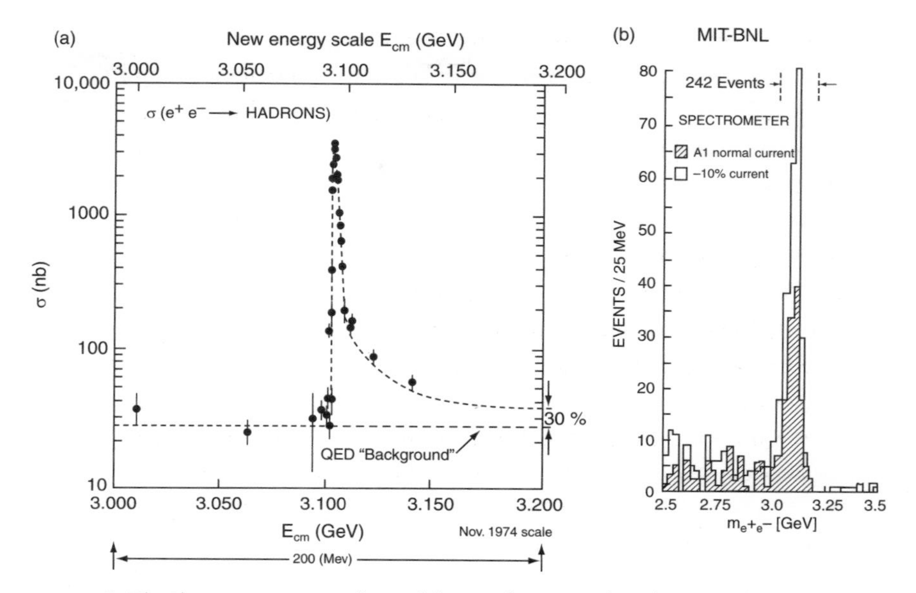

FIG 18 The J/ψ, a spin-1 meson formed from a charm quark and an anti-charm quark, was discovered simultaneously by research teams led by Richter at SLAC and Ting at Brookhaven. (a), the conspicuous resonance at 3.1 GeV from electron-positron annihilation experiments at SLAC. (b), the same resonance was revealed in high-energy proton-proton collision experiments at Brookhaven. Adapted from Hoddeson, *et al.*, *The Rise of the Standard Model: Particle Physics in the 1960s and 1970s*, Cambridge University Press, 1997, pp. 62 and 64.

It quickly emerged that both groups had made the same discovery. 'Sam! It's the same thing! It has to be right!' Richter declared.

What they had discovered was a spin-1 meson formed from a charm quark and an anti-charm quark. Glashow called it 'charmonium' by analogy with positronium, a version of the hydrogen atom in which the proton is replaced by a positron. The SPEAR researchers went on to find charmonium's first excited state at 3.7 GeV which they named the ψ' (psi-prime). Ting's gamble on finding a whole family of charmed particles had not paid off. It had very nearly backfired.

There followed something of a hiatus over priority. There was a Nobel Prize at stake, and as the discoveries at SLAC began to dominate discussion, Ting was concerned that he and the Brookhaven discovery were being edged out. In an issue of *Science* magazine published on 5 September 1975, Ting drew on laboratory notebooks to set out a clear timeline for the Brookhaven discovery. He also hinted that it was actually the rumours of the discovery at Brookhaven, as disseminated by Schwartz and others, that had led the SPEAR researchers to go back and look at 3.1 GeV. The Brookhaven team continued to talk about the J particle, the SPEAR team talked about the ψ. For either to adopt the other name would have been to concede priority.

Although Schwartz had mentioned his ten dollar bet with Ting to a Stanford colleague, there was no evidence that the vague rumours emanating from Brookhaven had influenced events at SLAC. It was accepted that this was an independent discovery by two laboratories. Both Richter and Ting shared the 1976 Nobel Prize for physics. Today, the particle is still called the J/ψ.

Samios and Palmer showed their single charmed baryon event at a meeting in March 1975, but no further evidence for 'naked charm'—particles containing either a single charm quark or an anti-charm quark—was forthcoming. At a conference in April 1976 Glashow urged Goldhaber to intensify the search. Glashow had predicted that a charmed neutral D meson, consisting of a charm quark and an anti-up quark, should be found at an energy of 1.95 GeV. In early May 1976, Goldhaber found it at 1.87 GeV.

At a conference on meson spectroscopy later that year, the organizer Roy Weinstein repeated Glashow's hat speech. Charm had indeed been found by 'outlanders'. Weinstein distributed small Mexican hats made of candy.

The meson spectroscopists proceeded to eat them.

28

Intermediate Vector Bosons

Geneva, January/June 1983

The discovery of the J/ψ was a triumph for theoretical physics. It also served to neaten the structure of the fundamental particles—the foundation of what was fast becoming the 'Standard Model' of particle physics.

There were now two 'generations' of fundamental particles, each consisting of two leptons and two quarks and the particles responsible for carrying forces between them. The electron, electron neutrino, up quark, and down quark form the first generation. The muon, muon neutrino, strange quark, and charm quark form the second generation, differentiated from the first by their masses. The photon carries the electromagnetic force, the W and Z particles the weak nuclear force, and eight coloured gluons carry the strong nuclear force between the coloured quarks.[1]

If the symmetry was pleasing, the pleasure was short-lived. By the spring of 1977, the evidence for an even heavier version of the electron—called the tau lepton—was overwhelming. Martin Perl and his colleagues at SLAC had for many years been studying anomalous electron–muon pair production in electron–positron annihilation experiments using SPEAR. They had convinced themselves that tau–anti-tau pairs were also

[1] The gluons are represented in terms of superpositions of various colour–anti-colour combinations, such as (red–anti-blue + blue–anti-red), (blue–anti-green - green–anti-blue), etc. These combinations result from the SU(3) symmetry of QCD. If it was simply a case of taking all the different possible colour–anti-colour combinations then there would be nine gluons instead of eight.

being produced in the annihilation, with each separately decaying into electrons and anti-muons or muons and positrons. It was not what physicists really wanted to hear, however, and their results were greeted with scepticism. When independent corroboration of the results became available from the Deutsches Elektronen Synchrotron (DESY) in Hamburg, there could be no denying the existence of the tau.

A tau lepton demanded a tau neutrino and, inevitably, speculation mounted that there are actually three generations of leptons and quarks. Leon Lederman made up for missing the J/ψ by finding the upsilon (Υ) and its first and second excited states at Fermilab in August 1977.[2] *This is a meson consisting of what had by then come to be known as a bottom quark and its anti-quark. With a mass of about 4.2 GeV, the bottom quark is a heavier, third-generation version of the down and strange quarks with a charge of −⅓. It was assumed that the final member of the third generation—the top quark—was heavier still and would be found as soon as colliders capable of the requisite collision energies could be built.*[3]

Although it had made something of a surprise appearance, the third generation of leptons and quarks was readily absorbed into the Standard Model. At a symposium organized at Fermilab in August 1979, evidence was presented for the appearance of quark and gluon 'jets' produced in electron–positron annihilation experiments. These are directed sprays of hadrons produced from the formation of a quark–anti-quark pair in which an energetic gluon is also 'liberated' from one of the quarks. Such tell-tale 'three-jet events' provide the most striking evidence yet found for both quarks and gluons.[4]

The top quark was still missing, as was direct evidence for the W and Z particles, the intermediate vector bosons of the weak force. As the Standard Model became the new orthodoxy, Glashow, Weinberg, and Salam learned that they had been awarded the 1979 Nobel Prize for physics.

Faith in the essential correctness of the Standard Model was growing. Yet it could be argued that the evidence for the W and Z intermediate vector

[2] Lederman and his colleagues thought they had found the Υ in the form of a 'bump' at an energy of 6 GeV and published their results prematurely. When the bump later disappeared, the phantom particle became known as the 'oops-Leon'. The Υ subsequently appeared with an energy of 9.5 GeV. See Lederman, p. 321.

[3] There were attempts to name the new quarks 'truth' and 'beauty' but the more prosaic 'top' and 'bottom' won out.

[4] Three-jet events were first identified by a research team at DESY and corroborated by an international research team headed by Ting at CERN.

bosons of the weak force was still rather circumstantial. These force carriers were predicted by the renormalizable, spontaneously broken SU(2) × U(1) quantum field theory of the weak and electromagnetic forces, augmented by a Higgs field which gave the particles their mass.

Indirect evidence for the existence of the W and Z particles had been obtained through observation of weak-force decays and of weak neutral currents. None of these observations made any sense without these particles. They just *had* to be there. But, in experimental physics, seeing is believing. In order to bolt the SU(2) × U(1) electro-weak theory firmly into its place within the Standard Model it would be first necessary to see the W and Z particles directly. To put this in perspective: there had been no Nobel Prize for finding weak neutral currents. There would certainly be a Nobel Prize for whoever could find the intermediate vector bosons.

In his Nobel Prize lecture, Weinberg had explained that the electroweak theory predicted masses for the W and Z particles which depend on a weak force 'mixing angle', θ_W. For the $W^\pm$ particles, Weinberg predicted masses equal to about 40 GeV/sinθ_W. For the Z^0, the mass was predicted to be about 80 GeV/sin2θ_W. Later in his lecture he referred to experimental estimates for $\sin^2\theta_W$ of 0.23±0.01, corresponding to a mixing angle of about 29°. This figure puts the masses of the $W^\pm$ particles at about 83 GeV and the Z^0 at 94 GeV.[5] In mass terms, the W and Z particles are about as heavy as the nucleus of a strontium atom.

CERN's Super Proton Synchrotron (SPS) was a 6.9 kilometre circumference proton accelerator commissioned in June 1976. It was originally specified as a 300 GeV accelerator but constructed to generate particle energies of 400 GeV. A month before its commissioning, it had been outpaced by the proton accelerator at Fermilab, which had reached 500 GeV.

But the problem was that these were accelerators, not colliders. Despite such high accelerated particle energies, smashing particles into stationary targets results in substantial energy wastage. Only particles with

[5] The mixing angle θ_W (also referred to as the Weinberg angle) relates the masses of the W and Z particles according to the expression $\cos\theta_W = M_W/M_Z$, where M_W is the mass of the $W^\pm$ particles and M_Z is the mass of the Z^0. The mixing angle varies with collision energy. An angle of 29° suggests this ratio is about 0.875.

considerably lower energies can be created and observed this way. To reach the energies of the W and Z particles would require an accelerator considerably larger than any yet built. Or a collider.

The world's first hadron collider, called the Intersecting Storage Rings (ISR), had been constructed at CERN and was in operation from 1971.[6] This was a proton–proton collider. However, its peak beam energy was insufficient to reach the W and Z particles.

In April 1976 a study group was assembled at CERN to report on the next major construction project: the Large Electron–Positron (LEP) collider. This was to be built in a 27-kilometre circular tunnel passing beneath the Swiss-French border near Geneva. It would use the SPS to accelerate electrons and positrons to speeds close to that of light before injecting them into the collider ring. The initial design energy was 45 GeV for each particle beam which, when combined, would produce head-on collision energies of 90 GeV, just within reach of the W and Z particles.

There was one problem. This was a long-term project. The LEP was not expected to be operational until 1989. The CERN physicists didn't have the patience to wait. 'The pressure to discover the W and Z was so strong,' recalled CERN physicist Pierre Darriulat, 'that the long design, development and construction time of the LEP project left most of us, even the most patient among us, unsatisfied. A quick (and hopefully not dirty) look at the new bosons would have been highly welcome.'

Arguments were advanced for the construction of various kinds of proton–proton collider, but these were rejected by CERN managers fearful of delays to the LEP project.

What the CERN physicists needed was to figure out how to stretch the existing facilities—such as the SPS—into the all-important energy regime. This could be done by colliding protons and anti-protons. However, the mechanism by which anti-protons are produced, colliding high-energy protons with stationary targets, produce particles with a large spread of energies. A storage ring can accept particles only with a narrow range of energies around the ring's design value. The result would be an

[6] The decision to build the ISR had been approved in December 1965, in Victor Weisskopf's last Council Session as CERN Director-General.

accelerated anti-proton beam with low intensity or *luminosity* (a measure of the number of collisions that the beam can produce).

To make a beam of anti-protons sufficiently luminous for proton–anti-proton collider experiments would require that the anti-proton energies be somehow 'gathered' and concentrated around the desired beam energy.

Fortunately, CERN physicist Simon van der Meer had figured out how to do precisely this. Van der Meer was an accelerator theorist, primarily concerned with the practical application of theoretical principles to the design and operation of particle accelerators and colliders. He had performed some experiments using the ISR in 1968 and published an internal report four years later. The reason for his tardiness was simple: the physics he was pursuing seemed vaguely mad. In his report he wrote: 'The idea seemed too far-fetched at the time to justify publication.'

It is possible to illustrate van der Meer's idea by recalling James Clerk Maxwell's 1867 thought experiment concerning the second law of thermodynamics. A gas at room temperature consists of many billions of atoms moving through space, colliding with each other and with the walls of the container with a broad range of speeds. The average speed is a measure of the energy and temperature of the gas.

Suppose, Maxwell suggested, that we fit a partition in the container which has a small hole covered by a shutter. The container now has two compartments, which we call A and B. Now further suppose that an experimenter with sufficiently sharpened faculties (called Maxwell's *demon*) is able to sense when an atom moving with a high speed in compartment A is heading towards the hole in the partition. He opens the shutter and the atom passes through to compartment B. Similarly, he can sense when an atom moving with a low speed in compartment B is heading towards the hole. He opens the shutter and the atom passes through to compartment A.

After some time, all the high-speed atoms have passed into compartment B, and all the low-speed atoms have passed into compartment A. Without expending any work, the demon has raised the temperature of the gas in compartment B and lowered it in compartment A. This is in apparent violation of the second law of thermodynamics, which states that the entropy (the amount of 'disorder') of the gas in an isolated system cannot spontaneously reduce.

The violation is apparent, not real. In 'sensing' the speeds of the atoms in the gas, the demon must have expended energy, leading to entropy increases in some larger system. It can be argued that when the demon and his measuring instruments are included, then the entropy of the total system does indeed increase in accordance with the second law.

Van der Meer needed to do something rather similar. If it could be possible to 'sense' anti-proton energies distant from the desired beam energy and somehow 'nudge' them back towards this energy, then the result would be a concentration of anti-proton energies and an increase in the luminosity of the beam. His 1968 experiments had hinted that this could be done.

He called it 'stochastic cooling'. Pick-up electrodes positioned in one part of the beam detect anti-protons whose energies deviate from the desired beam energy. A signal is then sent to the other side of the beam storage ring. The size of this signal is proportional to the extent of the deviation. The signal is amplified and applied to a 'kicker' which applies an electric field to the same section of the beam and deflects the offending particles towards the centre of the beam orbit. By repeating this process many millions of times, the beam gradually converges on the desired beam energy.

The word stochastic simply means 'random', and the cooling refers not to the temperature of the beam but to the random motions and the energy spread of the particles contained within it.

In 1974 van der Meer carried out some tests of stochastic cooling using the ISR. The results were not substantial, but they were sufficient to suggest that the principle worked. Two years later a small storage ring was converted and used for an Initial Cooling Experiment, which yielded results in 1977 and 1978 that were greatly encouraging. It seemed that by using the stochastic cooling technique it would be possible to create anti-proton beams of sufficient luminosity to perform proton–anti-proton collision experiments in the SPS. No new collider would be needed.

Proton and anti-proton beam energies of 270 GeV would combine in the SPS to produce collisions with a total energy of 540 GeV, well in excess of the energies required to reveal the W and Z particles.

Carlo Rubbia had begun to champion the proton–anti-proton collider experiments at CERN in 1976. When the Initial Cooling Experiment was shown to be successful in June 1978, Rubbia was given the go-ahead to form a collaborative team of physicists to design the elaborate detector facility that would be required to prove the existence of the W and Z particles. As this was to be constructed in a large excavated area on the SPS the collaboration was called Underground Area 1, or UA1. The team would grow to include some 130 physicists.

This was still a long-shot, and the potential for disruption to the LEP project remained. It was not an easy decision for the CERN Research Board. But the Board decided to take a gamble. America had dominated high-energy physics since the end of the Second World War, taking full advantage of its lead during the slow recovery of post-war European physics. American particle physicists had notched up a string of prestigious discoveries. Although CERN physicists had been involved in the discovery of weak neutral currents, new particles had so far eluded them. Besides, Rubbia, notoriously difficult to work with,[7] would have simply taken his proposal elsewhere if he hadn't secured the go-ahead from CERN. 'Most likely, if CERN hadn't bought Carlo [Rubbia]'s idea, he would have sold it to Fermilab,' Darriulat explained.

Six months after the decision a second, independent collaboration—UA2—was formed under Darriulat's leadership. This would be a smaller collaboration, consisting of some 50 physicists, designed to provide friendly competition with UA1. The UA2 detector facility would be less elaborate (it would not be able to detect muons, for example), but would nevertheless be able to provide independent corroboration of the UA1 findings.

Protons were first accelerated to 26 GeV in CERN's proton synchrotron and used to produce anti-protons from a stationary copper target.[8] The anti-protons were then injected into a purpose-built anti-proton accumulator (AA), one batch with energies around 3.4 GeV every few seconds. The anti-protons were accumulated in the AA for a day or so, and the stochastic cooling technique applied to converge the range of

[7] Martinus Veltman wrote, of Rubbia: 'When he was Director of CERN, he changed secretaries at the rate of one every three weeks. This is less than the average survival time of a sailor on a submarine or destroyer in World War II . . .'. See Veltman, p. 74.

[8] The CERN physicists had estimated that about a million protons hitting the target would be required to produce two anti-protons, an estimate that proved to be a factor of two too high.

anti-proton energies, before being re-injected into the proton synchrotron to be accelerated to 26 GeV.

The anti-protons were then passed to the SPS ring, where they joined protons travelling in the opposite direction. The protons and anti-protons were then accelerated to 270 GeV, arranged in bunches each a few nanoseconds (a few billionths of a second) in duration. The bunches of protons and anti-protons would collide at six points around the ring. The UA1 and UA2 detector facilities were positioned at two of these points.

Anti-protons were first injected into the AA in July 1980. In July 1981 the SPS accelerated the anti-protons to 270 GeV for the first time, and the first collisions were registered. The UA1 and UA2 detector facilities were designed to be rolled away from the ring to allow the SPS to be switched back to operation as an accelerator with fixed targets, as the search for the W and Z particles alternated with more conventional experiments.

The first opportunity to perform 'full-blooded' proton–anti-proton collision experiments therefore came in the spring of 1982. But the UA1 detector facility was brought down by a contaminated compressed air supply. There was no choice but to dismantle UA1 and clean its delicate components, a task that would take many months.

As a consequence, two separate proton–anti-proton runs were merged into a single run commencing in October 1982. UA1 and UA2 began logging data. Not every collision was to be captured. It was anticipated that collisions producing the W and Z particles would be very rare, so both detector facilities were set so that they would respond only to selected collisions meeting pre-programmed criteria. The collider would produce several thousand collisions per second over a period of two months. Only a handful of W- and Z-producing events were expected.

The detector facilities were programmed to identify events involving the ejection of high-energy electrons or positrons at large angles to the beam direction. Electrons carrying energies up to about half the mass of the W would be the signature of the decay of W^- particles. High-energy positrons would likewise signal the decay of W^+ particles. Measured energy imbalances (differences between the energies of the particles going into the collision versus those coming out) would signal the concomitant production of anti-neutrinos and neutrinos, which could not be detected directly.

Preliminary results were presented at a workshop on the physics of proton–anti-proton collisions in Rome in early January 1983. Rubbia, uncharacteristically nervous, made the announcement. From the several thousand million collisions that had been observed, UA1 had identified six events that were candidates for W-particle decays. UA2 had identified four candidates. Though somewhat tentative, Rubbia was convinced: 'They look like Ws, they smell like Ws, they must be Ws.' 'His talk was spectacular,' wrote Lederman. 'He had all the goods and the showmanship to display them with a passionate logic.'

On 20 January, CERN physicists packed into the auditorium to hear two seminars delivered by Rubbia for UA1 and Luigi Di Lella for UA2. A press conference was called on 25 January. The UA2 collaboration preferred to reserve judgement, but judgement was soon forthcoming. The W particles had been found, with energies close to the predicted 80 GeV. The UA1 collaboration published its results in the 24 February 1983 issue of *Physics Letters*. The UA2 collaboration published its results in the same journal less than a month later.

It had always been understood that the Z^0 would be a little more difficult to hunt down. When the proton–anti-proton collider experiments were resumed in April 1983, the CERN physicists pushed the SPS even harder. At least the signature of the Z^0 decay would be easier to identify—electron–positron or muon–anti-muon pairs observed to carry never-before seen energies.

The UA1 discovery of the Z^0, with a mass around 95 GeV, was announced on 1 June 1983 and published in *Physics Letters* on 7 July. This was based on the observation of five events—four producing electron–positron pairs and one producing a muon pair. The UA2 collaboration had accumulated a few candidate events by this time but preferred to wait for results from a further experimental run before going public. UA2 eventually reported eight events producing electron–positron pairs. They published their results in *Physics Letters* on 15 September 1983.

By the end of 1983, UA1 and UA2 between them had logged about a hundred $W^{\pm}$ events and a dozen Z^0 events, with masses around 81 GeV and 93 GeV, respectively.

It had been a long journey, one that was, arguably, begun with Yang and Mills' seminal 1954 work on an SU(2) quantum field theory of the strong force.[9] This was the theory which predicted massless bosons that had so irked Pauli. In 1957 Schwinger had speculated that the weak nuclear force is mediated by three field particles, and his student Glashow had subsequently reached for an SU(2) Yang–Mills field theory.

Much water had since been carried under the bridge. The massless bosons had acquired mass through spontaneous symmetry-breaking and the Higgs mechanism. The resulting theory had been shown to be renormalizable. And now the intermediate vector bosons had been found, precisely where they had been expected.

Rubbia and van der Meer shared the 1984 Nobel Prize for physics.

[9] Actually, in 1938 Oskar Klein had suggested that the weak force might be mediated by 'electrically charged photons'.

29

The Standard Model

Geneva, September 2003

The character of particle physics had changed. Many particle physicists would come to regard the 1960s and 1970s as a golden age of their science. Back then, the commissioning of a new accelerator or collider was herald to the beginning of an exciting journey into an unfamiliar and largely uncharted territory, populated by particles that were, at best, only vaguely hinted at by the theorists. Particle discoveries were often surprising, if not bewildering. They spoke of what we didn't yet understand.

Now, it seemed, the task of the particle physicist was simply to confirm the existence of particles that everybody knew must be there. The purpose of a new accelerator or collider was to verify what physicists already believed to be true. This was now a journey into increasingly familiar and charted territory. There were still surprises, still much that the physicists didn't yet understand, but the kinds of questions that were within the reach of the experimentalists had to a large extent already been answered.

Flush with success from their search for the W and Z particles, CERN physicists pushed to find the top quark. They failed, concluding that the mass of the top quark must be greater than 41 GeV, almost ten times the mass of its third-generation partner, the bottom quark.[1] *By the early 1990s, competition between CERN and the Fermilab Tevatron, a new proton–anti-proton collider capable of reaching collision energies of 1.8 TeV (almost*

[1] Observations of Z^0 particle decays had revealed no evidence for creation of top–anti-top pairs, suggesting that the mass of the top quark must be greater than a little less than half the mass of the Z^0.

two thousand billion electron volts) had helped to push the lower mass limit of the top quark to 77 GeV. And then on to 91 GeV, close to the mass of the Z^0. CERN physicists could not stretch the utility of their proton–anti-proton collider any further than a total collision energy of 620 GeV. They dropped out of the race.

The discovery of the top quark was eventually announced at Fermilab on 2 March 1995, by two competing research teams each consisting of about 400 physicists.[2] *Its mass was an astonishing 175 GeV, equivalent to the mass of a rhenium nucleus and almost 40 times larger than the mass of the bottom quark. It was identified through its decay products. Energetic protons and anti-protons collide to produce a top–anti-top pair. Each of these particles decays into a bottom quark and a W particle. One W particle decays into a muon and a muon anti-neutrino. The other decays into an up and a down quark. The end result is a collision which produces a muon, a muon anti-neutrino, and four quark jets.*

Aside from the Higgs boson, the only particle that remained to be discovered was the tau neutrino. Its discovery was announced at Fermilab five years later, on 20 July 2000.

With the discoveries of the top quark and the tau neutrino, the Standard Model was virtually complete. Physicists faced the unprecedented situation that there were now no experimental data that did not conform to the predictions of the quantum field theories that formed the basis of the Standard Model.

From the very beginning of its history, the development of quantum physics had been driven by disconcerting and inexplicable experimental results. Planck had experienced the most strenuous work of his life forging a theoretical basis for his empirical radiation formula. Stopped in a street in Copenhagen, Pauli had explained that the source of his misery was the anomalous Zeeman effect. As the number and types of 'elementary' particles had exploded in the 1950s and 1960s, quantum field theorists occupied an unfashionable backwater. At almost every stage, experiment had tended to outpace theory, leaving the theorists scrambling for explanations.

Now that had all changed. There were now no disconcerting or inexplicable experimental results. Theory had triumphed. And yet, it was also painfully apparent that we were far from the end of physics.

The Standard Model comprises three types of Yang–Mills quantum field theories in the gauge group SU(3) × SU(2) × U(1). It describes the

[2] The first few pages of the papers reporting these results consist just of a long list of names.

interactions of three generations of matter particles through three kinds of force, mediated by a collection of field particles or 'force carriers'.

The everyday material substances with which we are most familiar are composed of atoms. Atoms consist of a central nucleus of protons and neutrons surrounded by ghost-like electron wave-particles. Protons and neutrons are in their turn composed of up and down quarks. These quarks, the electron, and electron neutrino are all spin-½ fermions and together they form the first generation of matter particles in the Standard Model. They are all that is needed to describe everything we can experience in the material world around us.

Each quark flavour (up, down) comes in three different colour varieties—red, green, and blue. The quarks are bound inside protons and

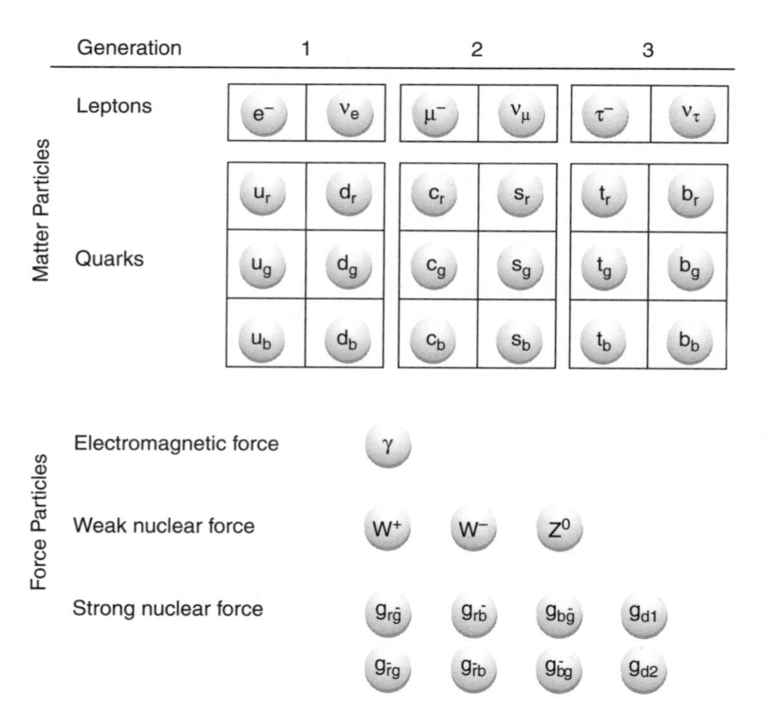

FIG 19 The Standard Model of particle physics describes the interactions of three generations of matter particles through three kinds of force, mediated by a collection of field particles or 'force carriers'.

neutrons by the strong colour force, which increases in strength as the quarks are pulled apart. In consequence, the quarks are permanently 'confined', forbidden from appearing without a chaperone. The colour force is carried by coloured spin-1 gluons, which interact with the quarks and with themselves.

The up and down quarks pair with anti-quarks to form the spin-0 pions. The positive pion, π^+, is formed from the combination of an up quark and an anti-down quark ($u\bar{d}$). The negative pion, π^-, is formed from a down quark and an anti-up quark ($d\bar{u}$). The neutral pion, π^0, is a superposition of $d\bar{d}$ and $u\bar{u}$ combinations. These are relatively low-mass particles (which is why they were discovered in the 1940s and 1950s) and can be thought of as pseudo Nambu–Goldstone bosons. They are responsible for mediating the interaction between protons and neutrons, binding them together in the nucleus.

There are two further generations of matter particles, following the pattern established by the first generation and differing only in the particle masses. The second generation consists of the strange and charm quarks and the muon and muon-neutrino. The third generation consists of the top and bottom quarks and the tau and tau-neutrino.

There are only three generations. Detailed studies of Z^0 decays using the Large Electron–Positron (LEP) collider at CERN revealed that there can be no more than three different kinds of neutrino. If there was a fourth, or a fifth, further decay routes would be open to the Z^0 which would affect its measured lifetime. These measurements do not rule out a radically different generation of matter particles consisting of a very heavy neutrino,[3] but there is no evidence for such a particle. It is therefore concluded that there are just three neutrinos and, by inference, three particle generations.[4]

The strong nuclear force acts on the quark colours. The weak nuclear force, mediated by the spin-1 W and Z particles, acts on the quark flavours. In beta radioactive decay, one of the more familiar manifestations of the weak force, a down quark in a neutron is transformed into an up

[3] The neutrino mass in such a generation would have to be heavier than half the mass of the Z^0.

[4] The number of neutrino species was determined to be 2.985 ± 0.008. See Cashmore *et al.*, p. 81.

quark, turning the neutron into a proton, with the emission of a W^- particle. The W^- particle then decays into an electron and an anti-neutrino.

Mixing via weak-force interactions allows transitions between up and strange quarks, and between down and charm quarks. This mixing is characterized by another angle, known as the Cabibbo angle, named for Italian physicist Nicola Cabibbo, with a measured value of about 13°. Further mixing allows transitions between up and bottom, down and top, charm and bottom, and strange and top quarks. This is a generalization of Cabibbo mixing, called CKM mixing after Cabibbo and Japanese physicists Makoto Kobayashi and Toshihide Maskawa. The CKM 'matrix' is characterized by three angles. The measured mixing angle between first- and third-generation quarks is about 0.2°. The measured angle for mixing second- and third-generation quarks is about 2.4°. A fourth 'angle' reflects the relative phase of the coupling between the quarks and is related to CP violation in weak-force decays.

Finally, the electromagnetic force, experienced between electrically charged particles, is mediated by massless spin-1 gauge bosons. These are the photons, first discovered by Planck in 1900 and championed by Einstein in his 'miracle year', five years later.

Lurking somewhat mysteriously beneath this formalism is the Higgs field, which pervades the vacuum and fills the universe. Interactions between massless particles and the Higgs field (or Higgs 'condensate') endow the particles with mass. The amount of mass acquired reflects the extent of the coupling between the particles and the field. The particle of the Higgs field is the spin-0 Higgs boson, which has been elevated in the Standard Model to the status of a 'God particle', responsible for the masses of all the particles.[5]

'The Higgs particle itself has never been detected,' wrote 't Hooft in 1995, 'but its *field* is being felt everywhere. If the Higgs were not there, our model would have so much symmetry that all particles would look alike; there would be too little *differentiation*.'

The Standard Model is a triumph of theoretical and experimental physics. 't Hooft summarized the theory as follows:

[5] It should be noted, however, that interaction with the Higgs field is not the only source of what we recognize to be mass. In fact, 99 per cent of the mass of protons and neutrons is derived from the energy of the gluon fields that bind their constituent quarks together. In their turn, protons and neutrons account for 99 per cent of the mass of every atom.

> This is a mathematical description of all known particles and all known forces between them, enabling us to explain all of the behaviour of these particles... As far as we know, there is no single physical phenomenon that cannot be regarded as some consequence of the Standard Model, and yet its basic formulae are not terribly complicated. We do admit that the model is not absolutely perfect... however, the degree of perfection reached is quite impressive.

But the Standard Model does not represent a return to triumphalism of the kind characterized by Kelvin's possibly apocryphal speech to the British Association for the Advancement of Science in 1900. Despite the model's obvious and unmitigated successes, its deep flaws have been painfully apparent since its inception in the late 1970s.

The Standard Model has to accommodate a rather alarming number of 'fundamental' particles. Six quark flavours multiplied by three possible colour values gives 18 different types of quark. Add the leptons—the electron, muon, and tau, and their neutrinos, and we have 24 fermions. Then there are the anti-particles of all these, making 48, to which we need to add the field quanta: the photon, the W^+, W^-, and Z^0 particles and eight different types of gluon, making 60 in total. All of these particles have been 'observed' so to speak, but we should also add the so far unobserved Higgs boson, making 61 particles in all. This hardly seems the stuff of a fundamental theory.

These 61 particles are connected together in a framework that requires 20 parameters that cannot be derived from theory but must be obtained by measurement. As Lederman put it in 1993:

> The idea is that twenty or so numbers must be specified in order to begin the universe. What are these numbers (or parameters, as they are called in the physics world)? Well, we need twelve numbers to specify the masses of the quarks and leptons. We need three numbers to specify the strengths of the forces... We need some numbers to show how one force relates to another. Then we need a number for how the CP-violation enters, and a mass for the Higgs particle, and a few other handy items.

The pattern of particle masses is particularly troublesome. For the quarks, we must distinguish between the mass of a 'naked' quark and a quark 'dressed' in a covering of gluons whose energy contributes to the quark mass. The mass of a naked up quark has been determined to be between 1.5 and 3.3 MeV; the naked down quark mass between 3.5 and 6.0 MeV.[6] When we compare these figures with the measured mass of a proton (uud) of 939 MeV we get some idea of the scale of the mass contributed by the binding energy of the gluons.

The charm quark has a 'running' mass of 1270 MeV, the strange quark a mass of 104 MeV.[7] The top quark has a running mass of 171,200 MeV, the bottom quark a mass of 4200 MeV. This looks like a pretty random pattern of values, with no rhyme or reason. The masses of the leptons are likewise unfathomable.[8]

The mixing angles, which determine how strongly the weak and electromagnetic forces act on both the quarks and the leptons, must also be determined by experiment. The masses and mixing angles are the results of the interactions of the quarks and leptons with the Higgs field. The inability of theory to predict these values from first principles may therefore reflect the simple fact that the properties of the Higgs field and the precise nature of the symmetry-breaking mechanism are not properly accounted for in the Standard Model.

The Standard Model is not a theory in which the strong, weak, and electromagnetic forces can be said to be properly unified. And, of course, the model makes no attempt to accommodate the fourth known force of nature: gravity. This is a force so weak that it is of no consequence in the interactions between the fundamental particles. But it is always attractive, proportional to the particle mass, and cumulative. Scale up the quarks into nucleons, nucleons into atoms, atoms into molecules, molecules into solid matter, matter into planetary bodies, and the force of gravity becomes irresistible.

[6] These quark mass data are taken from C. Amsler *et al.*, *Physics Letters B*, **667**, 2008, p. 1.

[7] The 'running' mass reflects the fact that mass values vary when measured at different energy scales.

[8] Although the neutrinos were thought to be massless, recent observations of the flux of neutrinos from the sun suggest that they oscillate between the types, an electron neutrino turning into a muon or tau neutrino, etc. This is possible only if the neutrinos possess very small masses. This can all be accommodated in the Standard Model but at the cost of introducing further parameters.

The gravitational field implies a quantum field theory of the gravitational force, and a force carrier, called the graviton. To account for the properties of the gravitational force, the graviton would need to be a spin-2 field boson. Familiar (and by now tried and trusted) approaches to a quantum field theory of gravity lead to a result which is non-renormalizable.

Impasse.

With no disconcerting or inexplicable experimental results to pursue and no theoretical predictions within reach of experiment, there was now no guidance on how the Standard Model should be developed. Or, indeed, no guidance on whether the Standard Model should be abandoned in favour of an altogether different approach.

Although the mass of the Higgs boson could not be predicted, physicists at CERN and Fermilab were able to place lower and upper bounds on its value. Approval was given in the late 1980s for the construction of a new facility, the Superconducting Supercollider (SSC), in Waxahachie, Texas. This was to be an 87-kilometre circumference ring producing a collision energy of 40 TeV, potentially capable of bringing the Higgs boson within reach of experimental physics. On giving his approval, US President Ronald Reagan had urged his Cabinet secretaries to 'Throw deep.' Construction began in 1991. The project was directed by Roy Schwitters.

Budget estimates mushroomed, from $4.4 billion in 1987 to $12 billion in 1993. Despite efforts by President Bill Clinton to garner support for the project, it was cancelled by Congress. Nearly 23 kilometers of tunnel had been excavated and nearly $2 billion had been spent.

Two years later a more modest project was approved to construct the Large Hadron Collider (LHC) at CERN, using the 27-kilometre tunnel that houses the LEP. Rubbia declared that CERN would 'pave the LEP tunnel with superconducting magnets.' The LHC is expected to eventually produce a collision energy of 14 TeV, less than half the energy that could have been achieved with the SSC, but nevertheless with a theoretical capability of generating one Higgs boson every couple of hours.

On the morning of Tuesday, 16 September 2003 a small but highly distinguished international group of physicists gathered in CERN's large auditorium to celebrate the double anniversary of the discoveries of the weak neutral currents (1973) and the W and Z particles (1983). After a brief welcome by Maiani, then CERN Director General, Weinberg stood to describe the twists and turns that had led ultimately to the creation of the Standard Model of particle physics. He concluded as follows:

> Well, those were great days. The 1960s and 1970s were a time when experimentalists and theorists were really interested in what each other had to say, and made great discoveries through their mutual interchange. We have not seen such great days in elementary particle physics since that time, but I expect that we will see good times return again in a few years, with the beginning of a new generation of experiments at this laboratory.

PART VI

Quantum *Reality*

30

Hidden Variables

Princeton, Spring 1951

David Bohm was arrested on 4 December 1950 and charged with contempt of court.

He had joined the Communist Party in November 1942 and had been part of a close-knit group of radical young physicists studying at Berkeley under Oppenheimer. In March 1943 one among this group—Joseph Weinberg—had been caught betraying atomic secrets by an illegal FBI bug planted in the home of Steve Nelson, a key figure in the Communist Party apparatus in the Bay Area of San Francisco. This evidence was inadmissible in court and, in an attempt to expose Weinberg's betrayal by more legal means, the House Un-American Activities Committee (HUAC) had called Bohm to testify in May 1949.

Einstein had advised Bohm to refuse, suggesting that he 'may have to sit for a while', meaning that the penalty for his silence might be a short spell in prison. He had chosen to testify, but refused to divulge names at the hearing and at a subsequent HUAC hearing in June. Princeton University had expressed support and declared Bohm a 'thorough American'.

But events over the next twelve months would conspire to whip anti-Communist sentiment in America to fever pitch. Earlier testimony by Whitaker Chambers, a former Soviet agent and editorial staff member of Time *magazine, had revealed the names of two highly placed Communists in US President Harry Truman's administration—Alger Hiss at the State Department and Harry Dexter White at the Treasury. White died of a heart attack in August 1948. Hiss was convicted on two counts of perjury in January 1950 and sentenced to two concurrent five-year prison sentences. As the 'red scare' gathered momentum, in February 1950 Republican Senator Joseph McCarthy saw an opportunity to launch an anti-Communist crusade. It became a witch-hunt.*

HUAC had concluded in September 1949 that Weinberg and Bohm had been members of a Communist cell that had passed atomic secrets to the Soviets. After Bohm's arrest he was bailed, but now the Princeton University administrators withdrew their support. Bohm was suspended from his post for the duration of the trial.

Bohm came to trial on 31 May 1951. He was acquitted (as was Weinberg a few years later). But Princeton University did not renew Bohm's contract when it expired the following month. John Wheeler, who had helped to bring Bohm to Princeton, later summarized the prevailing sentiment: 'I found it hard to accept Bohm's decision to shield those who adhered to Communist ideology at a time when the Soviet Union was suppressing its own people and threatening world peace.' Einstein wanted to offer Bohm a position at the Institute for Advanced Study but Oppenheimer, fearing for his own position as Director of the Institute, vetoed the move.

With his life turning upside-down, it was hardly possible for Bohm to concentrate on his physics. He had just finished writing a book, simply titled Quantum Theory, *and was correcting the proofs. But he admitted that it was 'hard to concern myself with getting all these [mathematical] formulas correct.'*

The book was published in February 1951. It received many favourable reviews. On the question of interpretation, Bohm had stuck fairly closely to the orthodoxy of the Copenhagen school, though he was closer in spirit to Pauli than to Bohr. Einstein welcomed the book, claiming it was the clearest presentation of the Copenhagen interpretation he had ever read. But, of course, this didn't mean that Einstein accepted what Bohm had written. Einstein asked for an opportunity to explain his objections, and invited Bohm to visit him.

In 1935 Einstein, Podolsky, and Rosen had asserted that quantum theory is incomplete. They had left open the question of whether or not the theory could be somehow 'completed', declaring only that this should be possible in principle. The simplest way to complete the theory and restore causality and determinism is to invoke *hidden variables* of some form.

Einstein himself had toyed with just such an approach in May 1927. This was a modification of quantum theory that combined classical wave and particle descriptions, with the wavefunction of Schrödinger's wave mechanics taking the role of a 'guiding field' (Führungsfeld), guiding the physically real point-particles.

In this scheme, the wavefunction is responsible for all the wave-like effects, such as diffraction and interference, but the particles maintain their integrity as local, physically real entities. Instead of waves *or* particles, as the Copenhagen interpretation demanded, Einstein's hidden variables version of quantum theory was constructed from waves *and* particles.

But Einstein had lost his enthusiasm for this approach within a matter of weeks of formulating it. The guiding field was capable of exerting 'spooky' non-local influences. When de Broglie had stood to present his 'double solution' on the second day of the fifth Solvay conference later in October that year, Einstein had already rejected this approach. He had remained silent through de Broglie's presentation.[1]

This experience probably led Einstein to conclude that his initial belief—that quantum theory could be 'completed' through a more direct fusion of classical wave and particle concepts—was misguided. He subsequently expressed the opinion that a complete theory could only emerge from a much more radical revision of the entire theoretical structure. One such possibility was an elusive grand unified field theory, the search for which took up most of Einstein's intellectual energy in the last decades of his life.

In his book *Quantum Theory*, Bohm appeared to have accepted Bohr's response to the EPR argument as having settled the matter in favour of the Copenhagen interpretation. He wrote: 'EPR's criticism has, in fact, been shown to be unjustified, and based on assumptions concerning the nature of matter which implicitly contradict the quantum theory at the outset.' Despite what he had written, this conclusion had left him with a distinct feeling of dissatisfaction.

In his description of the EPR argument, Bohm had pushed the *gedankenexperiment* into a different domain of applicability. He considered a molecule consisting of two atoms in a quantum state in which the total electron spin angular momentum is zero. A simple example would be a hydrogen molecule, H_2, with its two electrons spin-paired in the lowest (so-called 'ground') electronic state.

[1] See Chapter 13, pp. 116–117.

Suppose that this molecule is broken apart in a process that preserves the total angular momentum to produce two equivalent atomic fragments. The hydrogen molecule is split into two hydrogen atoms. These atoms move apart but the spin orientations of the electrons in the individual atoms remain opposed—one spin-up and one spin-down.

The spins of the atoms are therefore *correlated*. Measurement of the spin of one atom (say atom A) in some arbitrary laboratory frame allows us to predict, with certainty, the direction of the spin of atom B in the same frame. We might be tempted to conclude that the spins of the two atoms are determined by the nature of the initial molecular quantum state and the method by which the molecule is fragmented. The atoms move away from each other with their spins fixed in unknown but opposite orientations and the measurement merely tells us what these orientations are.

But this is not how quantum theory deals with the situation. The two atoms are instead described by a single wavefunction until the moment of measurement. The atoms are entangled. If we choose to measure the component of the spin of atom A along the laboratory z-axis, for example, our observation that the wavefunction collapses into a state in which atom A has its spin orientation aligned in the $+z$-direction (say) means that atom B *must* have its spin orientation aligned in the $-z$-direction.

However, what if we choose, instead, to measure the x- or y-components of the spin of atom A? No matter which component is measured, the physics demand that the spins of the atoms must still be correlated, and so the opposite results must always be obtained for atom B. If we accept the definition of physical reality offered by Einstein, Podolsky, and Rosen, then we must conclude that *all* components of the spin of atom B are elements of reality, since it appears that we can predict them with certainty without in any way disturbing B.

However, the wavefunction specifies only one spin component, associated with the magnetic spin quantum number m_s. This is because the operators corresponding to the three components of the spin orientation in Cartesian (x, y, z) coordinates do not commute (the components are complementary observables). So, either the wavefunction is incomplete, or Einstein, Podolsky, and Rosen's definition of physical reality is inapplicable.

The Copenhagen interpretation says that no spin component of atom B 'exists' until a measurement is made on atom A. The result we obtain for B will depend on how we *choose* to set up our instrument to make the measurements on A, no matter how far away B is at the time.

This was a singular re-imagining of Einstein, Podolsky, and Rosen's original argument. It made completely transparent the nature of quantum entanglement and its implications for non-local, 'spooky' action at a distance. The measurement of the spin orientation of an atom is much more practicable than the measurement of its position or momentum. It opened the possibility that further elaborations of Bohm's version of the *gedankenexperiment* might be carried out in the laboratory, and not just in the mind.

Responding to Einstein's suggestion, Bohm met with him at Princeton sometime in the spring of 1951. The doubts over the interpretation of quantum theory that had begun to creep into Bohm's mind as he had worked on his book now crystallized into a sharply defined problem. As Einstein explained the basis for his own misgivings, most probably in terms clearer than those in the original Einstein, Podolsky, Rosen paper, Bohm acknowledged the need to change his position.

'This encounter had a strong effect on the direction of my research,' Bohm later wrote, 'Because I then became seriously interested in whether a deterministic extension of quantum theory could be found.'[2] Bohm was more committed to notions of causality and determinism than perhaps he had realized. These notions also lay at the heart of his Marxist ideology. In Marx's dialectical materialism, social change is *caused* by competing social forces and cannot be left to chance. The Copenhagen interpretation, by contrast, appears to leave everything to the roll of a dice.

Spurred by his discussion with Einstein, Bohm now reflected more deeply on the question of interpretation. At issue was the Copenhagen school's outright denial that individual quantum systems could be described objectively. He wrote:

[2] According to Basil Hiley, one of Bohm's long-term collaborators, Bohm said of his meeting with Einstein: 'After I finished [*Quantum Theory*] I felt strongly that there was something seriously wrong. Quantum theory had no place in it for an adequate notion of an individual actuality. My discussions with Einstein clarified and reinforced my opinion and encouraged me to look again.' Quoted by Basil Hiley, personal communication to the author, 1 June 2009.

> The usual interpretation of the quantum theory is self-consistent, but it involves an assumption that cannot be tested experimentally, *viz.*, that the most complete possible specification of an individual system is in terms of a wave function that determines only probable results of actual measurement processes. The only way of investigating the truth of this assumption is by trying to find some other interpretation of the quantum theory in terms of at present 'hidden' variables... the mere possibility of such an interpretation proves that it is not necessary for us to give up a precise, rational, and objective description of individual systems at a quantum level of accuracy.

Bohm was not specifically seeking a new theory or a return to simple classical physics. Rather, he acknowledged that quantum theory was constructed on a set of assumptions of which the most important, concerning the 'completeness' of the theory, is not subject to experimental test. The Copenhagen interpretation is founded on this completeness postulate and, rather than accept it at face value as the Copenhagen school demanded, Bohm wanted to explore the possibility that other descriptions and hence other interpretations are conceivable *in principle*.

As it appears that the only things we can detect are whole particles, which form diffraction or interference patterns only when many such particles have been detected, the suggestion that the particles are real entities that follow precisely defined trajectories is very compelling. Were there other ways in which the particles' motions might be predetermined?

Perhaps. If Schrödinger's wavefunction was reinterpreted as describing an objectively real wave-like field, this could serve to guide the motion of objectively real particles. Bohm now reworked Schrödinger's wave equation into a form resembling a fundamental dynamical equation in classical physics that is actually a statement of Newton's second law of motion and which is therefore much more closely associated with a particle interpretation. Bohm simply assumed that the wavefunction of the field can be written in a form containing real amplitude and phase functions. In itself, the assumption of a specific form for the wavefunction represents no radical departure from conventional quantum theory. However, Bohm now assumed the existence of a real particle, following a real trajectory through space, its motion embedded in the field and tied to or guided by the phase function through the imposition of a 'guidance condition'.

Every particle in every field therefore possesses a precisely defined position *and* a momentum, and follows trajectories determined by their respective phase functions. The equation of motion is then found to depend not only on the classical potential energy, but also on a second, so-called *quantum potential.*

The quantum potential is intrinsically non-classical and is alone responsible for the introduction of quantum effects in what would otherwise be an entirely classical description. Take out the quantum potential or allow it to decline to zero and Bohm's equations revert to the classical equations of Newtonian mechanics.

True to its nature, the quantum potential has some peculiar properties. It can exert effects on the particle in regions of space where the classical potential disappears. This contrasts markedly with the effects exerted by classical potentials (such as a Newtonian gravitational field), which tend to fall off with distance. A particle moving in a region of space in which no classical potential is present can therefore still be influenced by the quantum potential and some of the cherished notions of classical physics—such as straight-line motion in the absence of a (classical) force—must be abandoned.

The particle position and its trajectory are 'defined' at all times during its motion, and it is therefore not necessary in principle to resort to probabilities. When we consider a large number of particles all describable in terms of the same wavefunction, the above reasoning can still be applied. There is nothing in principle preventing us from following the trajectories of each particle. However, in practice we do not usually have access to a complete specification of all the particle initial conditions and, just as in Boltzmann's statistical mechanics, we resort to classical probabilities as a practical necessity.

This contrasts strongly with the notion of quantum probability. In conventional quantum theory, the wavefunction is really a calculation tool for probabilities interpreted as the relative frequencies of possible outcomes of repeated measurements on a collection of identically prepared systems. These outcomes are not determined until a measurement is made. In Bohm's theory, the particle motions are predetermined and we calculate probabilities because we are ignorant of the initial conditions of all the particles in the collection. These probabilities refer to individual states of individual particles—their positions and their trajectories—not measurement outcomes. Measurement therefore has no mystical role: the measurements merely tell us the actual states or positions of the particles or their actual trajectories through an apparatus, which are determined all along.

The probability is still related to the amplitude of the wavefunction, as in Born's original prescription, but this does not mean that the wavefunction has only a statistical significance. On the contrary, it is assumed that the wavefunction has a very strong physical significance—it determines the shape of the quantum potential.

This is a hidden variable theory. The hidden variable is not the guiding field—that is revealed in the properties and behaviour of the quantum wavefunction. It is actually the particle *positions* that are hidden. Bohm's theory reintroduces the classical concept of causality—classical particles are directed along classical trajectories dictated by guiding wave fields. But change the wave field simply by changing the measuring apparatus in some way, and the particles are obliged to respond instantaneously. The hidden variables are said to be 'non-local'. In this sense, at least, there is no conflict with Bohr's insistence on the primacy of the measuring apparatus: 'In this point,' Bohm wrote, 'We are in agreement with Bohr, who repeatedly stresses the fundamental role of the measuring apparatus as an inseparable part of the observed system.'

The theory restored causality and determinism, and eliminated the need to invoke a collapse of the wavefunction. But it had not eliminated non-local influences and 'spooky' action at a distance, and so appeared to be incompatible with special relativity.

In truth, Bohm had actually rediscovered and extended de Broglie's theory of the 'double solution'.[3] He drafted two papers on his hidden

[3] For this reason, Bohm's redevelopment is often referred to as de Broglie–Bohm theory. Nathan Rosen had also attempted a very similar approach in 1945 but had not pursued it further. See Jammer, *The Philosophy of Quantum Mechanics*, p. 285. Note that this is a non-local hidden variable *theory* which is capable, in principle, of making predictions which differ from conventional quantum theory. Subsequent non-local hidden variable *interpretations* of quantum theory make predictions which cannot be distinguished from those of conventional quantum theory.

variable theory and its application to the hydrogen atom and in July 1951 he submitted these to the journal *Physical Review*. His conversion had been swift. His book, overtly supportive of the Copenhagen view, had been published only four months previously.

He sent pre-prints to de Broglie, Bohr, Pauli, and Einstein. From de Broglie he learned for the first time all about the double solution and why de Broglie had abandoned it shortly after the fifth Solvay conference in 1927. From Pauli he received objections concerning the implications of the theory for multi-particle systems, but Bohm was confident that he could deal with these.

Unable to find a position in America or Britain, in October 1951 Bohm left Princeton and went into exile at the University of São Paulo in Brazil. Both Einstein and Oppenheimer had provided letters of recommendation and Bohm received support from Abrahão de Moraes, the head of the physics department. As his plane taxied towards the runway, it was announced that the pilot had been asked to return to the terminal. An irregularity had been discovered with the passport belonging to one of the passengers. Bohm feared a second arrest, but it turned out that the irregularity concerned a different passenger, who was removed from the flight.

Bohm's two papers were published in January 1952. He observed the reaction from exile in Brazil. Feynman was supportive. But Oppenheimer reflected the mood of the majority by declaring Bohm's work to be 'juvenile deviationism', urging that '... if we cannot disprove Bohm, then we must agree to ignore him.'[4]

Not surprisingly, Einstein was not enamoured of the approach Bohm had taken. In a letter to Born he wrote:

> Have you noticed that Bohm believes (as de Broglie did, by the way, 25 years ago) that he is able to interpret the quantum theory in deterministic terms? That way seems too cheap to me.

[4] Bohm died in 1992, but his vision for a causal form for quantum theory lives on. For example, Basil Hiley has recently extended Bohm's original theoretical structure using a Clifford algebra approach to accommodate relativistic effects.

31

Bertlmann's Socks

Boston, September 1964

Bohm did not adapt well to exile. He grew restless and in 1955 relocated to the Technion, the Israel Institute of Technology in Haifa.[1] *It was here that he met Yakir Aharonov, an outstanding 22-year old undergraduate student who had already established a reputation as a maverick. Together they worked on a further elaboration of Bohm's version of the Einstein, Podolsky, Rosen* gedankenexperiment *which they submitted for publication in May 1957.*

Bohm also published another book—essentially a manifesto for his deterministic programme—titled Causality and Chance in Modern Physics. *De Broglie supplied a foreword. In the summer of 1957 Bohm moved once more, to a research associate position at Bristol University in England. From Bristol, he subsequently took a professorship at the University of London's Birkbeck College, where he remained for the rest of his life.*

Whilst his was not exactly a lone voice, Bohm had recruited few followers to his cause. Although Bohm's deterministic theory yielded all the predictions of non-relativistic quantum mechanics, the majority of the physics community had moved on. Quantum electrodynamics had triumphed in 1949. Physicists had turned their attentions to the search for a quantum theory of the nuclear forces. In 1954 Yang and Mills had published their 'beautiful idea'. Gell-Mann and Zweig had introduced the idea of a triplet of fractionally charged constituents of mesons and baryons—quarks or aces—in 1963.

[1] The American authorities had withdrawn Bohm's passport. To travel to Israel he first had to become a Brazilian citizen.

There seemed little to be gained by raking over the ashes of old philosophical conundrums. Besides, hadn't Bohr already set the record straight on the Copenhagen interpretation in 1935? Hadn't von Neumann proved that all hidden variable theories are impossible in principle?

But the effort to interpret quantum theory and what it had to say about the nature of physical reality was about to take a surprising turn. Belfast-born CERN physicist John Bell had read Bohm's 1952 papers with great interest. The collapse of the wavefunction and the rather arbitrary boundary drawn by Bohr between the quantum objects of measurement and the classical measuring device seemed to Bell to be at best a confidence-trick, at worst a fraud. 'A theory founded in this way on arguments of manifestly approximate character,' he wrote some years later, 'however good the approximation, is surely of provisional nature.'

In 1952 he had conceived some ideas concerning hidden variable theories that were to form the basis for a paper he wrote twelve years later, whilst on leave from CERN at the Stanford Linear Accelerator Center.[2] *In the intervening years Bell had developed deep reservations about the relevance of von Neumann's impossibility proof. In his paper he argued that the proof hinged on a critical assumption that, whilst valid, applied to a situation in which two complementary physical quantities are measured simultaneously. But such measurements require completely incompatible measuring devices (or incompatible configurations of a single measuring device), which means that such measurements just cannot be made simultaneously. He concluded that the proof is, in fact, irrelevant.*

This meant that all varieties of hidden variables theories—local and non-local—were 'fair game' once again. But as Bell probed further, he made a discovery that was at once both simple and profound. In Quantum Theory, *Bohm had asserted that: '. . . no theory of mechanically determined hidden variables can lead to* all *of the results of the quantum theory.'*

Bell now discovered just how right this assertion was. What he found was that the choice between conventional quantum theory and adaptations based on local hidden variables was not after all just a matter of philosophical disposition. It was a matter of correctness.

In their 1957 paper, Bohm and Aharonov had pushed the EPR *gedankenexperiment* even closer to practical realization. Indeed, the purpose of the paper

[2] Due to a mix-up, this paper, published in *Reviews of Modern Physics*, did not appear until 1966.

was to claim that experiments capable of measuring non-local correlations between distant quantum particles had already been carried out.[3]

After visiting SLAC in 1964, Bell spent some time at the University of Wisconsin at Madison and thence moved on to Brandeis University near Boston. It was during this visit that he had an insight that was completely to transform questions about the nature of reality at the quantum level. He derived what was to become known as *Bell's inequality*: 'Probably I got that equation into my head and out on to paper within about one week-end. But in the previous weeks I had been thinking intensely all around these questions. And in the previous years it had been at the back of my head continually.'

Bell built his derivation on the version of the EPR experiment as elaborated by Bohm and Aharonov. After fragmenting the hydrogen molecule, the two hydrogen atom fragments move apart in opposite directions.[4] Hydrogen atom A moves to the left, atom B moves to the right. Bell now imagined that two magnets would be placed left and right, designed to determine the spin orientation of each atom using the Stern–Gerlach effect.

Atoms passed between the poles of the magnet are either deflected upwards, in the direction of the north pole (spin-up) or downward in the direction of the south pole (spin-down). Because of the correlation established between atom A and atom B, if the fields of both magnets are aligned, then opposite results are expected. If A is found to be in a spin-up orientation, then B will be found in a spin-down orientation, and *vice versa*.

Bell then turned his attention to an extremely simple example of a local hidden variable theory, one that would not only restore causality and determinism and eliminate the collapse of the wavefunction, as Bohm had done in 1952, but would eliminate spooky action at a distance as well. Any such theory is characterized as *locally real*. The hydrogen atoms in the

[3] We will examine these claims in the next chapter.

[4] Neither Bohm and Aharonov nor Bell were specific about the nature of the molecule involved, specifying only that it possess a total spin of zero. I will nevertheless continue to refer to the example of a hydrogen molecule.

above example are assumed to possess some variable or variables which predetermine the outcomes of the subsequent spin measurements. These variables are not obliged to be fixed at the moment the hydrogen molecule is fragmented, although it helps us to understand what needs to happen in this simple example if we assume that they are.

Imagine, then, that hidden within each atom there exists a tiny subatomic dial. The dial has a pointer. We assume that this pointer can point in any direction on the dial and, at the moment of fragmentation, the pointer directions within each fragmented atom are randomly fixed, such that they take up any direction in the entire 360° range on the face of the dial. However, whatever pointer direction is (randomly) fixed for atom A, the pointer direction for atom B must be firmly fixed in the opposite (180°) direction, and *vice versa*. We assume that the physics of the fragmentation process demands this (we can think of it as a law of aligned pointers).[5]

The atoms move towards the poles of their respective magnets. For the sake of simplicity, we assume that if the pointer lies at any angle in the top half of the dial face, this will determine that the spin is 'up' and the atom is deflected upwards towards the north pole of the magnet. Likewise, if the pointer lies anywhere in the bottom half of the dial face, the atom is detected in a spin-down orientation and is deflected downwards towards the south pole.

This is a local hidden variable theory. Whatever they are and however they are supposed to work, the pointers predetermine the outcomes of the spin measurements in a way which eliminates the need for mysterious influences travelling over long distances faster than light-speed. It is also a relatively commonsense theory: the outcomes of the experiments are predetermined as soon as the hydrogen molecule is fragmented. The measurements simply tell us what these outcomes are.

What happens in the situation where both magnetic fields are aligned, both with their north poles lying in the same direction? If the pointer for atom A lies anywhere in the top half of its dial face, the result is spin-up, which we designate as a '+' result. The pointer direction for B *must* lie in the bottom half of its dial face, predetermining a spin-down result which we

[5] It is actually the law of conservation of angular momentum.

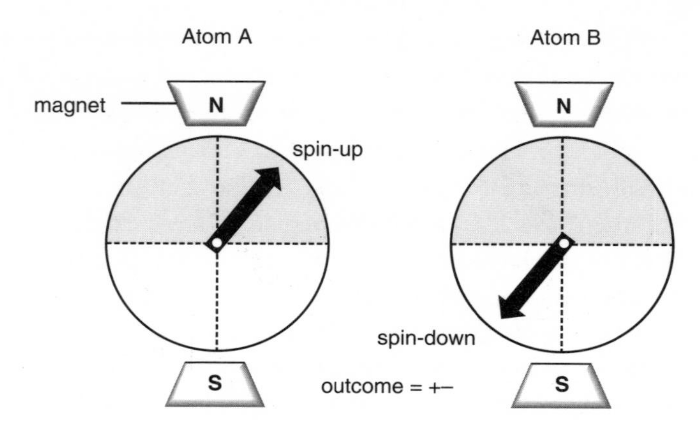

FIG 20 In this simple example of local hidden variables, we suppose that each fragmented atom contains a sub-atomic dial. The pointer on the dial pre-determines the spin orientation of the electron in the atom. The physics of the fragmentation process ensures that the pointers of each atom must be oriented in opposite directions. A pointer pointing anywhere in the top half of the dial face will lead to detection as spin-up. A pointer pointing anywhere in the bottom half of the dial face will lead to detection as spin-down. For the two pointer orientations shown, the outcome is a '+−' result.

designate as a '−' result. The probability of obtaining this combined result, which we denote P_{+-}, is simply the probability that the pointer direction for A lies in the top half of the dial face. This is obviously 50 per cent, the proportion of the area of the top half to the total area of the dial face. By the same argument, the probability P_{-+} is also 50 per cent.

Now, Bell speculated, what happens if we *rotate* the orientation of one of the magnets relative to the other? The diagram in Figure 21 shows the effects of rotating the magnet used to detect the spin of atom B through angles of 45°, 90°, and 180°. Changing the orientation of this magnet changes the orientation of what will be recognized as the top half of the dial face. The probability P_{+-} now depends on the *overlap* of the 'top' halves of the dial faces of the two atoms. This overlap shrinks as we increase the angle between the axes of the two magnets, with P_{+-} falling to zero when the magnets are aligned in opposite directions (180°). What happens is that as P_{+-} declines, the probability of obtaining the results in which both spins are measured as 'up' increases. We denote this probability as P_{++}.

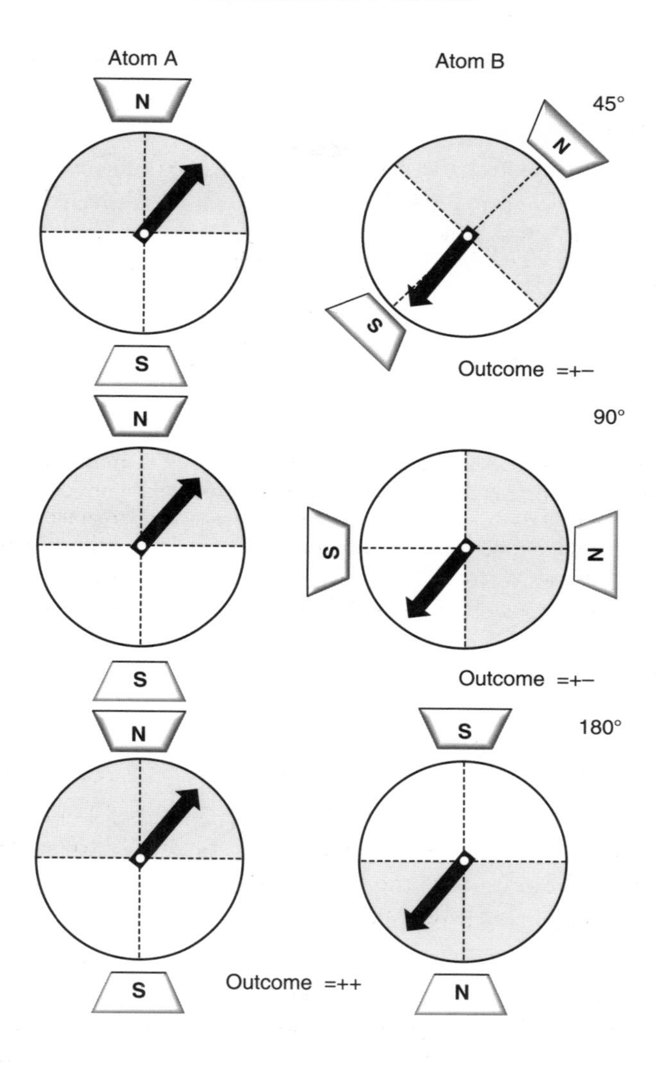

FIG 21 As the second magnet is rotated, so does the area of what will be recognised as the 'top' half of the dial face. The probability of obtaining the outcome '+–' declines as, for fixed orientations of the pointers, this depends on the overlap between the areas of the top half of the dial face for atom A (shaded area) and the top half of the dial face for atom B. This overlap declines as we rotate the second magnet through 45°, 90° and 180°. By the time we reach 180°, there is no overlap at all. The probability for achieving a '+–' result is zero. Instead, the outcome is a '++' result—both atoms are recorded to have spin-up orientations.

The conventional quantum theory prediction for P_{+-} is $\frac{1}{2}\cos^2(\varphi/2)$, where φ is the angle between the axes of the magnets.[6] It is possible to show that the simple hidden variable theory used by Bell predicts results for P_{+-} in agreement with quantum theory only for the angles 0°, 90°, and 180°, which give P_{+-} = 50 per cent, 25 per cent, and 0 per cent, respectively. The predictions diverge for all other angles, with maximum divergence at an angle of about 40°, for which the local hidden variable theory predicts P_{+-} = 38.9 per cent, compared to the quantum theory prediction P_{+-} = 44.2 per cent.

Perhaps we shouldn't be too unhappy with this result. This is an extremely crude local hidden variable theory. Surely it is not beyond the bounds of possibility that with a little ingenuity we should be able to construct a more sophisticated local hidden variable theory, one which accounts for *all* the predictions of quantum theory?

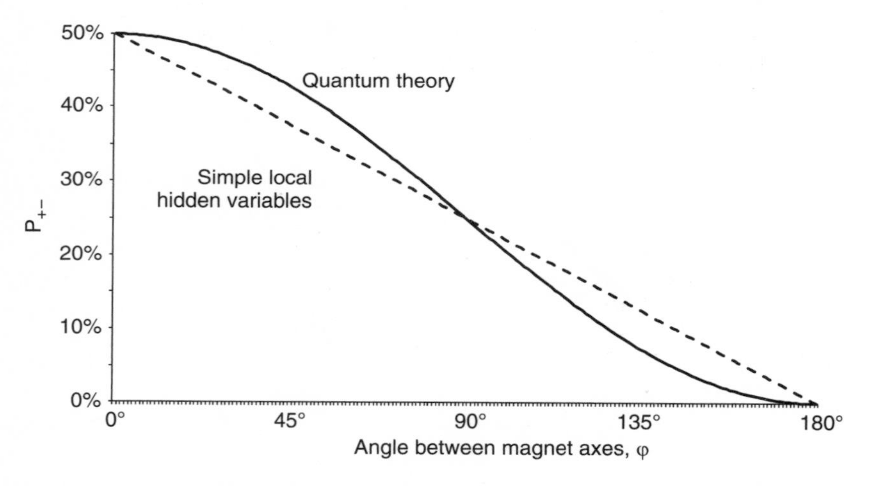

FIG 22 The simple local hidden variables theory used by Bell in 1964 predicts results for P_{+-} in agreement with quantum theory only for the angles 0°, 90° and 180°, which give P_{+-} = 50 per cent, 25 per cent and 0 per cent, respectively. The predictions diverge for all other angles, with maximum divergence at angles of 40° and 140°.

[6] I will use this result without presenting a derivation. Readers interested to understand where this comes from should consult John Bell, *Journal de Physique Colloque* C2. suppl. au numero 3. Tome 42, 1981, pp. 41–61.

The answer is no, we can't. And this is what Bell discovered.

Some seventeen years after his original discovery, Bell derived much the same result through the agency of a character called Dr Bertlmann. This was a real person (presumably with an unusual dress-sense) used by Bell to highlight the almost trivial nature of the inequality that bears his name. He opened his discussion with these words:

> The philosopher in the street, who has not suffered a course in quantum mechanics, is quite unimpressed by Einstein–Podolsky–Rosen correlations. He can point to many examples of similar correlations in everyday life. The case of Bertlmann's socks is often cited. Dr Bertlmann likes to wear two socks of different colours. Which colour he will have on a given foot on a given day is quite unpredictable. But when you see that the first sock is pink you can be already sure that the second sock will not be pink. Observation of the first, and experience of Bertlmann, gives immediate information about the second. There is no accounting for tastes, but apart from that there is no mystery here. And is not this [Einstein–Podolsky–Rosen] business just the same?

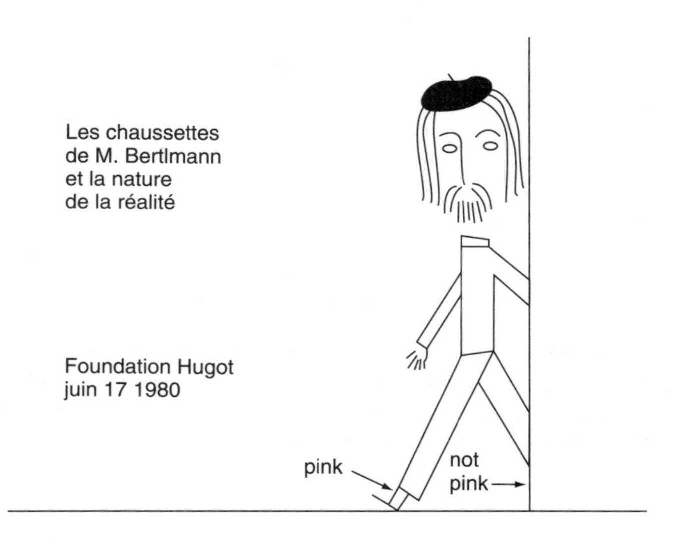

FIG 23 Bertlmann's socks and the nature of reality. Reprinted with permission from *Journal de Physique (Paris), Colloque C2*, (suppl. au numero 3), **42** (1981) C2 41–61. www.journaldephysique.org

Let's take a closer look at the physical characteristics and behaviour of these infamous socks. We want to hypothesize how these socks will stand up to the rigours of prolonged washing at three different temperatures. We subject Bertlmann's left socks (socks A) to three different tests. These are: washing for one hour at temperature *a*, washing for one hour at temperature *b*, and washing for one hour at temperature *c*. For the moment, we will defer our choice of specific temperatures.

We measure the numbers of socks that survive intact (which, for reasons of our own, we call a '+' result) or are destroyed (a '−' result) by prolonged washing at these different temperatures.[7] Having some grounding in theoretical physics, we know that we can discover some simple relationships between the numbers of socks that survive or are destroyed in these tests without actually having to perform the tests using real socks and real washing machines.

We denote the number of socks that survive (+) at temperature *a* and are destroyed (−) at temperature *b* as $n(a_+,b_-)$. We can write this number as the sum of two subsets. In one of these subsets, the sock survives at temperature *a*, is destroyed at *b*, and survives at *c*, which we write as $n(a_+,b_-,c_+)$. In the second subset the sock survives at *a*, is destroyed at *b*, and is destroyed at *c*, which we write as $n(a_+,b_-,c_-)$. Put simply, any sock which is counted in either one of these subsets will also count towards the set $n(a_+,b_-)$. Similarly, $n(b_+,c_-)$ is the sum of $n(a_+,b_+,c_-)$ and $n(a_-,b_+,c_-)$. If we now add these two results together, we get:

$$n(a_+,b_-) + n(b_+,c_-) = n(a_+,b_-,c_+) + n(a_+,b_-,c_-) + n(a_+,b_+,c_-) + n(a_-,b_+,c_-).$$

But the sum of the subsets $n(a_+,b_-,c_-)$ and $n(a_+,b_+,c_-)$ is simply the number $n(a_+,c_-)$. We conclude from this that the sum of $n(a_+,b_-)$ and $n(b_+,c_-)$ must be greater than or equal to the number $n(a_+,c_-)$.[8]

We sit back with a sense of satisfaction. But then we notice the error in our logic. Of course, if the sock is destroyed by washing at temperature *b* then it is simply not available for a further test at temperature *c*. And if it

[7] This derivation is based on an example originally used by Bell in 1981.

[8] We say greater than or equal to because the subsets 'left over' – $n(a_+,b_-,c_+)$ and $n(a_-,b_+,c_-)$ – might or might not both be zero.

survives washing at temperatures a or b then it might not necessarily give the result at c that could be expected of a brand new sock.

Hmm...

But then we remember that Bertlmann's socks always come in *pairs*. We assume that, apart from differences in colour, the physical characteristics and behaviour of each sock in a pair are identical. A test performed on the left sock (sock A) can be used to predict what the result of the same test would be if it was performed on the right sock (sock B), even though we might actually perform a different test on B. We must further assume that *whatever test we choose to perform on A in no way affects the outcome of any other test we might perform on B*. But this seems so obviously valid that it's hardly worth a second thought.

We now perform three experiments on three samples containing the same total number of pairs of socks. In the first experiment, for each pair, sock A is washed at temperature a and sock B is washed at b. The number of pairs of socks for which A survives and B is destroyed is denoted $N_{+-}(a,b)$. This must be equal to the purely hypothetical number of individual socks which survive at a and are destroyed at b. In other words, although it is sock B that we have washed at temperature b, we assume that we would have got the same result if we had somehow been able to do this test on sock A. Consequently, $N_{+-}(a,b)$ must be equal to $n(a_+,b_-)$.

In the second experiment, for each pair, sock A is washed at temperature b and sock B is washed at c. We use the same kind of reasoning to deduce that $N_{+-}(b,c)$ is equal to $n(b_+,c_-)$. Finally, in our third experiment, for each pair, sock A is washed at temperature a and sock B is washed at c, for which it follows that $N_{+-}(a,c)$ must be equal to $n(a_+,c_-)$.

It is now clear where this is leading. We can make use of the result we obtained above and conclude that the sum of $N_{+-}(a,b)$ and $N_{+-}(b,c)$ must be greater than or equal to $N_{+-}(a,c)$.

We can now generalize this result for any collection of pairs of socks. By dividing each number by the total number of pairs (which was the same for each experiment) we can determine the relative frequencies with which each joint result was obtained. We can identify these relative frequencies with probabilities for obtaining these results in future experiments yet to be performed. We are led to the inescapable conclusion that the sum of the probabilities $P_{+-}(a,b)$ and $P_{+-}(b,c)$ *must be greater than or equal to* the probability $P_{+-}(a,c)$. This is one form of Bell's inequality.

This result has nothing whatsoever to do with quantum physics, the Copenhagen interpretation, or hidden variables. It is a consequence of the simple fact that any pair of socks producing an up-down (+−) result for the combination (a,c) must in principle have been capable of producing an up-down result either for (a,b) or (b,c).

Now follow the above arguments through once more, replacing socks with hydrogen atoms, pairs of socks with pairs of correlated atoms, washing machines with magnets, and temperatures with magnet orientations, and we will arrive again at Bell's inequality.

So, what's the big deal? The big deal is that the conventional quantum theoretical predictions for the probabilities are given by relations of the kind $P_{+-}(a,b) = \frac{1}{2}\cos^2(\varphi/2)$, where a and b are now orientations of the magnets and the angle $\varphi = (b - a)$. We are entirely free to choose any three orientations of the magnets we like. If we set $a = 0°$, $b = 135°$, and $c = 270°$ we find that Bell's inequality insists that $\frac{1}{2}\cos^2(67\frac{1}{2}°) + \frac{1}{2}\cos^2(67\frac{1}{2}°)$ must be greater than or equal to $\frac{1}{2}\cos^2(135°)$.

This seems reasonable. Until we do the maths and realize that this implies that 0.146 (a combined probability of 14.6 per cent) must be greater than or equal to 0.250 (25 per cent).

The conclusion is inescapable. Quantum theory predicts results that violate Bell's inequality.

The most important assumption we made in the reasoning which led to the inequality was that the hidden variables which determine the spin orientations of the atoms are locally real. We assumed that it is possible to carry out measurements on atom A without in any way disturbing atom B. This result is therefore quite independent of the nature of the local hidden variable theory itself. Bell concluded that it is possible to find measurement configurations for which *quantum theory is incompatible with any local hidden variable theory and hence local reality*. This is Bell's theorem. 'If the [hidden variable] extension is local it will not agree with quantum mechanics, and if it agrees with quantum mechanics it will not be local. This is what the theorem says.'

He concluded his 1964 paper as follows:

16. David Gross.

17. Frank Wilczek.

18. David Politzer.

Setting out to prove that local gauge field theories could not be asymptotically free, in 1973 American theorists Gross and Wilczek found the exact opposite. Harvard graduate student David Politzer proved the same result.

19. English physicist Peter Higgs developed a mechanism for spontaneous symmetry-breaking in quantum field theories in 1964.

20. In 1971 Dutch theorist Martinus Veltman told his young graduate student Gerard 't Hooft that '... we must have ... at least one renormalisable theory with massive charged vector bosons.'

21. 'I can do that,' said 't Hooft, pictured here on the left with Chen Ning Yang.

22. In experiments culminating in the 'November revolution' of 1974, Samuel Chao Chung Ting's group (pictured) found evidence for a new particle, which they called the 'J', a meson formed from a charm-anti-charm pair.

23. On 11 November 1974 Ting and Burton Richter (pictured) realised that their laboratories had simultaneously discovered the same particle, the J/Ψ.

24. Accelerator theorist Simon van der Meer (pictured here on the extreme right) experimented with a technique he called stochastic cooling. Carlo Rubbia (extreme left) and his team at CERN used the technique to reveal the W and Z particles in 1982/83.

25. Influenced by Einstein, American theorist David Bohm developed a variation of the Einstein, Podolsky, Rosen argument that brought the experiment a step closer to practical realisation.

26. Irish physicist John Bell derived a general result – called Bell's inequality – which shows that the predictions of quantum theory are incompatible with those of all classes of local hidden variable theories. With this result, a direct test of the nature of reality at the quantum level became possible.

27. French physicist Alain Aspect and his team published experimental results in 1981 and 1982 which showed that quantum reality is non-local.

28. British physicist Antony Leggett derived a new inequality related to a broad class of 'crypto' non-local hidden variable theories. The inequality provided a direct test not just of locality, but of realism too.

29. Leggett's inequality was put to the test by Markus Aspelmeyer and Anton Zeilinger (pictured) in 2006. The results suggest that the properties we ascribe to quantum objects are 'real' only in the context of measurement.

30. A discussion between John Wheeler (pictured) and Bryce DeWitt at Raleigh-Durham airport in 1965 led to the first formal equation of quantum gravity.

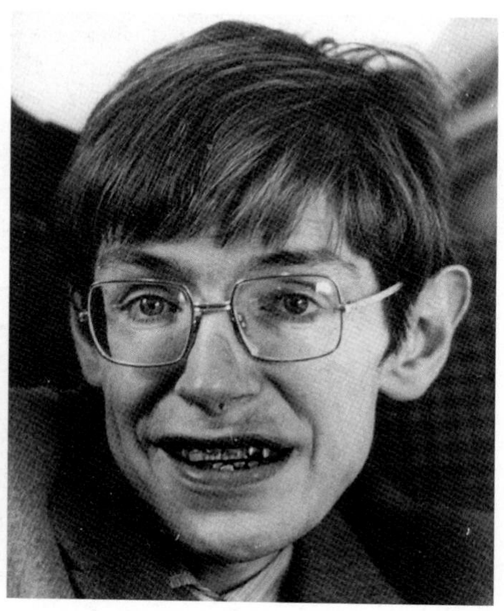

31. In 1974 English physicist Stephen Hawking applied quantum theory in the curved space-time around a black hole. He found that a black hole should consequently radiate energy.

32. American physicist John Schwarz began collaborating with David Gross on superstring theory in the late 1960s. With French theorist Joël Scherk, he found that superstring theory predicted a particle with the properties of the graviton.

33. As superstring theory began to dominate theoretical physics, in the late 1980s Italian physicist Carlo Rovelli (pictured left) and American Lee Smolin revitalised the canonical approach through their work on loop quantum gravity.

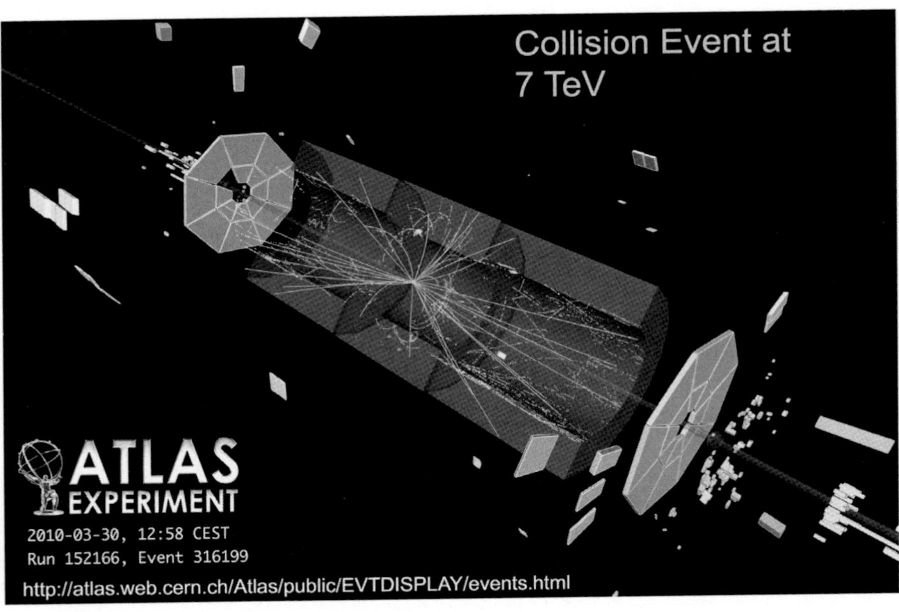

34. CERN's Large Hadron Collider (LHC) frustratingly broke down just a few weeks after it was first switched on in September 2008. The LHC was re-started over a year later and produced the first 7 TeV proton-proton collisions on 30 March 2010.

> In a theory in which parameters are added to quantum mechanics to determine the results of individual measurements, without changing the statistical predictions, there must be a mechanism whereby the setting of one measuring device can influence the reading of another instrument, however remote. Moreover, the signal involved must propagate instantaneously, so that such a theory could not be [consistent with special relativity].

Unless, of course, as Einstein, Podolsky, and Rosen had asserted in 1935, quantum theory is incomplete and reality is local. In this case Bell's inequality provides a straightforward test. If experiments like the ones described could actually be carried out in the laboratory, the results might allow us to determine which was correct: quantum theory or an extension based on local hidden variables.

32

The Aspect Experiments

Paris, September 1982

The work of Bohm and Aharonov in 1957 and of Bell in 1964 left physicists confronted with a tantalizing question. The possibility of actually carrying out experiments on correlated pairs of quantum particles now seemed very real. Bell's theorem suggested that it should be possible to perform experiments which would either prove that reality at the quantum level is inherently non-local and 'spooky', or that quantum theory is in some way demonstrably incomplete. Either way, what might have been dismissed as so much pointless philosophizing had now been turned into a direct experimental test.

The correlated particles did not necessarily need to be atoms. In 1946 John Wheeler at Princeton University had proposed studies on pairs of photons produced by electron–positron annihilation. Instead of spin orientations, the directions of polarization of the photons were to be measured.[1] *Experiments of this kind had actually been carried out in 1949 by Chien-Shiung Wu and Irving Shaknov, and these had confirmed that the photons so produced were correlated, and hence 'entangled'. Bohm and Aharonov believed that these experiments demonstrated polarization correlations as predicted by quantum theory, but they did not provide a direct test based on Bell's inequality.*

[1] In fact, the origin of light polarization lies in the spin properties of photons. Photons are bosons with spin quantum number 1. In principle this gives rise to three different magnetic spin quantum numbers and hence three possible spin orientations. However, one of these, corresponding to the magnetic spin quantum number $m_s = 0$, is forbidden in relativistic quantum theory for particles which travel at the speed of light. The other two, with $m_s = +1$ and $m_s = -1$, correspond to left and right circularly polarized light, respectively.

But there were other sources of entangled photons that were actually easier to generate in laboratory experiments. Photons emitted in rapid succession from electronically excited calcium atoms proved to be one of the most attractive sources of correlated quantum particles. Carl Kocher and Eugene Commins at the University of California at Berkeley had used this source in 1966, although they, too, had not set out explicitly to test Bell's inequality.

The first such direct tests were performed in 1972, by Stuart Freedman and John Clauser at Berkeley using an extension of the Kocher–Commins experimental design. These experiments produced the violations of Bell's inequality predicted by quantum theory but, because of the need to extrapolate the data and make some further assumptions, they left unsatisfactory 'loopholes'. It could be argued that the case against local hidden variables was still not yet proven.

In the meantime, particle physicists continued to dig deeper into the nucleus. They were largely unconcerned with what quantum theory might ultimately have to say about the nature of the reality that they were now revealing. The deep inelastic scattering experiments in 1968, the observation of weak neutral currents in 1973, and the November revolution in 1974 all conspired to put quarks and the intermediate vector bosons firmly on the particle map. By the end of the 1970s, the Standard Model was accepted as the theory of all the forces of nature except gravity.

Despite these successes, a small international community of physicists continued to focus attention on quantum theory's foundations and its philosophical problems. There were further reports of experimental results on entangled pairs of photons, but the first comprehensive experiments designed specifically to test a generalized form of Bell's inequality were those performed in the early 1980s by Alain Aspect and his colleagues Philippe Grangier, Gérard Roger, and Jean Dalibard, at the Institute for Theoretical and Applied Optics at the University of Paris in Orsay.

French physicist Alain Aspect had studied the philosophical problems of quantum theory and the Einstein, Podolsky, Rosen *gedankenexperiment* in the early 1970s, whilst doing three years' voluntary service in Cameroon. He was strongly influenced by Bell's papers. He concluded that the experimental tests performed up to that time had fallen short of the ideal, and he set himself the challenge of perfecting an apparatus that would provide the ultimate test of Bell's inequality.

Specifically, he wanted to build an apparatus that would allow him to change the orientation of the measurement and detection devices *after* the entangled particles had been created and were 'on their way' to their respective detectors. Bohm and Aharonov had suggested this possibility in 1957, and Bell had restated the importance of such a test in his 1964 paper.

This, Aspect decided, would be the subject of his 'thèse d'état'. He persuaded Christian Imbert, a young professor at the Institute for Theoretical and Applied Optics in Paris, to act as his thesis advisor, and proceeded to seek funding and assemble equipment in the basement of the Institute.

Imbert recommended that he talk to Bell, and in 1975 Aspect travelled to CERN in Geneva to discuss his proposed experiments. He nervously explained what he was intending to do, as Bell listened in silence.

'Are you tenured?' Bell finally asked, concerned that experiments designed to answer esoteric questions concerning the nature of reality at the quantum level were hardly the stuff on which solid academic careers could be built.

Aspect explained that he was still only a graduate student. He had yet to secure his doctorate, although he held a permanent position at the Institute.

'You must be a very courageous graduate student…,' Bell replied.

Bell agreed that the experiments that Aspect was proposing would represent a substantial improvement on previous studies, and offered his encouragement.

Aspect published details of his proposed experiments in the journal *Physical Review* in October 1976. He settled on excited calcium atoms as the source of entangled photons. In the lowest energy 'ground' electronic state of the calcium atom, the outermost atomic orbit is spherical and filled with two spin-paired electrons. If one of these electrons absorbs a photon of the right wavelength, then the electron may be promoted to a higher-energy, dumbbell-shaped orbit.[2] In this process, the photon that is absorbed imparts a quantum of angular momentum to the electron in the atom, and this appears as orbital angular momentum of the excited electron.[3]

[2] The atomic orbitals in question are 4s and 4p, respectively.

[3] This additional angular momentum cannot appear in the spin of the electron, since this is fixed at $s = ½$.

Now suppose that it is possible to excite a second electron (the one 'left behind' in the ground-state orbit) also into this same dumbbell-shaped orbit, but in a way that maintains the alignment of the electron spins. In other words, we create a doubly excited state in which the electron spins remain paired. This doubly excited state undergoes a rapid cascade emission, producing two photons in quick succession as the electrons return to the ground-state orbit. To conserve angular momentum in this process, the two photons must be emitted in opposite states of circular polarization. The photons are entangled, in precisely the way that Einstein, Podolsky, and Rosen and, subsequently, Bohm and Bell had anticipated.

In fact, the two emitted photons have wavelengths in the visible region. One photon is green, the second is blue.

Aspect and his colleagues would use two high-power lasers to produce the excited calcium atoms, which were formed in an atomic 'beam', produced by passing gaseous calcium from a high-temperature oven through a tiny hole into a vacuum chamber. Subsequent collimation of the atoms entering the sample chamber would provide a well-defined beam of atoms. The low density of atoms at the point of intersection with the laser beams would ensure that the calcium atoms did not collide with each other or with the walls of the chamber before absorbing and subsequently emitting photons.

The physicists would monitor the light emitted in opposite directions from the atomic beam source, using coloured filters to isolate the green photons (which we will call photons A) on the left and the blue photons (photons B) on the right. The photons would then be passed into an arrangement consisting of two polarization analysers, four photomultipliers to amplify the signals from the detected photons, and electronic devices designed to detect and record coincident signals from the photomultipliers.

The analysers transmitted light polarized parallel to the plane of incidence (vertical polarization), and reflected light polarized perpendicular to this plane (horizontal polarization).[4] Detecting these linear polarization states is more straightforward than detecting circular polarization

[4] Polaroid polarized sunglasses reduce glare by filtering scattered light with random polarization to produce linearly polarized light.

states and changing from one to the other does not affect the basis of the correlation between photons A and B. The correlation is such that if the optical axes of both analysers are aligned in the same direction, then if photon A is detected in a state of vertical polarization, photon B must also be detected in a state of vertical polarization.[5]

Each polarization analyser was to be mounted on a platform which allowed it to be rotated about its optical axis. Experiments could therefore be performed for different relative orientations of the two analysers, which would be placed about 13 metres apart. The electronics would be set to look for coincidences in the arrival and detection of the photons A and B within a time window just 20 billionths of a second wide. Any kind of 'spooky' signal passed between the photons, 'informing' photon B of the fate of photon A, for example, would therefore need to travel the 13 metres between the detectors within this narrow time window. In fact, it would take about twice this amount of time for a signal moving at the speed of light to cover this distance.

The practical realization of these experiments took another five years. Aspect, Grangier, and Roger published their first definitive results in the journal *Physical Review Letters* in August 1981.

The physicists performed four sets of measurements with four different orientations of the two analysers. This allowed them to test a generalized form of Bell's inequality.[6] For the specific combination of analyser orientations chosen, the generalized form of the inequality demands a value less than or equal to 2. Quantum theory predicts a value of 2.828.[7] The

[5] This is a little different from the situation where we measured the spin orientations of entangled atoms, which were correlated up–down. It results from the simple fact that left and right circular polarization is defined for photons moving *towards* their respective detectors. If photon A moves to the left in a state of left circular polarization, then photon B must move towards the right in a state of right circular polarization. But the right-hand detector sees right circular (clockwise) rotation as left circular (counter-clockwise) rotation. This translates into the expectation that the photons will ultimately be detected either as both vertically or both horizontally polarized.

[6] This was developed in 1969 by John Clauser, Michael Horne, Abner Shimony, and Richard Holt. It involves extension to a fourth arrangement of the analysers but does not depend on the assumption of 'perfect' correlation between the entangled particles. It is therefore valid for the situation in which the experimental instruments are 'imperfect'.

[7] The quantum theory prediction is actually 2 multiplied by the square-root of 2.

physicists obtained the result 2.697 ± 0.015, a violation of Bell's inequality by 83 per cent of the maximum possible predicted by quantum theory.

The quantum theory prediction of 2.828 assumes that the experimental apparatus performs 'perfectly'. But real experimental apparatus is imperfect. Not all the photons could be detected. The analysers would occasionally 'leak', with a few vertically polarized photons being reflected and horizontally polarized photons transmitted. All these instrumental deficiencies serve to reduce the measured extent of the correlation below the quantum theory predictions. If we think of entanglement as a kind of quantum interference between distant particles, then the practical limitations of experimental instruments in the real world tend to reduce the extent of quantum interference that can be observed. Physicists refer to this as a reduced 'visibility' of the full extent of quantum entanglement.

By taking the instrumental deficiencies into account, the physicists obtained a modified quantum theory prediction of 2.70 ± 0.05, in excellent agreement with experiment. And well in excess of the limit set by Bell's inequality.

The physicists concluded:

> ...our results, in excellent agreement with quantum mechanical predictions, are to a high statistical accuracy a strong evidence against the whole class of realistic local theories; furthermore, no effect of the distance between measurements on the correlations was observed.

These results appeared to confirm that the photons remain mysteriously bound to one another, sharing a single wavefunction, until the moment of measurement. At that moment the wavefunction collapses and the photons are 'localized' in polarization states that are correlated to an extent that simply cannot be accounted for in any local hidden variable theory. Measuring the polarization of photon A does affect the result that will be obtained for photon B, and *vice versa*, even though the photons are so far apart that any communication between them would have to travel faster than the speed of light. Nature, it seems, is inherently non-local, and 'spooky'.

Instead of ending the matter decisively, the Aspect experiments were sufficiently perturbing and provocative to encourage even more speculation and debate. Diehard proponents of local hidden variable theories could point to the fact that the polarization analysers in these first Aspect experiments were set in position *before* the correlated photons were emitted by the calcium atoms. Could it be that the photons were somehow influenced in advance by the way the apparatus was set up? Though seemingly improbable, this is not impossible. This is referred to as the 'locality loophole'.

But Aspect had intended to meet this challenge from the very beginning, and with his colleagues Jean Dalibard and Gérard Roger he sought to close this loophole just a year later. The physicists modified their original experimental arrangement to include devices which could switch the paths of the photons, directing each of them towards two differently orientated analysers. This prevented the photons from 'knowing' in advance along which path they would be travelling, and hence through which analyser they would eventually pass. The results were equivalent to changing the relative orientations of the two analysers *while the photons were in flight.*

The physicists obtained the result 2.404 ± 0.080, once again in clear violation of the generalized form of Bell's inequality. Taking account of instrument deficiencies allowed them to obtain a modified quantum theory prediction of 2.448, once again in excellent agreement with experiment.[8] They submitted a paper on their results to *Physical Review Letters* in September 1982. It was published in December.

Ever more sophisticated experiments performed since have not altered the basic conclusions of the Aspect experiments. Much more efficient sources of entangled photons were subsequently obtained from something called type-I and type-II parametric down-conversion. This involves the conversion of photons from an intense source, such as a laser, into pairs of photons of longer wavelength by passing them through certain types of crystalline materials. The polarization states of such photons are entangled and, because they are generated in fixed spatial directions, no photon pairs are 'lost' as a result of being emitted in the 'wrong' directions.

Using the calcium atom source, Aspect and his colleagues were able to measure up to about 40 detection coincidences per second, each

[8] It could still be argued that the switching between the photon paths was not completely random. The locality loophole was closed decisively in experiments reported in December 1998 by Anton Zeilinger and his colleagues at the University of Innsbruck.

corresponding to detection of a correlated photon pair. Using type-II parametric down-conversion, coincidence rates as high as 360,800 per second have been recorded. In experiments using this method reported in 1998, a result for the generalized form of Bell's inequality of 2.6979 ± 0.0034 was obtained.

In August 1998 a research group from the University of Geneva reported the results of experiments demonstrating a clear violation of the generalized form of Bell's inequality using entangled photons detected at observer stations located in Bellevue and Bernex, two small Swiss villages outside Geneva almost 11 kilometres apart.[9] From these experiments it is possible to estimate that any 'spooky' action at a distance would need to propagate from one detector to another with a speed at least 20,000 times that of light. In more recent experiments, entangled photons have been measured at La Palma and Tenerife in the Canary Islands, at separation distances of 144 kilometres.[10]

A further loophole, called the 'efficiency loophole', was closed in 2001. In all the experiments described above the numbers of photon pairs detected were considerably smaller than the numbers of pairs generated. Why does this represent a loophole? Because if only a very small proportion of the pairs that are generated are ultimately detected, then we are required to assume that the detected pairs represent a true statistical sample (called a 'fair' sample) of the properties and behaviour of the total number of pairs of particles.

With more than a little ingenuity, it is possible to devise a local hidden variable theory which, because of data rejection, predicts the same measurement outcomes as quantum theory. Although such theories demand what seems like some kind of grand conspiracy on the part of nature, physicists were determined to close the efficiency loophole as well.

In February 2001, physicists from the National Institute of Standards and Technology in Boulder, Colorado, and the Department of Physics at the University of Michigan published the results of experiments they had

[9] In fact, these photons were entangled in energy and time, mirroring the position–momentum entanglement originally envisaged by Einstein, Podolsky, and Rosen.

[10] In May 2009, reports appeared confirming that entangled photons had been bounced off an orbiting satellite back to a receiver station on earth.

conducted on entangled states created from positively charged beryllium ions. When they tested the generalized form of Bell's inequality, they obtained the result 2.25 ± 0.03.

In these experiments it was not necessary to assume the fair sampling hypothesis: *all* the entangled ion pairs created were detected and contributed to the measured results. To salvage local hidden variable theories from this experimental evidence we would need to invoke a very grand conspiracy indeed, one that somehow exploits both locality and efficiency loopholes at the same time, but not independently.

This brings to mind another of Einstein's famous quotations: 'The Lord is subtle, but he is not malicious.'[11]

In case there should be any remaining doubts about the validity of Bell's theorem and Bell's inequality, in 2000 Anton Zeilinger and his team at the University of Innsbruck created entangled *triplets* of photons. These three-photon entangled states are known as Greenberger–Horne–Zeilinger (GHZ) states, after physicists Daniel Greenberger, Michael Horne, and Zeilinger.

The Innsbruck physicists made measurements of various combinations of linear and circular polarization states of the photons. The GHZ states have properties such that the polarization combinations predicted by local hidden variable theories are the exact opposite of the combinations predicted by quantum theory. The experimental results confirmed the predictions of quantum theory, once again ruling out all classes of locally real theories.

It seems therefore that we can establish the 'reality' of the state of one quantum particle by choosing what kind of measurement to perform on another, no matter how far apart these particles are at the time of measurement. In 1935, Einstein, Podolsky, and Rosen were convinced that invoking such 'spooky' action at a distance is unnecessary and that quantum theory is incomplete, though they were not explicit about the way in which the theory might be 'completed'.

These experimental tests demonstrate that the very basis of the Einstein, Podolsky, Rosen argument is unfounded. Although the precise

[11] The phrase 'Raffiniert ist der Herr Gott. Aber Boshaft ist Er Nicht' is carved in stone above the fireplace in a room in Fine Hall, Princeton University, in memory of Einstein.

nature of the 'spookiness' remains a subject for debate, the simple fact is that any theory which attempts to describe physical reality as local is doomed to fail.

Reality at the quantum level is determinedly non-local.

The reality advocated by the proponents of hidden variable theories does not have to be a local reality, as was demonstrated by Bohm in 1952. It is nevertheless clear that these experiments leave the realist with a lot of explaining to do. It seems we have no alternative but to change radically the way we think about reality and abandon our natural, but naïve, presumptions. Aspect agreed:

> But of course we already know that, since we knew quantum mechanics looks like a good theory, and that quantum mechanics is not compatible with the naïve image of reality. But here we have shown that in this kind of very unusual situation quantum mechanics works very well, and so this must convince us that truly we must change the old picture of the world.

But these experiments did not spell the end of the story. Commenting on the Aspect experiments in 1985, Bell observed:

> It is a very important experiment, and perhaps it marks the point where one should stop and think for a time, but I certainly hope it is not the end. I think that the probing of what quantum mechanics means must continue, and in fact it will continue, whether we agree or not that it is worth while, because many people are sufficiently fascinated and perturbed by this that it will go on.

Bell was to be proved right. Now fully awake to the evidence threatening to undermine completely our attempts to understand reality at the quantum level, more and more physicists turned to questions concerning fundamental quantum theory. The theory would be tested in ways that could not have been dreamt of during the great debate between Bohr and Einstein, the debate which had established the supremacy of the Copenhagen orthodoxy over all other rivals.

33

The Quantum Eraser

Baltimore, January 1999

CERN physicists reported the discovery of the W and Z bosons in 1983 and pushed hard to find the top quark. However, there did not appear to be much doubt that the few remaining matter particles required to complete the three generations of the Standard Model and the particles carrying forces between them would be found. It was only a matter of time.

But the noise from the community of physicists concerned with the foundations of quantum theory was growing. This was noise about the very meaning of the theory on which the Standard Model was constructed. Aspect and his colleagues had shown that quantum reality is non-local, but what did this mean? Was it possible to probe more deeply into the essence of quantum theory's interpretation, perhaps to penetrate to the very meaning of complementarity itself?

In his classic Lectures on Physics, *first published in 1965, Richard Feynman had introduced a hypothetical variant of the famous two-slit interference experiment, in this case using electrons instead of light. By introducing a light source behind the two slits he proposed to observe through which slit a particular electron had passed on its way to a detector. This is directly analogous to one of the early* gedankenexperiments *that Einstein had used in his great debate with Bohr.*[1]

In Feynman's version, a flash of light scattered by the electron as it emerges from one or other of the two slits reveals through which slit it has passed. Feynman goes on to conclude that the very act of trying to gain this kind of 'which-way' information must

[1] See Chapter 13.

prevent us from observing interference effects. And, as in Bohr's initial defence of quantum theory in the face of Einstein's early challenges, he cites the 'clumsiness' of the measuring instrument combined with the uncertainty principle as the mechanism by which we are prevented from observing simultaneous particle-like (which-way) and wave-like (interference) behaviour. Feynman wrote:

> *If an apparatus is capable of determining which [slit] the electron goes through, it cannot be so delicate that it does not disturb the [interference] pattern in an essential way. No one has ever found (or even thought of) a way around the uncertainty principle. So we must assume that it describes a basic characteristic of nature... if a way to 'beat' the uncertainty principle were ever discovered, quantum mechanics would give inconsistent results and would have to be discarded as a valid theory of nature.*

But, although Bohr had initially resorted to a defence based on the clumsiness of the measurement, he was eventually forced to abandon this approach in the face of EPR's 'bolt from the blue'. He had to do this for the simple reason that such a defence requires an almost classical realist conception of the initial measurement interaction, a conception entirely consistent with EPR's 'reasonable' definition.

Bohr had no option but to deny the validity of this definition and he therefore shifted his defence to a much more subtle (and rather vague) argument to the effect that the complementary wave–particle nature of quantum particles essentially precludes simultaneous observation of wave-like and particle-like behaviour. This incompatibility, he claimed, derives from the limits placed on our ability to acquire deeper knowledge of the quantum world using classical-scale measuring instruments.

From Bohr's shifting position we can infer that complementarity *prevents us from watching the electrons in the way that Feynman proposed, not clumsiness of the kind that he associated with the uncertainty principle. It suggests that if we could ever find a way of 'beating' the uncertainty principle as Feynman had speculated, then the results of any such experiments would still remain consistent with quantum theory's predictions.*

The essential duality of wave and particle behaviour—complementarity—would ensure this. This duality is what Feynman described as the central mystery at the heart of quantum theory.

The prospect of 'beating' the uncertainty principle (strictly speaking, 'beating' the clumsiness constraint) in the way Feynman described seems very slight, if not negligible. At the level of individual quantum interactions it would appear that we will always be defeated by the essential parity in scale and energy between the initial interaction required for measurement and the measured object itself. To make the kinds of measurements that Feynman believed were impossible, we would need to probe a quantum system in a way that does not impart to it what we might refer to simply as a classical 'kick', an uncontrollable transfer of momentum in the classical sense.

In 1982 theorists Marlan Scully and Kai Drühl, at the Max Planck Institute for Quantum Optics near Munich and the Institute for Modern Optics at the University of New Mexico, believed they had found a way.

They conceived a thought experiment involving the interference of photons emitted from two different atoms. The atoms act as distinct sources of light in much the same way as the two slits in the classic Young experiment act as sources of waves that expand beyond the slits, going on to overlap and interfere. We will call these atoms A and B.

In this thought experiment, each atom is irradiated with a pulse of laser light and an electron is excited to a higher-energy orbit. In the situation where both atoms subsequently emit a photon and return directly to the ground state, the photons cannot be distinguished and we have no way of knowing which photon has come from which atom. In other words, we cannot gain which-way information and we would therefore anticipate that the photons should exhibit an interference pattern. Scully and Drühl showed that this is, indeed, what quantum theory predicts: terms appear in the equations characteristic of interference effects.

Now suppose that there is an alternative way in which the excited atoms can decay, perhaps via an intermediate electron orbit lower in energy than the initial excited orbit created by the laser pulse, but higher in energy than the ground state. We arrange the experiment so that we detect only photons derived from the transition from the initial excited state to the intermediate state. At first sight, this looks no different to the situation we considered above, and we might therefore be tempted to conclude that the interference pattern should remain. Observation of interference would be behaviour characteristic of the wave description.

But it seems that we can now also gather which-way information characteristic of the particle description. By discovering whether the atoms are in the ground or intermediate states *after* emission of the photons, we can in principle tell which atom the detected photon came from. For example, if we find that atom A is in the intermediate state and atom B is in the ground state, we know immediately that the photon came from atom A and we therefore know its trajectory from atom A to the detector (let's call this path A).

This gives us which-way information, equivalent in Feynman's version of the two-slit experiment to discovering which slit the electron went through. And as we can do this without in any way disturbing the photons after they have been emitted, it really does seem as though we can measure both the wave-like and particle-like behaviours of the photons simultaneously.

The Copenhagen interpretation categorically denies that we can do this. So, if we cannot invoke the 'clumsiness' defence, what is it in this experiment that prevents us from observing both particle-like and wave-like behaviour simultaneously? How, we might wonder, would Bohr have responded to such a challenge? Commenting on a different (though entirely equivalent) thought experiment almost ten years later, Scully

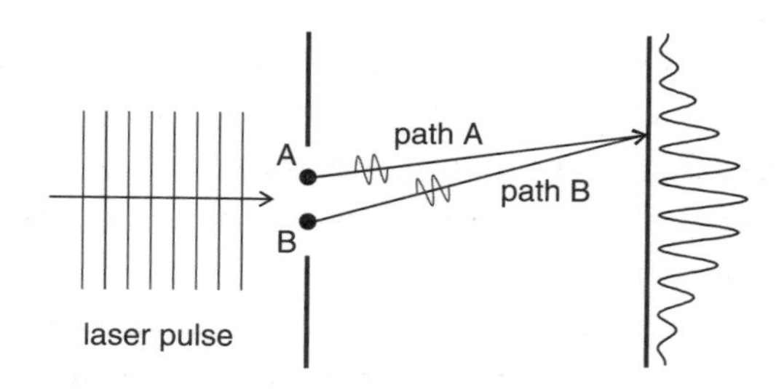

FIG 24 Scully and Drühl replaced the two slits of the classic Young's experiment with two atoms, A and B. The atoms are irradiated with a pulse of laser light and an electron in one or other is, with equal probability, excited to a higher-energy orbit. The excited atom subsequently emits light and the emitted photon goes on to be detected. As both atoms may absorb and emit light with equal probability, the emitted photons from both atoms should produce an interference pattern, characteristic of wave behaviour. However, if the emission of a photon leaves the atom in an intermediate state, it may be possible to discover which of the two atoms is left in this state and therefore which atom emitted the photon. This gives 'which way' information characteristic of particles.

and his colleagues Berthold-Georg Englert and Herbert Walther at the Max Planck Institute for Quantum Optics and the University of Munich observed that:

> ...Bohr would not have been distressed by the outcome of these considerations, as the wave-like (interference) phenomenon is lost as soon as one is able to tell which path the [photon] traversed. Quantum mechanics contains a built-in safeguard such that the loss of coherence in measurements on quantum systems can always be traced to correlations between the measuring apparatus and the system being observed.

The safeguard these physicists refer to arises when the entire system is considered from the perspective of quantum theory. Even though the photons are left to propagate towards the detector unaffected by a 'clumsy' measurement, they have become inexorably entangled with the tell-tale quantum states of the atoms they have left behind. It is this entanglement that destroys the interference.

By introducing the final quantum states of the atoms into the equation, the terms responsible for wave interference disappear.[2] Simple mathematics demonstrates that we can't have it both ways. If we can discover which-way information—even in principle—the interference terms disappear and we are denied the possibility of observing interference effects. If we are prevented from discovering which atom the photon came from, then the interference terms are preserved but we can obviously no longer gain which-way information.

If we force the photon to reveal which path it followed or which slit it went through, quantum theory denies us the possibility of observing interference effects through a mechanism that has nothing whatsoever to do with the clumsiness of the measurement. If wave–particle duality is the central mystery at the heart of quantum theory, then complementarity—not the uncertainty principle—is the mechanism by which it operates.

[2] The interference terms are multiplied by the 'overlap integrals' of the final quantum states of the two atoms, and as these integrals are zero the terms disappear. See Scully and Drühl, *Physical Review A*, **25**, 1982, p. 2209.

And, perhaps for the first time, we have here a basis from which we can begin to understand how complementarity actually *works*.

But now here's another thought. What if we set up the experiment in a way which gives us the opportunity to gain which-way information, but we *choose not to look*? By choosing not to look, can we expect the interference pattern to be restored? What if we wait until the photons have passed through the apparatus and have been detected and *then* choose whether or not we want to look to see which way they had gone? Is it really possible that we can switch the interference pattern on and off by choosing whether to look or not *after* the photons have been detected?

The answers to these questions are quite sophisticated. According to Scully and Drühl's original analysis of this hypothetical quantum 'eraser' experiment, the interference pattern does indeed come back if we choose not to look, but the mechanism by which it returns is very subtle.

In the situation where the atoms possess an intermediate electron orbit, we saw that this gives us the possibility of discovering which-way information just by looking to see which state the atom was in after emission of a photon. But now instead of looking to see what final state each atom is in, we hit the atoms with another pulse of laser light which excites any atom present in the intermediate state to another, higher-energy state. This decays rapidly back to the ground state with the emission of another photon, which we will call a ϕ-photon to distinguish it from the photons that go on to interfere (or not). By doing this we lose any opportunity to measure the final states of the atoms and so we erase the which-way information from the quantum system.

The interference pattern does not (yet) come back.

To see the interference pattern we must look closely at the ϕ-photons. Suppose we trap these in elliptically shaped optical cavities, one with atom A at one of its focal points and the second with atom B at one of its focal points. The cavity material allows the laser light and the interference photons to pass through unhindered. The cavities are joined together at their second focal points by material designed to detect a ϕ-photon and this detector is isolated by means of a mechanical shutter.

So, what happens now? Let's assume that the first laser pulse excites both atoms A and B. One of the atoms decays to the intermediate state, emitting a photon which goes on to be detected. The second atom decays directly to the ground state. We now excite both atoms using a second laser pulse, erasing the information about their final states and leaving a ϕ-photon in one of the optical cavities. We have no way of knowing which atom has emitted the ϕ-photon and so we have no way of knowing which cavity the ϕ-photon is in.

The correct quantum-theoretical description of this situation is in terms of two 'partial waves', one in each cavity, representing a 50:50 chance of finding the ϕ-photon in one or the other.

We wait until the interference photon has been detected and open the mechanical shutter. The ϕ-photon partial waves are reflected by the cavity walls towards the detector. The partial waves now combine. If they combine constructively at the point where we have placed the detector, we expect to record detection of a single ϕ-photon. Alternatively, if the partial waves combine destructively, the ϕ-photon goes undetected. The probability of detecting the ϕ-photon and not detecting it is again 50:50.

Of course, by opening the shutters we have denied ourselves the possibility of finding out which cavity the ϕ-photon was in and so gaining which-way information. The interference pattern does indeed return, but to see it we must correlate detection of the interference photon with detection (or non-detection) of the ϕ-photon.

Assume we could in some way code each spot recorded by detection of an interference photon. If the spot is produced by a photon which is correlated with detection of a ϕ-photon we code it red. If the spot is produced by a photon which is correlated by non-detection of a ϕ-photon, we code it blue. If we then look at the resulting pattern of spots through different coloured filters, we would see an interference pattern formed by the red spots and a second interference pattern formed by the blue spots with one displaced from the other.

If we describe the pattern formed by the red spots as interference fringes, then the pattern formed by the blue spots are 'anti-fringes', where the peak of a fringe coincides with the trough of an anti-fringe. If we do not look at the pattern through coloured filters and therefore do not differentiate between red and blue, we cannot see any interference pattern at all (in fact, we get a scatter pattern comparable to that associated with which-way detection).

FIG 25 By coding photons that are detected coincidently with ϕ-photons as red, we recover an interference pattern that looks something like (a). Coding photons detected coincidentally with non-detection of ϕ-photons as blue yields an interference pattern given by (b). If we do not discriminate between red and blue photons we get the sum of interference fringes and anti-fringes, which yields what looks like a scatter pattern, (c).

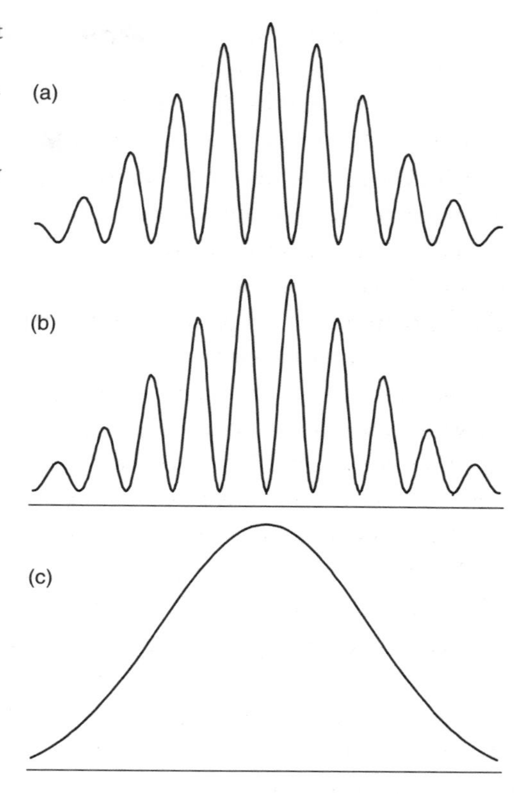

These ideas are fascinating, and quantum theory again seems to have all the answers. But can these ideas be transformed into real experiments? Recognizing the enormity of the challenge to experimentalists, Scully raised the stakes: 'Realizing quantum erasure in an experiment, however, has been difficult for many reasons (even though Scully offered a pizza for a convincing demonstration).'

The experiments were tricky, but they were done some years later. The thought experiment outlined by Scully and Drühl was judged to be too difficult. But Thomas Herzog, Paul Kwiat, Harald Wienfurter, and Anton Zeilinger at the University of Innsbruck in Austria used more familiar sources of pairs of entangled photons in an entirely analogous quantum eraser experiment. They reported their results in a paper submitted to *Physical Review Letters* in July 1995.

In fact, the physicists produced two pairs of vertically polarized photons using type-I parametric down-conversion. One pair, consisting of photons with wavelengths in the red and near-infrared regions of the spectrum, was produced in a first pass of laser light through the crystalline material. We will call this path A. The second photon pair, with identical wavelengths, was produced by reflecting the laser light back through the crystal for a second pass. We call this path B. The pair generated by the first pass was also reflected back into the crystal, with the end result that the photon pairs created by direct (first pass) or reflected (second pass) laser beams could no longer be distinguished.

The result was interference, detectable as 'fringes' in the rates of detection both the red and near-infrared photons as a function of the differences in the paths they had taken through the apparatus. In effect, the two sources of photon pairs (first pass and second pass) acted like two slits in the classical interference experiment.

With this arrangement, interference could be observed so long as no attempt was made to identify which path the photons were following, and hence identifying their source—first pass or second pass (and, by analogy, identifying which 'slit' they had passed through).

However, which-way information could be simply obtained by, for example, changing the polarization of the first-pass near-infrared photons from vertical to horizontal and placing a polarizing analyser in front of the detector. The nature of the photon polarization then revealed what kind of photon it was—first pass or second pass—and therefore what path it had taken—path A or path B. Obtaining such which-way information for the near-infrared photons implied equivalent which-way information for the red photons, too. The end result was that interference in both red and near-infrared photon detection rates disappeared.

This which-way information could then be erased simply by rotating the analyser to an angle of 45°, preventing the possibility of learning the polarization orientation of the near-infrared photons and hence identifying which path they had taken. This arrangement was sufficient to restore interference in the near-infrared photon detection rate and in the rate of red and near-infrared photon coincidences, but not the red photon detection rate.

The physicists now rotated the polarization of the first-pass red photons from vertical to horizontal and introduced an analyser which could be used to tell whether the red photons were first or second pass. This allowed the possibility of regaining which-way information for the near infrared photons, too. The interference duly disappeared.

But now, the physicists could play the same game once more. Rotating the analyser through 45° again erased the which-way information. But this time interference was not restored in the detection rates of either the red or near-infrared photons. Interference *did* return if detection of a near-infrared photon was *correlated* with detection of a red photon with the polarizer orientated at either +45° or −45°, analogous to correlation with red or blue spots in the original thought experiment. This interference was revealed in the coincidence detection rates, which formed fringes (+45°) and anti-fringes (−45°).

The physicists concluded:

> The use of mutually exclusive settings of the experimental apparatus implies the complementarity between complete path information and the occurrence of interference. In conclusion, our results corroborate Bohr's view that the whole experimental setup determines the possible experimental predictions.

History does not record if Herzog, Kwiat, Weinfurter, and Zeilinger claimed their free pizza.

In truth, the Innsbruck physicists had not quite replicated all the facets of the original thought experiment. Most importantly, they had not demonstrated the possibility of delaying the choice between observing which-way and interference phenomena until after the measurements had been made. Such a delayed-choice quantum eraser experiment was demonstrated by Scully and Yoon-Ho Kim, Rong Yu, Sergei Kulik, and Yanhua Shih at the University of Maryland in Baltimore. They submitted a paper describing their results to *Physical Review Letters* in January 1999. It was published in 2000. Their results confirmed the bizarre nature of the quantum world. The interference pattern can indeed be switched on and off by choosing whether or not to look at the which-way information *after the experiment is over.*

Further experiments reported in 2002 used a real-life two-slit arrangement to perform both quantum eraser and delayed-choice quantum eraser experiments. Quantum eraser experiments with kaons were reported in 2004.

There is of course a direct relationship between complementarity and quantum non-locality. Interference effects characteristic of waves are the direct manifestation of non-local behaviour. These effects appear in the mathematical structure of quantum entanglement—in this case entanglement of the states responsible for interference with the states used to register which-way information. It is simply not possible to design an apparatus to reveal which-way information which does not result in entanglement of this kind. And these states cannot be disentangled without forcing the system to reveal one type of behaviour or the other. They cannot be disentangled to reveal both types of behaviour simultaneously.

This has nothing to do with our ability to conceive an apparatus which avoids 'clumsy' measurements. An apparatus which simultaneously reveals both wave-like and particle-like behaviour is simply inconceivable. This is the essence of complementarity.

34

Lab Cats

Stony Brook/Delft, July 2000

The experiments designed to test Bell's inequality and to explore the nature of complementarity served only to demonstrate the extraordinarily counter-intuitive behaviour of the quantum world. There was now much experimental evidence to make the realist feel distinctly uncomfortable. And there was more discomfort yet to come.

Bohr had insisted on a clear distinction between the microscopic world of quantum particles and the macroscopic world of classical measuring devices. In the context of von Neumann's collapse postulate, it is at this boundary between the microscopic and macroscopic that we might expect to find the elusive collapse of the wavefunction. But Bohr was never clear about where this distinction should be drawn. Under Einstein's influence, Schrödinger had used his cat paradox to turn this lack of distinction into a reductio ad absurdum.

Many years later, John Bell wrote about the 'shifty split' between the measured quantum object and classical perceiving subject:

> *What exactly qualifies some physical systems to play the role of 'measurer'? Was the wavefunction of the world waiting to jump for thousands of years until a single-celled living creature appeared? Or did it have to wait a little longer, for some better qualified system... with a PhD?*

Bohr believed that the answer lay in the irreversibility of the act of measurement and John Wheeler subsequently wrote more specifically about an 'irreversible act of amplification'. We gain information about the quantum world only when we can amplify elementary quantum events, like the absorption of photons, and turn them into perceptible signals, such as the deflection of a pointer on a scale or the appearance of a visible spot on a piece of photographic film.

In 1970 the physicist Dieter Zeh noted that the interaction of a wavefunction with a measuring apparatus and its environment might lead to rapid, irreversible decoupling or 'dephasing' of its components.[1] *This 'decoherence', as it came to be known, would work in such a way that any interference terms in the superposition are destroyed and we are consequently prevented from observing interference amplified to the level of macroscopic objects. Schrödinger's cat is spared at least the discomfort of being both dead and alive because decoherence destroys the superposition long before it can be amplified to macroscopic dimensions.*

Decoherence is a kind of quantum 'friction'. The wavefunction does not collapse (or decohere) instantaneously but over a finite—if very short—time. Clearly, for individual photons, electrons, and atoms moving through the vacuum, the decoherence times are long. But when large numbers of particles are involved, as in the interaction of photons, electrons, and atoms with a macroscopic measuring apparatus, the decoherence time becomes extremely short. For all practical purposes the collapse can be assumed to be essentially instantaneous.

But, as the saying goes, there is no such thing as a free lunch. Whilst this is all intuitively appealing, decoherence cannot explain why the wavefunction decoheres only into components that will be recognized as measurement outcomes, and not as awkward and embarrassing superposition states. Nor can it be used to predict specific measurement outcomes. Decoherence changes the basis of the prediction from a superposition state consisting of 'this' and 'that' to distinct components: 'this' or 'that', but leaves us dealing with the same quantum probabilities. As Bell complained:

> *The idea that elimination of coherence, in one way or another, implies the replacement of 'and' by 'or', is a very common one among solvers of the 'measurement problem'. It has always puzzled me.*

Of course, in the 75 years since Schrödinger published details of his cat paradox, nobody has ever reported seeing a cat in any kind of superposition state. Decoherence theory suggests that such superpositions are suppressed in a kind of 'quantum censorship'. But this just begs another question.

[1] We can think of the term 'dephasing' as implying a loss of the alignment between the peaks and the troughs (the phase) of all the components of the wavefunction under measurement. Note that the *idea* of decoherence predates Zeh's important contributions.

If cats are indeed too large to sustain quantum superpositions, is it nevertheless possible to find other kinds of macroscopic objects that can? Is it possible to create laboratory analogues of Schrödinger's infamous cat?

Such considerations lead directly to the notion of macroscopic quantum states and the possibility of their superposition, entanglement, and interference. It helps to begin with some definitions.

In the early 1980s British-born physicist Tony Leggett sought to define what is meant by 'macroscopic realism'. We can think of this simply in terms of a macroscopic object (such as a cat) which can be found in at least two distinct states (dead/alive). To be macroscopically real, it must be possible to determine through a measurement that, at any particular time, the object is in one or other of these states. By definition, the measurement must have no effect on the state itself.

Leggett defined 'distinctness' based on the difference of one or more 'extensive' physical properties of the object in question, such as its total charge, magnetic moment, position, momentum, etc. He called this the *extensive difference.*

But extensive difference alone is not sufficient to define distinctness. The experiments described in previous chapters involving measurements on correlated particles can be considered 'macroscopic' if the particles are separated by a large distance at the time of measurement. Clearly, such experiments do not reflect a Schrödinger cat-type situation. It is therefore necessary to introduce a further parameter, which Leggett called 'disconnectivity'.

The disconnectivity is not well defined in a quantitative sense, but can be thought of as a measure of the degree of quantum entanglement of the systems. If the systems consist of just one particle each, as in the original Einstein, Podolsky, and Rosen thought experiment, then their disconnectivity is correspondingly small. If, however, the systems consist of many particles (dust grains, water drops, golf balls, or cats) then the disconnectivity is correspondingly large. Complex systems with large extensive difference can be expected to possess disconnectivity of the order of the number of constituents in the systems.

Armed with these definitions, it is possible to search for systems exhibiting both high extensive difference and disconnectivity, and discover what kind of reality prevails: a quantum reality of superpositions of macroscopic states or macroscopic realism?

One major step in this direction was taken by Anton Zeilinger and his team at the University of Vienna in 1999. These physicists reported the diffraction and interference of a beam of buckminsterfullerene molecules, each consisting of 60 carbon atoms in a soccer-ball structure, in an elaborate version of the classic Young experiment.[2] These experiments were subsequently repeated with fullerene molecules consisting of 70 carbon atoms. By Leggett's reckoning, the extensive difference in these experiments was about a million, and the disconnectivity over a thousand.[3]

These are fascinating experiments, begging all kinds of questions about what happens to the mass of these molecules as they undergo wave-like behaviour. But they still take us only a relatively small step towards the macroscopic world of everyday objects. To go further down this path, we need to return to the properties of superconductors.

Electrons are fermions and obey the Pauli exclusion principle. However, when considered as though they are a single entity, two electrons of opposite spin and momentum have no net spin and, under the right conditions, they can collectively form a boson. Like other bosons (such as photons), these pairs of electrons can 'condense' into a single quantum state. When a large number of pairs so condense in a superconductor, the result is a macroscopic quantum state extending over large distances. We can think of a macroscopic quantum state as a state with some property of macroscopic dimensions, such as electrical conductivity, whose behaviour is governed by quantum, rather than classical, mechanics.[4]

As described in Chapter 23, this condensed state lies lower in energy than the normal electrical conduction band of the superconducting material and

[2] For more details on the discovery of buckminsterfullerene and its aftermath, see Jim Baggott, *Perfect Symmetry: The Accidental Discovery of Buckminsterfullerene*, Oxford University Press, 1992.

[3] Actually the disconnectivity for C_{60} is 60 carbon atoms multiplied by 18 (six protons plus six neutrons plus six electrons) = 1080.

[4] Another example of a macroscopic quantum state is superfluid helium. Because helium consists of an even number of fermions it can act like a boson. It also undergoes 'Bose condensation' at very low temperatures. At temperatures below about −270 °C, liquid helium loses virtually all of its viscosity and can 'creep' as a thin film up the sides of a beaker.

the paired electrons experience no resistance. The distance between each electron in a pair is quite large, and so many such pairs overlap within the metal lattice. The wavefunctions of the pairs likewise overlap and their peaks and troughs line up just like light waves in a laser beam. The result can be a macroscopic number of electrons (about 100 billion billion) moving through a metal lattice with their individual wavefunctions locked in step.

Imagine that an external magnetic field is applied to a superconducting ring, which is then cooled to its superconducting temperature. The current which flows in the surface of the ring forces the magnetic field to flow outside the body of the material.[5] The total field is just the sum of the applied field and the field induced by the current flowing in the surface of the ring. If the applied field is removed, the current continues to circulate (because the electrons feel no resistance) and an amount of magnetic flux is 'trapped'.

According to the quantum theory of superconductivity, this trapped flux is quantized: only integer multiples of the so-called superconducting flux quantum are allowed.[6] These different flux states therefore represent the quantum states of an object of macroscopic dimensions. Such superconducting rings are usually about a centimetre or so in diameter. The existence of these states has been confirmed by experiment.

In a superconducting ring of uniform thickness, the quantized magnetic flux states do not interact. The quantum state of the ring can only be changed by warming it up, changing the applied external field, and then cooling it down to its superconducting temperature once more. However, the mixing of the flux states becomes possible if the ring contains a *Josephson junction*. This is essentially a small region of the ring in which an insulator has been inserted but which is narrow enough to allow *quantum tunnelling* of pairs of electrons from one side to the other.[7]

[5] This is the Meissner effect. See Chapter 23, p. 230.

[6] The flux quantum is given by $h/2e$, where h is Planck's constant and e is the charge of the electron.

[7] In quantum tunnelling the wavefunction of a quantum particle can pass through a narrow energy barrier, allowing a small proportion of its amplitude to appear on the other side. Although this is often referred to as 'quantum' tunnelling, it is actually a wave phenomenon which can be exhibited by classical waves (such as sound waves). Quantum tunnelling boggles the mind when we use the wave amplitude as a measure of the probability of finding the associated particle, as it means that an electron (for example) has a finite probability of passing through a barrier in a way that would be classically impossible.

A variety of quantum interference effects then becomes possible and the rings are generally called superconducting quantum interference devices (SQUIDs, for short), first invented in 1964. These devices are incredibly sensitive, and are used to measure magnetic field strengths in a variety of medical applications. The smallest change in magnetic flux that can be detected in one second using a typical SQUID ring corresponds roughly to the energy required to raise a single electron one millimetre in the earth's gravitational field. More sensitive devices approach the limits imposed by the uncertainty principle.

Interestingly, this sensitivity is a feature of the Josephson junction, not the ring. This means that a macroscopic variable (such as a measurable flux of electrons travelling around the ring) can be controlled by a microscopic amount of energy in a manner that has little to do with the physical size of the ring itself. The question then becomes: is it possible to create superpositions of distinct macroscopic states of such devices—for example, states in which large numbers of electrons are flowing in *opposite* directions around the ring?

How would interference between such macroscopic states be manifested? Clearly this is not going to be as simple as observing visible interference fringes. This is a bit like trying to imagine how interference would be manifested between the macroscopic states of alive and dead cats.

Let's think instead about the macroscopic quantum states created in a superconducting ring in which electrons flow anticlockwise and clockwise around the ring. These states sit in separate, distinct, potential energy 'wells'. The bottom of each well represents the lowest energy configuration of each state. To cross from one well to another (i.e. to reverse the direction of flow of the electrons around the ring) we would have to raise the energy of the ring by warming it up, change the applied field, and then cool it back down to its superconducting temperature.

Now suppose that we insert a Josephson junction. The effect of the junction is to allow the anticlockwise and the clockwise states to combine in a superposition. In fact, where the energies of the two states are equal or near-equal, the states mix together to form a slightly lower energy anticlockwise + clockwise combination, and a slightly higher energy anticlockwise − clockwise combination. Spectroscopists refer to this kind of effect as an 'avoided crossing'. Instead of crossing, two new states are formed in this energy region with characteristics of mixtures of the original states.[8]

The amount of splitting between these mixed states is given by the so-called tunnel splitting which is a characteristic of the Josephson junction. This kind of splitting between quantum states is fairly common in atomic and molecular spectroscopy, but it is perhaps novel to consider its effects in quantum states involving large numbers of electrons moving in a metal ring of macroscopic dimensions.

The experimental demonstration of the possibility of superpositions of distinct macroscopic states—laboratory versions of Schrödinger cat states—then comes down to this: what happens in the energy region where the anticlockwise and the clockwise states are of equal or near-equal energy?

In practice, the experiments require the application of a bias to the junction in the form of an externally applied magnetic flux through the ring. As the bias is varied, microwave radiation is used to probe the size of the splitting between the states, and hence the extent of the interference between them. A splitting of the order of magnitude of the bias itself implies that the anticlockwise and clockwise states remain distinct macroscopic states and there is no interference. A splitting of the order of the square-root of the sum of the squares of the bias and the tunnel splitting implies that a superposition of macroscopic quantum states has been created.

The results of two kinds of experiments were reported in 2000 by two groups of researchers. Jonathan Friedman, Vijay Patel, W. Chen, S.K. Tolpygo, and J.E. Lukens at the State University of New York, Stony Brook, used a SQUID ring measuring 140 microns in diameter.[9] In a paper submitted to the journal *Nature* in April 2000 they reported results confirming the formation of macroscopic superpositions:

> The quantum dynamics of the SQUID is determined by the flux through the loop, a collective coordinate representing the motion of approximately [a billion] Cooper pairs acting in tandem.

[8] This is the same kind of mixing that produces the kaon states K_1^0 and K_2^0 from K^0 and $\bar{K}^0$ see p. 248.

[9] A micron is a millionth of a metre.

The paper was published in July. That same month, a second group based at the Delft Institute for Micro Electronics and Submicron Technology and at MIT submitted a similar paper to the American journal *Science*. The approaches of the two groups were somewhat different. The Stony Brook group used a SQUID ring with a single Josephson junction and performed measurements on excited levels of the anticlockwise and clockwise states. The Delft/MIT group used a 5-micron diameter SQUID ring with three Josephson junctions and measured the splitting between superpositions of the ground levels of the macroscopic quantum states.

But the results were very similar. The Delft/MIT group claimed their results indicated:

> ...symmetric [anticlockwise + clockwise] and anti-symmetric [anticlockwise − clockwise] quantum superpositions of macroscopic states. The two classical states have persistent currents of 0.5 microampere and correspond to the center-of-mass motion of millions of Cooper pairs.

Just stop and think for a moment. In a quantum superposition consisting of a symmetric combination of a macroscopic state in which a billion electron pairs are moving anticlockwise around the ring and another macroscopic state in which a billion electron pairs are moving clockwise around the ring, in just what direction are the electrons *actually* moving?

Leggett estimated that in these experiments both the extensive difference and disconnectivity are of the order of 10 billion.[10]

Of course, millions or even billions of pairs of electrons do not represent a cat-sized object. But this was never quite the point. The Copenhagen interpretation of quantum theory had always been stubbornly silent on the question of where the distinction between the microscopic and macroscopic worlds should be drawn. Whilst experiments showing non-local behaviour between pairs of correlated photons, electrons, atoms, or ions might be disconcerting, they leave the quantum weirdness firmly in the

[10] Leggett now feels this figure is somewhat overstated, and suggests instead a figure of 1 to 10 million. Tony Leggett, personal communication to the author, 4 August 2009.

quantum realm. Quantum superpositions involving billions of particles start to bring the weirdness into a realm we can see with the naked eye.

Writing about these experiments some years later, Leggett saw no reason to suppose that they represented fundamental limits to the size of macroscopic quantum superpositions that could in principle be created in the laboratory:

> ...we can say that on a logarithmic scale we have come about 40% of the way from the atomic level to that of the everyday world...there seems no obvious *a priori* objection to extending the SQUID experiments from rings of the size of these used in the existing experiments (a few microns) to much larger sizes, perhaps of the order of 1 [centimetre].

In a recent review article published in the journal *Nature*, Vlatko Vedral at the Univeristy of Leeds notes:

> There are many open questions regarding entanglement. Here I have stated that, in theory, entanglement can exist in arbitrarily large and hot systems. But how true is this in practice? Another question is whether the entanglement of massless bodies fundamentally differs from that of massive ones. Furthermore, does macroscopic entanglement also occur in living systems and, if so, is it used by these systems?

This remains a very active area of research. In May 2008 Markus Aspelmeyer and Zeilinger commented briefly on research that was ongoing in Vienna with the aim to demonstrate quantum interference effects in heavier and larger systems. The main goal of this work is to demonstrate interference for small viruses or bacteria with nanometre (billionth of a metre) dimensions.[11]

There is more to this search for macroscopic quantum phenomena than questions concerning the nature of reality and the transition from the quantum to classical realms. Quantum states that represent opposites, such as spin-up and spin-down, vertical and horizontal, clockwise and anticlockwise, can in principle represent information in the form of

[11] See Markus Aspelmeyer and Anton Zeilinger, *Physics World*, July 2008.

quantum binary digits, or *qubits*. A computer constructed from macroscopic qubits which makes use of quantum entanglement between these qubits can in theory perform operations much faster than a classical computer.[12]

Aside from quantum computers based on SQUIDs, other possibilities include computers based on entangled trapped ions, so-called qubit 'clusters' involving four-photon GHZ states, quantum 'dots', and many others.

When Einstein and Schrödinger developed their now infamous thought experiments, they had sought to use entanglement and 'spooky' action at a distance to undermine the foundations of the interpretation of quantum theory laid down by the Copenhagen school. It would have been quite impossible to imagine that their arguments would become the basis of an entire new quantum technology.

Those realists intent on standing firm in the face of the onslaught from experiments proving quantum non-locality and macroscopic quantum superpositions could still find some solace. Reality might indeed be non-local, they now had to admit, but that doesn't necessarily mean that quantum particles do not possess defined properties until they are measured.

Does it?

[12] This means that a quantum computer could in principle be used to crack the encryption systems used for most Internet transactions, which are based on factoring large prime numbers. Not to worry, though, as Internet security could perhaps be restored using cryptographic systems based on quantum entanglement!

35

The Persistent Illusion

Vienna, December 2006

In their 1935 paper, Einstein, Podolsky, and Rosen had set forth what they judged to be a 'reasonable' definition of reality. At its heart was the entirely commonsense assumption that quantum particles have the properties we ascribe to them independent of their measurement. An electron has a spin-up orientation: the nature of its interaction with a Stern–Gerlach magnet is determined by this property and its subsequent path through the apparatus reveals what this property is. Likewise, a photon has a vertical polarization: its passage through a polarization analyser simply tells us what it is.

We have to accept that this simplistic local realism must give way in quantum mechanics to something rather 'spookier'. The experimental tests of Bell's theorem demonstrate quite unequivocally that the spin orientation of an electron or the polarization state of a photon can be affected by what happens to another particle an arbitrarily long distance away. We have no idea how this kind of non-local influence might be propagated, or even if 'propagated' is the right word to use. But there is now plenty of evidence to suggest that, however it works, such non-local influences cannot be used to transmit meaningful information (messages) faster than the speed of light. In this sense, at least, quantum theory and special relativity coexist, if not peacefully then with only the grumbling characteristic of two old but venerable theories that don't quite get along.

In the face of the repeated and consistent experimental violation of Bell's inequality, we trade off the assumption of locality so that we may preserve some semblance of realism. But realism—the assumption that particles have the properties we assign to them even when we don't look—is also an assumption. It is an assumption that has been challenged by philosophers down the centuries. These philosophers have continued to ask impertinent, brain-teasing questions such as: If a tree falls in a forest with nobody around to hear, does it make a sound?

Bell had not trusted the Copenhagen interpretation of quantum theory and, like Einstein, felt that the theory was incomplete. Leggett, too, was distrustful. In 1976 he had left his post at the University of Sussex in England to take up a temporary teaching exchange at the University of Science and Technology in Kumasi, Ghana. Lacking access to the most up-to-date research papers at the University's library, he had applied himself instead to a problem that had attracted his attention some years earlier. Taking his cue from Bohm, Leggett examined the predictions of a general class of non-local hidden variable theories.

On his return to England, he drafted a paper presenting his ideas but did not get around to submitting it for publication. He put the manuscript in a drawer and forgot all about it. It was only with the advent nearly thirty years later of reliable sources of entangled photons of the kind that had been used to test Bell's inequality that he decided to take the manuscript out of the drawer and dust it off.

He had in the meantime moved to a professorship at the University of Illinois at Urbana-Champaign, and was awarded a share (with Alexei Abrikosov and Vitaly Ginzburg) in the 2003 Nobel Prize for physics for his work on superconductors and superfluidity. His paper on non-local hidden variable theories was published in the journal Foundations of Physics *in October 2003.*

At a scientific conference in Minnesota the following May, Leggett shared his results with German physicist Markus Aspelmeyer, who worked alongside Anton Zeilinger's group at the University of Vienna and the Institute for Quantum Optics and Quantum Information. In his paper Leggett had derived a new set of inequalities that could be used to test not just locality, but realism too. The inequalities relate to a broad general class of non-local hidden variable theories. Here, once again, was another straightforward test. If quantum theory failed, this would be direct evidence that it was not complete. If it passed, then our preciously held assumptions about the nature of reality at the quantum level could no longer be revised, they would have to be abandoned.

The experiments didn't seem too difficult.

Leggett had looked at it this way. In a cascade emission process of the kind that had been used in the Aspect experiments, we assume that the properties of the photons are governed by some, possibly very complex, set of hidden variables. These hidden variables possess unique values that determine the states of the photons and their subsequent interactions with their respective measuring devices. We further assume that the photons are emitted with a statistical distribution of variables and that the form of this distribution is determined only by the physics of the emission process and not by the way the measuring apparatus is set up.

The outcomes of the measurements to be made on the correlated photons are determined by the settings of the polarization analysers and the values of the hidden variables. So far, so familiar.

Local hidden variable theories are characterized by two further assumptions. In the first, we assume (as did Einstein, Podolsky, and Rosen) that the outcome of the measurement on photon A can in no way be affected by the *outcome* of the measurement on the distant photon, B, and *vice versa*. In the second, we assume that the outcome of the measurement on photon A can in no way be affected by the *setting* of the polarization analyser used to perform the measurement on B, and *vice versa*.

One or other of these assumptions is invalidated by the experimental demonstration of the violation of Bell's inequality. In particular, experiments designed to close the 'locality loophole' show that despite randomly changing the settings of the analysers whilst the photons are in flight towards them, the correlations predicted by quantum theory are nevertheless maintained. This kind of result does not allow us to conclude which assumption is invalidated, since changing the analyser setting will obviously also change the outcome of the subsequent measurement.

Leggett chose to relax the setting restriction. In other words, he chose to allow the behaviour of the photons and the outcomes of subsequent measurements to be influenced by the way the measursing devices are set up. He understood well enough that allowing the outcome of a measurement to be affected by the setting of a distant analyser was, to many physicists, highly counter-intuitive:

> But in physics we are normally accustomed to require some positive reason before we accept a particular part of the environment as relevant to the outcome of an experiment. Now the [distant] polarizer...is nothing more than (e.g.) a calcite crystal, and nothing in our experience of physics indicates that the orientation of distant calcite crystals is either more or less likely to affect the outcome of an experiment than, say, the position of the keys in the experimenter's pocket or the time shown by the clock on the wall.

But in the face of the experimental evidence, something had to give. In choosing to relax the setting restriction, Leggett was allowing for some unspecified non-local influence of the analyser settings to affect the outcomes of measurements on distant particles. In this sense he was being consistent with Bohr's response to Einstein, Podolsky, and Rosen's original challenge:[1]

> ...there is in a case like that just considered no question of a mechanical disturbance of the system under investigation during the last critical stage of the measuring procedure. But even at this stage there is essentially the question of *an influence on the very conditions which define the possible types of predictions regarding the future behaviour of the system.*

By preserving the outcome restriction, Leggett was defining a class of non-local hidden variable theories in which the individual particles possess defined properties before the act of measurement. What is actually measured will of course depend on the measurement settings that the particles encounter, and changing these settings does affect the behaviour of distant particles. But maintaining this restriction meant that there was no way that a measurement made on photon A could affect the outcome of a simultaneous or subsequent measurement on photon B, and *vice versa*. Leggett referred to this broad class of theories as 'crypto' non-local hidden variable theories.[2]

Leggett went on to show that relaxing the setting restriction is in itself insufficient to reproduce all the results of quantum theory. Just as Bell had done in 1964, Leggett now derived inequalities that were obeyed by such hidden variable theories but which for certain combinations of measurement settings were predicted by quantum theory to be violated. At stake then, was the rather simple question of whether or not quantum

[1] See Chapter 16, p. 146–147.

[2] Leggett not only sought to preserve the outcome restriction, but also to ensure that sub-ensembles of polarized photons would obey Malus' law, i.e. they would interact with polarization analysers in a manner consistent with quantum theory predictions (Tony Leggett, personal communication to the author, 4 August 2009). Note that Bohm's non-local hidden variable theory does not belong in this class. The effects not only of changing analyser settings but also of measurement outcomes affect the shape of the quantum potential and consequently influence instantaneously the behaviour of distant particles. The outcome restriction is not relaxed in this case.

particles have the properties we assign to them *before the act of measurement*. Put another way, here was an opportunity to test whether quantum particles have 'real' properties *before* they are measured.

The experiments implied by Leggett's work were not so very different from the experiments that had been performed to test Bell's inequality. There were some subtleties and some challenges, however. All the experiments conducted to this time had measured the polarization states of photons in a single basis (such as vertical/horizontal) using analyser orientations varied within a single plane. Testing the Leggett inequalities would require measurements in two bases (linear and elliptical polarization) and in two planes, such as the *xz*- and *yz*-planes. This was why the results of the earlier tests of Bell's inequality could not be used to determine the validity or otherwise of the Leggett inequalities.

By allowing non-local influences due to the settings of the measurement devices, the predictions of the general class of non-local hidden variable theories considered by Leggett were, not surprisingly, so much closer to the predictions of quantum theory. It was possible to discern a pattern. Imposing both the outcome and setting restrictions (local reality) results in a relatively large difference between the predictions of quantum theory and local hidden variables. Lifting the setting restriction (non-local reality) closes this gap but does not eliminate it. Relaxing the outcome restriction closes the gap completely: it is impossible to differentiate between quantum theory and this class of non-local hidden variable interpretations of quantum theory using experiment.

To discriminate between quantum theory and the crypto non-local hidden variable theories studied by Leggett would require the highest quality entangled photons and the highest 'visibility'. These tests would make the earlier experiments look like a walk in the park.

Aspelmeyer returned to Vienna and, with Leggett's help, worked together with a local theorist to re-derive Leggett's inequality for a specific experimental arrangement that could be realized in the laboratory. Aspelmeyer's student Simon Gröblacher carried out these experiments over one weekend. The physicists took the results to Zeilinger, who agreed to repeat them.

The two groups submitted their joint results for publication in the British journal *Nature* in December 2006. They were published in April 2007.

For the kind of arrangement used in these experiments, the Leggett inequality demands that the value of the experimental correlation is restricted to be less than or equal to the value given by $4 - (4/\pi)|\sin(\varphi/2)|$, where φ is the angle difference between two analyser orientations in both the *xz*- and *yz*-planes. Quantum theory predicts the value $|2(\cos\varphi + 1)|$. The maximum divergence between these two sets of predictions occurs for $\varphi = 18.8°$. At this angle, the general class of crypto non-local hidden variable theories studied by Leggett predicts the value 3.792. Quantum theory predicts 3.893, a difference of less than 3 per cent.

Nevertheless, the results were once again unequivocal. Violations of the Leggett inequality were observed for φ between 4° and 36°, with a maximum violation around 20°. For this angle the experimental value for the correlation was found to be 3.8521 ± 0.0227, compared with the Leggett inequality prediction of 3.779 and the quantum theory prediction of 3.879. This is a violation of the Leggett inequality by more than three standard deviations.

The Vienna physicists ensured that the same arrangement yielded results that could be used in a simultaneous test of the generalized form of Bell's inequality. For the same difference angle φ, they obtained a correlation 2.178 ± 0.0199 compared to the Bell limit of 2, a violation of Bell's inequality by nine standard deviations.

The physicists concluded their paper as follows:

> We believe that our results lend strong support to the view that any future extension of quantum theory that is in agreement with experiments must abandon certain features of realistic descriptions.

Further reports followed hard on the heels of this paper. The Vienna group and a group from the Universities of Geneva and the National University of Singapore confirmed the violation of Leggett's inequality under less restrictive conditions, expanding the class of non-local hidden variable theories that could now be ruled out.

Zeilinger had not studied quantum theory as a student in Vienna in the 1960s and so came late to its paradoxes and conundrums. 'Quantum mechanics is very fundamental,' he explained, 'Probably even more fundamental than we appreciate. But to give up on realism altogether is certainly wrong. Going back to Einstein, to give up realism about the moon, that's ridiculous. But on the quantum level we do have to give up realism.'

What does it mean?

If a tree falls in the forest when there's nobody around to hear, we might hypothesize that it produces certain wave disturbances—compressions and rarefactions in the air around it. We might hypothesize that some of these disturbances will have characteristic audio frequencies. We might predict that if a measuring device (a human being or an audio tape recorder) was to be placed near the location of the tree then the result *will* be the hearing or recording of an audio signal. We might even call this signal a sound.

But we commonly use the word 'sound' in two senses. We sometimes refer to wave disturbances with frequencies in the audio range as *sound waves*, even though we might not actually hear them. In this sense, the word 'sound' is used to describe a physical phenomenon—the wave disturbance, with characteristic amplitude and frequency. Sound is also a human *experience*, the result of physical signals delivered by human sense organs (specifically the tympanic membrane) which are synthesized in the human mind as a form of perception.

Now, to a large extent, we can interpret the action of human sense organs in much the same way we interpret the actions of mechanical measuring devices. The human auditory apparatus simply translates one set of physical phenomena into another, leading eventually to stimulation of those parts of the brain cortex responsible for the perception of sound. It is here that the distinction comes. Everything to this point is explicable in terms of physics, but the process by which we turn electrical signals in the brain into human perception and experience in the mind remains, at present, unfathomable.

Philosophers have long argued that sound, colour, taste, smell, and touch are all *secondary qualities* which exist only in our minds. We have no basis for our commonsense assumption that these secondary qualities reflect or represent reality as it really is. Like the prisoners in Plato's cave, we may well live in a world of crude appearances of objects, which we mistake for the objects themselves.

If we interpret sound in terms of human experience rather than just physical phenomena, then when there is nobody around there is a sense in which the falling tree makes no sound at all.

And this is what the tests of Leggett's inequality suggest. We must now accept that the properties we ascribe to quantum particles, such as spin-up, vertically polarized, 'here' and 'there' are properties that have no meaning except in relation to a measuring device. We can no longer assume that the properties we measure necessarily reflect or represent the properties of the particles as they really are. As Heisenberg had earlier argued:[3]

> ...we have to remember that what we observe is not nature in itself but nature exposed to our method of questioning.

This does not mean that quantum particles are not real. What it does mean is that we can ascribe to them only an *empirical* reality. This is a reality that depends on our method of questioning, a reality that can be affected by the outcomes of measurements on distant particles.

Although we may speak of electron spin, position, orbital angular momentum, and so on, these are empirical properties we have assigned to an electron based on our experiences of them. Each property becomes 'real' only when the electron interacts with an instrument specifically designed to reveal that property. These concepts help us to correlate and describe our observations, but they have no meaning beyond their use as a means of connecting the object of our study with the measuring device we use to study it.

Before measurement, the properties of quantum particles such as electrons and photons are clearly constrained by the physics that produced them, but they are in a sense 'undetermined'. Their properties are 'vanilla': incipiently spin-up or spin-down, vertical or horizontal, but undetermined until the act of measurement. The nature of the interaction with a measuring device somehow 'determines' these properties, in a way that we may never hope to fathom.

[3] See Chapter 12, p. 109.

'Reality is merely an illusion,' Einstein once admitted, 'albeit a very persistent one.'[4]

Like Bell before him, Leggett had been encouraged by a general scepticism of the Copenhagen interpretation to develop a theorem which led ultimately to a decisive experimental test. Whilst he accepts that the experimental results mean that realism is no longer tenable, he still does not accept that quantum theory is complete.

'I'm in a small minority with that point of view,' he demurred, 'And I wouldn't stake my life on it.'

[4] Or, if you prefer: 'Reality is that which, when you stop believing in it, doesn't go away.' Philip K. Dick.

PART VII

Quantum *Cosmology*

36

The Wavefunction of the Universe

Princeton, July 1966

Einstein presented his general theory of relativity in a series of lectures delivered to the Prussian Academy of Sciences in Berlin, culminating in a final, triumphant lecture on 25 November 1915. Yet within a few short months he was telling the Academy that his new theory of gravitation might need to be modified:

> *Due to electron motion inside the atom, the latter should radiate gravitational, as well as electromagnetic energy, if only a negligible amount. Since nothing like this should happen in nature, the quantum theory should, it seems, modify not only Maxwell's electrodynamics but also the new theory of gravitation.*

Attempts to construct a quantum theory of gravity were begun in 1930 by Bohr's protégé, Léon Rosenfeld. He noted the problems with divergences that would plague quantum field theory for many years to come. Soviet physicist Matvei Bronstein later pictured the challenge in the form of a cube whose faces are distinguished by three fundamental physical constants. These are the speed of light, c, Planck's constant divided by 2π, $\hbar$ and Newton's gravitational constant, G.[1] *Proceeding from the top-left-back corner to Newton's theory of gravitation at the top-right-back corner requires the introduction of G. Introducing a*

[1] This is now known as the Bronstein cube. Strictly speaking, as pulling the physical theories back to the top-left corner (which we can denote as 'Galilean physics') requires the assumption $\hbar = 0$, $G = 0$, and $c = \infty$, it might be better to think of the constants in terms of the group $\hbar$, G, and $1/c$.

finite speed of light c takes us to general relativity at the top-right-front corner. To get to a quantum version of general relativity at the bottom-right-front corner requires the further introduction of ħ.

Just looking at this cube suggests at least two directions to a quantum version of general relativity. One can seek to 'quantize' general relativity, a step that Einstein himself tended to dismiss as 'childish'. A second direction involves starting with relativistic quantum field theory and making it conform to the general covariance requirements of general relativity.[2] *There is, of course, a third direction (not pictured), which is to start afresh.*[3]

Whichever direction we take, we run into a series of profound problems right at the outset. General relativity is about the motions of large-scale bodies such as planets, stars, solar systems, galaxies, and the entire universe within a four-dimensional space–time. In general relativity these motions are described by a set of complex mathematical equations known as Einstein's gravitational field equations. They are complex because the

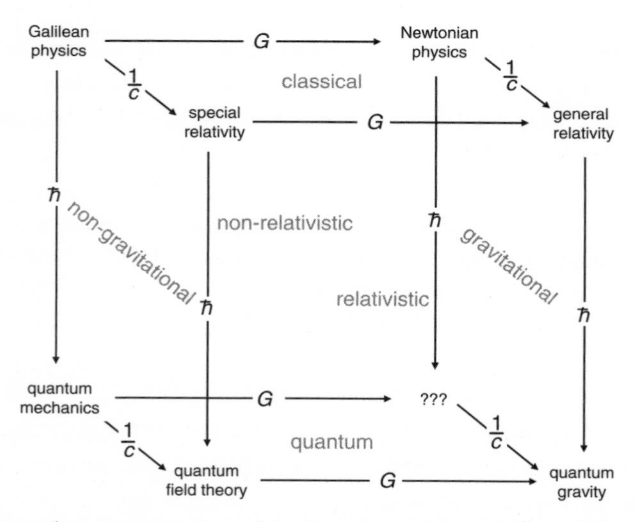

FIG 26 A modern interpretation of the Bronstein cube. Each face of the cube represents a domain of theoretical physics, and passing from one domain to another involves introducing the physical constants *c*, *ħ* and *G*. The ??? represents a quantized but non-relativistic version of Newtonian physics. Adapted from Roger Penrose, *The Large, the Small and the Human Mind*. Cambridge University Press, 1997, p. 91.

[2] General covariance means that the physical laws described by the theory are invariant to arbitrary coordinate transformations.

[3] Which acknowledges that if you want to get there, then you shouldn't start from here.

mass they consider distorts the geometry of space–time around it and the geometry of the space–time around it governs how the mass moves.

But space–time itself is contained entirely within general relativity—it is a fundamental dynamical variable of the theory. The theory itself constructs the framework within which mass moves and events happen. In this sense the theory is 'background independent'—it does not presuppose the existence of a background framework against which motions of large masses are to be registered. Quantum theory, in contrast, presumes precisely this. It is 'background dependent', requiring a classical 'container' of space and time within which the wavefunctions of quantum particles evolve.

Then we run into the uncertainty principle, which changes our understanding of the very meaning of 'empty' space. It is not empty at all. It is filled with virtual particles, flickering in and out of existence.[4] *General relativity assumes that space–time is certain, and not subject to quantum theory's characteristically probabilistic laws. In general relativity we can be certain that space–time is 'here' or 'there', curves this way or that way, at this rate or that rate.*

Early attempts to forge a quantum theory of gravity were premature and frustrated. Momentum began to pick up again in the early 1950s.

In May 1952, John Wheeler at Princeton pulled a new bound notebook from his shelf and labelled it 'Relativity I'. He was pleased to learn that he had been given the go-ahead to teach a course on relativity, and thought to get into the subject properly by writing a book about it. 'That fall, fifteen graduate students enrolled in my course,' Wheeler explained. 'It was the first time that a relativity course had been offered at Princeton—and together we worked our way through the subject, trying to get behind the mathematical formalism that had dominated the theory for decades, looking for real, tangible physics.'

In fact, when Einstein first developed his field equations in 1915 he considered them too complex to solve. Yet only a year later German physicist Karl Schwarzchild had produced a solution, which he discovered whilst serving in the German army on the Russian front. Schwarzchild's solutions predicted that a particularly massive body could collapse under

[4] The effect of the creation and destruction of these virtual particles in empty space can be demonstrated in the laboratory through something called the Casimir effect, discovered in 1948 by Dutch physicist Hendrik Casimir. Two closely spaced metal plates will actually be pushed closer together due to the fact that the pressure from virtual photons in the gap between the plates no longer balances the pressure of virtual photons outside the gap.

the influence of its own gravity to a so-called *singularity*. Such a singularity would distort the space–time around it so much that nothing, not even light, would escape its 'event horizon'. Wheeler was initially troubled by this idea, but grew to accept it. He called it a *black hole*.

The early predictions of general relativity are well known. The theory helped to account for a wobble in the orbit of the planet Mercury around the sun, a wobble that Newton's gravity couldn't explain. Einstein also predicted that light from a distant star passing by the sun would be bent in its path by the curvature of space–time. Such bending of starlight was proven by observations made during a solar eclipse in 1919, and helped to make Einstein a cultural icon of the twentieth century.

A few years after teaching his first courses on general relativity, Wheeler began to think about the potential implications of quantum effects. In seeking to base the search for a theory of quantum gravity on the fundamental concepts of geometry, he elucidated a theory of *quantum geometrodynamics*, an allusion to the parallels with quantum electrodynamics, or QED. In discussions with one of his students, American Charles Misner, he applied the logic of quantum uncertainty to space–time itself, replacing the perspective of a flat or gently curving space–time with the chaos of uncertainty and quantum fluctuations.

The uncertainty principle is a licence for the bizarre, and in the quantum domain the topology of space–time becomes twisted and tortured and riddled with bumps, lumps, and tunnels—'wormholes' connecting one part of space–time with another. Wheeler referred to this as quantum or space–time 'foam'.

In the absence of a fully fledged theory of quantum gravity these were simply ideas about what space–time might look like at extremely small distances, of the order of a millionth of a billionth of a billionth of a billionth of a centimetre (a distance called the Planck length) and in extremely small time intervals of the order of a tenth of a millionth of a trillionth of a trillionth of a trillionth of a second (called the Planck time).[5]

[5] The Planck length and Planck time can be calculated from the fundamental constants G, $\hbar$, and c. The Planck length is given by $\sqrt{G\hbar/c^3}$ and the Planck time is given by $\sqrt{G\hbar/c^5}$ (the Planck length divided by c).

In fact, on these scales the very concept of distance and time interval lose their meaning. There is no distance smaller than the Planck length, no time shorter than the Planck time.

Misner worked on aspects of classical general relativity under Wheeler's direction, publishing a long paper in the journal *Annals of Physics* in 1957 and intending to submit this paper as his Princeton PhD thesis. He then discovered that the key ideas in his work had been published already in 1925. Undaunted, he turned his attention to quantum gravity. He published details of an approach to the Feynman quantization of general relativity in the journal *Reviews of Modern Physics* later that same year.

It was in this paper that Misner suggested three lines of research on quantum gravity, much like three possible routes to the conquest of Everest. The quantization of general relativity could be approached by first recasting the theory into a form more familiar in quantum mechanics (this is a so-called constrained Hamiltonian formulation). This reformulation is a kind of stepping-stone. The second step is to apply a quantization technique. In the 1940s and early 1950s German theorist Peter Bergman at the Institute for Advanced Study and Dirac in Cambridge had worked on quantization techniques based on gauge symmetries. This became known as canonical quantization, and the line of research became known as the canonical approach to quantum gravity.[6]

An alternative quantization technique was to apply Feynman's path integral or sum over histories approach, a technique that Misner had outlined in his paper. The third route was to devise a quantum field theory of the gravitational field which conforms to the general covariance requirements of general relativity using a fictitious 'flat space'. Once constructed, it was assumed that the resulting field equations could be solved using perturbation techniques, in much the same way that the equations of QED had been solved, perhaps using Feynman diagrams. This is known as the covariant approach.

Although purportedly pursuing the same objective, the canonical and covariant approaches looked distinctly different. Not surprisingly, as it builds

[6] It involves structuring the equations to define classical variables (position, momentum) which are then replaced by their quantum-mechanical operator equivalents.

on general relativity, the canonical approach emphasizes the geometry of space–time and the (quantum) dynamics of objects moving within this geometry. The covariant approach emphasizes the quantum field and the graviton as force carrier. Although some theorists acknowledged the benefits of pursuing multiple lines of research, the great conceptual divisions and lack of common ground between these approaches was already ominous.

Work on the canonical approach had led to some puzzling results. General relativity demands that space and time be treated on an equal footing and there is no absolute space–time coordinate system—all coordinate systems are arbitrary. In general relativity there is no meaningful 'here' and 'there' or 'now' and 'then'. The theory rather deals with space–time *intervals*.

Differences in time enter the theory as ct, the time interval multiplied by the speed of light, which has the same units as a distance interval. Once entered, the time dimension becomes in principle indistinguishable from the three spatial dimensions in a four-dimensional space–time.[7]

However, what Dirac discovered is that in his constrained Hamiltonian reformulation of general relativity the dynamics are governed by only three of the four dimensions. 'This result,' he declared, 'has led me to doubt how fundamental the four-dimensional requirement in physics is.' The three dimensions, called a three-space, hold all the information about the geometrical relationships between masses and, to all intents and purposes, look like three spatial dimensions. Space–time had been unpicked, and although time had not exactly disappeared in this reformulation, it had become rather mysterious and elusive.[8] In fact, time had become the result of the changing geometrical relationships between material objects.

In 1961, Misner, together with Americans Richard Arnowitt and Stanley Deser, published a greatly simplified elaboration of the constrained

[7] In fact, space–time intervals are given by $\sqrt{d^2-(ct)^2}$, where d is the difference in spatial coordinates (simplified here to one dimension), t is the time difference, and c is the speed of light. This means that some intervals can be imaginary (i.e. they are multiplied by i, the square-root of -1), and this is a characteristic 'signature' of a time interval. For example, the space–time interval between your current position and this same position in five minutes' time is roughly 90i million kilometres.

[8] Of course, the nature of time has been the subject of intense debate among philosophers for centuries. Deep questions concerning the implications of both special and general relativity for our understanding of time had rumbled virtually from the moment the theories had been written down.

Hamiltonian formulation of general relativity, building on the earlier work of Dirac and Bergman. In the Arnowitt, Deser, and Misner (ADM) formulation, space–time is treated as a series of purely spatial 'hypersurfaces', which may individually be curved, connected together through their relationships one with another representing the evolution of the system in time.

The stage was set for a significant development in the search for a quantum theory of gravity.

Many of the possible solutions of Einstein's gravitational field equations describe a universe in which space–time itself is expanding. Einstein initially resisted the idea of an expanding universe and fudged his equations to produce static solutions, by introducing an arbitrary 'cosmological constant'. But the fact that our own universe is indeed expanding was suggested by American astronomer Edwin Hubble's observations in 1929. He noted that distant galaxies are all receding from us at rates that increase with their distance, much like dots scribbled on a deflated balloon will all move apart from each other as the balloon is inflated.

If the universe is expanding then this implies that it had an origin at some point in time in an infinitesimally small, immensely hot 'big bang'. Subsequent expansion and cooling would then give rise to the universe as we know it today.

The violent explosion of radiation some time after a hot big bang was predicted to have left a detectable signature. This radiation would pervade all of space, cooled to a few degrees or tens of degrees above absolute zero as the universe expanded. It would take the form of microwave radiation. Whilst working on radar at MIT in 1946, physicist Robert Dicke had set a limit on this background radiation at less than 20 degrees above absolute zero.[9]

A few years later George Gamow, Ralph Alpher, and Robert Herman had predicted a temperature of five degrees above absolute zero. In 1965, Dicke, together with Princeton physicists Jim Peebles, David Wilkinson, and Peter Roll, were building an apparatus to try to detect this microwave background radiation when they discovered that it had already been found. Arno Penzias and Robert Wilson at nearby Bell Laboratories

[9] Absolute zero (0 kelvin) is equal to −273.15 °C.

in Holmdel, New Jersey, had the previous year detected a persistent and annoying hiss of microwave radiation using a highly sensitive 20-foot horn antenna. It was coming uniformly from all directions in the sky. Try as they might, they could not eliminate what they took to be interference. They did not know what it was.

Penzias was subsequently advised of a preprint by Peebles concerning the microwave background radiation, and he and Wilson made the connection. The radiation they had discovered fit closely with the Princeton physicists' predictions, indicating a radiation temperature of about three degrees above absolute zero. Penzias and Wilson and the Princeton group published companion papers announcing the discovery in the *Astrophysical Journal Letters* in May 1965.

The big bang was a proven hypothesis. This meant that our universe had an origin about 12 or 13 billion years ago.[10] And now there could be no denying it. Although the universe is the ultimate macroscopic object, the first moments in the creation of all things—space, time, force, radiation, matter—were indisputably microscopic in scale. The origin of the universe was a quantum phenomenon, governed (for better or worse) by quantum laws. What was needed to understand these early moments was a *quantum cosmology*, of which a quantum theory of gravity would be a fundamental component.

American theorist Bryce DeWitt had studied under Schwinger at Harvard University, completing his doctorate in 1950. He had preferred to study at Harvard rather than Caltech because of his passion for rowing. He met and married Cecille Morette whilst working at the Institute for Advanced Study. After conducting research at a number of institutes overseas, accompanied by his wife, he returned to America in 1952 and worked first at the Lawrence Livermore National Laboratory before moving to the University of North Carolina at Chapel Hill in 1956.

When in 1965 Wheeler realized he had to make a short stopover at Raleigh-Durham airport in North Carolina, he called DeWitt and asked to meet him there. DeWitt arrived as Wheeler was waiting for

[10] More recent satellite observations of the microwave background radiation have allowed the origin of the universe to be set about 13.7 billion years ago, give or take a couple of hundred million years.

his connecting flight. They discussed recent work on the reformulation of general relativity and DeWitt mumbled something about doing for relativity what Schrödinger had done for the hydrogen atom in 1925. To DeWitt's surprise, Wheeler was immediately enthusiastic, declaring that the equation for quantum gravity had been found.

DeWitt subsequently spent a short time at the Institute for Advanced Study in Princeton, where he held further conversations with Wheeler and developed the first formal theory of quantum gravity. He produced a series of papers, submitted to *Physical Review* in July 1966 and eventually published in August 1967.[11] The first of these describe a canonical approach applied to a simple model universe based on the solutions of Einstein's gravitational field equations first derived in 1922 by Russian mathematician Alexander Friedmann. In a Friedmann universe, space–time is homogeneous and expanding uniformly in all directions. DeWitt filled this model universe with non-interacting material particles at rest.

As Dirac had observed, so DeWitt now noted that in his quantum wave equation of the universe, time had disappeared. The wavefunction, which DeWitt called 'the wavefunction of the universe', was found to depend only on the geometry of the resulting three-space. It was the wavefunction of a stationary universe with zero total energy. He wrote: '...one therefore comes to the conclusion that nothing ever happens in quantum gravidynamics, that the quantum theory can never yield anything but a static picture of the world.' His solution was to combine many different wavefunctions into a 'wavepacket' state, which would trace out a classical trajectory on the three-space.

DeWitt's result implies that time is 'phenomenological'. It suggests that our experience of time is not born of an intrinsic, fundamental element of reality called 'time'. What we experience is rather the changing geometry of the universe and the masses within it, which we synthesize in our minds and interpret as evolving instants of time.

DeWitt had no clear sense of how the wavefunction of the universe should be interpreted and, in mitigation, cited historical precedent.

[11] It seems that the delay was, in part, the result of difficulties concerning page charges.

Schrödinger had faced a similar challenge when confronted with the wavefunction of the electron in a hydrogen atom. In that case, the wavefunction, and specifically Born's probabilistic interpretation, had been subsumed into the Copenhagen interpretation of quantum theory. But, as DeWitt now argued, in a quantum theory of gravity the Copenhagen interpretation was of little help:

> The Copenhagen view depends on the assumed *a priori* existence of a classical level to which all questions of observation may ultimately be referred. Here, however, the whole universe is the object of inspection; there is no classical vantage point, and hence the interpretation question must be reargued from the beginning.

If it was assumed that everything there is, is inside the universe, then there could be no classical 'measuring device' sitting outside whose purpose was to collapse the wavefunction of the universe and make it 'real'.

DeWitt had in mind another, altogether more radical interpretation, devised by American physicist Hugh Everett III in 1957 and submitted as his Princeton PhD thesis under Wheeler. In Everett's 'relative state' formulation the pure Schrödinger wave mechanics is all that is needed to make a complete theory. The wavefunction obeys the deterministic, time-symmetric equations of motion at all times in all circumstances. There are no discontinuities, the wavefunction does not collapse and this process occupies no special place in the theory.

However, the apparent restoration of reality in Everett's formalism comes with a fairly large metaphysical trade-off. If there is no collapse, each term in a superposition is taken to be real, and this means that *all* experimental outcomes are therefore realized. But, of course, an individual observer only ever experiences one outcome: the electron is 'here', or 'there', spin-up, or spin-down. Everett wrote:

> Thus with each succeeding observation (or interaction), the observer state 'branches' into a number of different states. Each branch represents a different outcome of the measurement and the *corresponding* eigenstate for the [superposition]. All branches exist simultaneously in the superposition after any given sequence of observations.

The act of measurement causes the world to split into separate worlds, with different outcomes realized in different worlds. Looking back at the paradox of Schrödinger's cat, we can see that the difficulty is now resolved. The cat is not simultaneously alive and dead in one and the same world, it is alive in one world and dead in another.

DeWitt would later champion Everett's formulation as the 'many worlds' interpretation of quantum theory. In his 1967 paper on canonical quantum gravity he wrote:

> Everett's view of the world is a very natural one to adopt in the quantum theory of gravity, where one is accustomed to speak without embarrassment of the 'wave function of the universe'. It is possible that Everett's view is not only natural but essential.

Because his equation had been derived by combining Einstein's gravitational field equations with Schrödinger's wave mechanics, DeWitt called it the Einstein–Schrödinger equation. Everybody else called it the Wheeler–DeWitt equation.

For sure, it was an equation of limited validity and applicability and did not seem to have any immediate solutions. It was far from a fully fledged theory of quantum gravity, but it did serve to highlight the tremendous problems associated with the quantization of general relativity.

It also represented the beginning of formal attempts to construct a quantum cosmology.

37

Hawking Radiation

Oxford, February 1974

Whilst the Wheeler–DeWitt equation was to be a milestone in the search for a quantum theory of gravity, it quickly proved to be unsatisfactory. DeWitt came to call it 'that damned equation'.

The covariant approach didn't fare much better. In the early 1970s 't Hooft and Veltman studied the renormalizability of a quantum field theory of gravity. As a warm-up exercise, they turned their attention first to the renormalizability of Yang–Mills field theories. Early in 1971 Veltman told 't Hooft that they had to have at least one renormalizable field theory with massive charged vector bosons. 't Hooft told him 'I can do that.' Despite their subsequent success (they won the 1999 Nobel Prize for physics), the theorists concluded that covariant quantum gravity is plagued with unrenmoralizable divergences.

With the search for quantum gravity temporarily stalled, a new generation of cosmologists were instead exploring the various exotic objects predicted by general relativity. Gravity might be the weakest of nature's forces, but it is ultimately irresistible. Gravity binds together and compresses clouds of gas drifting in space. Compression of clouds with sufficient mass leads to the ignition of the nuclear fuel at their cores. Stars are born. Radiation pressure released by burning the nuclear fuel holds back further compression, and the star enters a period of relative stability.

However, as the fuel is expended, the force of gravity grips tighter. For any mass greater than about 1.4 times the mass of the sun,[1] *the force of gravity is ultimately overwhelming. It crushes the body of matter into the obscurity of a black hole.*

Black holes were compelling theoretical objects because they pushed physics to extraordinary limits and because, despite their exotic nature, it was possible that they played important roles in our own universe. In the late 1960s, a young Cambridge University physicist called Stephen Hawking produced a series of papers on black-hole physics in collaboration with mathematician Roger Penrose, then at Birkbeck College in London.

General relativity, they claimed, predicted that at the heart of a black hole there beats a singularity, a region of infinite density and space–time curvature where the laws of physics break down. Of course, what goes on in the region of a singularity is completely hidden from observation by the black hole's event horizon, a fact that Penrose elevated to the status of a principle, called the cosmic censorship hypothesis.

Together with Canadian Werner Israel, Australian Brandon Carter, and British physicist David Robinson, Hawking demonstrated that black holes are also one of the simplest objects in the universe. Their properties and behaviour depend only on their mass, angular momentum, and electric charge, a conjecture called the 'no hair' theorem.[2] *All other information about the matter inside the black hole is lost.*

In a moment of inspiration one night in November 1970, Hawking had realized that the properties of the event horizon meant that this could never shrink—the area of a black hole could never decrease. He later wrote:

> *If the rays of light that form the event horizon, the boundary of the black hole, can never approach each other, the area of the event horizon might stay the same or increase with time but it could never decrease—because that would mean that at least some of the rays of light in the boundary would have to be approaching each other.*

He called Penrose in a state of excitement the next morning. Penrose agreed. There is one other well-known physical property that can never decrease: entropy. The second law of thermodynamics demands that in a spontaneous change entropy always increases.

Could there be a connection between the area of a black hole and its entropy?

[1] This is known as the Chandrasekar limit, named for Indian physicist Subrahmanyan Chandrasekhar.

[2] Here 'hair' represents all other kinds of information other than mass, angular momentum, and electric charge, which is lost behind the black hole's event horizon.

Hawking began to show symptoms of amyotrophic lateral sclerosis (ALS), a type of motor neurone disease, in 1963, shortly after moving to Cambridge. He began to lose neuromuscular control. The initial prognosis was not good. He was given only a few years to live, and there seemed little point in completing his PhD.

But it turned out that he had a rare type of ALS which, though no less debilitating, would take much longer to rob him of control over his body. He married Jane Wilde in 1965 and, with her help, he recovered his sense of purpose and enthusiasm for physics. He secured his doctorate and became first a research fellow, then professorial fellow at Gonville and Caius College, Cambridge.

Though slow, his disease progressed inexorably. In 1965 he could walk with the help of a cane. By 1970 he needed a walking frame. By 1972 he was confined to a wheelchair. Hawking and his wife took on local government bureaucracy, occasionally winning small but important battles against a public infrastructure that did not look kindly on the disabled.

It was one of Wheeler's students, Jacob Bekenstein, who in his Princeton PhD thesis made the connection between the non-decreasing area of a black hole and its entropy. This appeared to resolve a problem with the status of black holes in the context of the second law of thermodynamics. If a black hole consumed material of high entropy—such as a quantity of gas, for example—then the entropy of the universe outside the black hole would decrease.

The entropy of the black hole must therefore increase to ensure that the overall change in entropy associated with the process was either zero or positive. But, according to the 'no hair' theorem, all information is lost inside the black hole. There appeared to be no way of knowing what the entropy of the black hole was, and therefore no way of knowing if the second law was obeyed or violated.

Hawking had argued in 1970 that in such a change the *area* of the black hole would increase. If, as Bekenstein was now suggesting, the area is directly related to the entropy of the black hole, then the second law could be saved.

But Hawking was irritated. Whilst this was a neat solution it carried with it a number of implications which Bekenstein hadn't addressed and which made the idea much less credible. For one thing, a body with entropy also has to have a temperature. And a body with a temperature has to emit radiation. For a perfect 'black' body this would be precisely the spectrum of radiation that Planck had studied in 1900. But how could a black hole, with properties and behaviour determined only by its mass, angular momentum, and electric charge, possess a temperature and *emit* radiation?

In August 1972 Hawking confronted Bekenstein at a month-long summer school on black-hole physics at Les Houches in the French Alps. During the summer school, Hawking worked together with Brandon Carter and Yale physicist Jim Bardeen and derived four definitive new laws that governed black-hole physics. They looked remarkably like the laws of thermodynamics. 'Each black-hole law, in fact, turned out to be identical to a thermodynamical law,' wrote American physicist Kip Thorne, 'if one only replaced the phrase "horizon area" by "entropy", and the phrase "horizon surface gravity" by "temperature".'

Bekenstein was more convinced than ever that he was on the right track, but when challenged on the question of temperature and black-hole radiation his response involved a certain amount of arm-waving.

The rest of the black-hole physics community was convinced that this was all a coincidence. Hawking, Carter, and Bardeen subsequently wrote a paper pointing out the apparently fatal flaws in Bekenstein's arguments. 'I must admit,' Hawking wrote, 'that in writing this paper I was motivated partly by irritation with Bekenstein, who, I felt, had misused my discovery of the increase of the area of the event horizon.'

But, despite its apparent absurdity, the idea of black-hole radiation had been the subject of some earlier discussions. In June 1971 Thorne had visited Soviet physicist Yakov Zeldovich in Moscow. Zeldovich had called Thorne to his apartment early one morning and confronted him with the proposal that a spinning black hole must radiate. 'That's one of the craziest things I've ever heard,' Thorne declared. 'How can you make such a crazy claim? Everyone knows that radiation can flow into a hole, but nothing, not even radiation, can come out.'

Zeldovich argued that a spinning metal sphere emits electromagnetic radiation and, by analogy, a spinning black hole should emit gravitational energy in the form of waves. Thorne was not aware that a spinning metal sphere should behave this way, and asked for the explanation. 'The metal sphere will radiate when electromagnetic *vacuum fluctuations* tickle it,' Zeldovich said. 'Similarly, a black hole will radiate when gravitational vacuum fluctuations graze its horizon.'

The basis for his argument was quantum electrodynamics. The vacuum fluctuations would produce virtual waves which would be accelerated as they skimmed around the event horizon, gaining energy from the black hole's rotation. This transfer of energy would promote the virtual wave to the status of a real wave, which would appear as radiation.

As the black hole radiated its rate of spin would slow. With its rotational energy drained the black hole would eventually stop spinning. And it would then stop radiating.

Thorne was not convinced, and they agreed to a bet. If subsequent theoretical developments showed that Zeldovich was right and spinning black holes could radiate, then Thorne would buy him a bottle of White Horse scotch whisky.

Zeldovich published his proposal, but the idea was largely forgotten. It was not discussed at the Les Houches summer school.

A year later, in August 1973, Hawking and his wife Jane attended a conference in Warsaw to celebrate the five-hundredth anniversary of the birth of Nicolaus Copernicus. In September they travelled on to Moscow to meet with Zeldovich and his research student, Alexei Starobinsky. Thorne joined them as a translator and guide.

Hawking now learned about Zeldovich's crazy proposal from Starobinsky. The Soviet physicists had proved to their own satisfaction that a spinning black hole should radiate. 'They convinced me that, according to the quantum mechanical uncertainty principle [responsible for the vacuum fluctuations], rotating black holes should create and emit particles,' Hawking later wrote. He was nevertheless sceptical about Zeldovich and Starobinsky's proof. He returned to Cambridge to think about it some more, determined to devise a better theoretical treatment.

In the meantime, Canadian William Unruh, another student of Wheeler's, and Don Page, a student of Thorne's, provided independent, though tentative, confirmation of Zeldovich's claim. It looked as though Thorne was going to have to hand over that bottle of whisky.

This was all about quantum effects in the microphysics of a large gravitational body whose properties are predicted by the macrophysics of general relativity. Obviously, what Hawking needed was a fully fledged quantum theory of gravity. In the absence of such a theory, alternative approximate approaches had to be tried. Hawking chose to retain an essentially classical, general relativistic description of the black hole itself and apply quantum field theory to the curved space–time around the event horizon.

He found some curious differences in the application of quantum field theory in curved compared to flat space–times. However, these paled into insignificance when Hawking realized just what he had discovered. Zeldovich had indeed been right: spinning black holes do radiate. Although Hawking felt that he now had a better mathematical basis for this conclusion, this was relatively old news. What was new was that when the black hole stopped spinning, it didn't stop radiating.

'I didn't want particles coming out...,' Hawking explained. 'I wasn't looking for them; I merely tripped over them. I was very sorry because it destroyed my framework, and I did my best to get rid of them. I was rather annoyed.'

'When I did the calculation, I found...that even nonrotating black holes should apparently create and emit particles at a steady rate. At first I thought that this emission indicated that one of the approximations I had used was not valid. I was afraid that if Bekenstein found out about it, he would use it as a further argument to support his ideas about the entropy of black holes.'

Hawking had found the phenomenon, which would come to be known as Hawking radiation, but he did not yet have an explanation. He presented some of his results at an informal seminar in Oxford in November 1973. He then went on to calculate the spectrum of this radiation, and discovered to his astonishment that it is identical to the spectrum of a black body. When combined in this way, two of the most successful theories of twentieth-century physics, whose predictions were bizarre and the subjects of endless debate, were now predicting behaviour no different from nineteenth-century physics.

Where did this radiation come from?

It comes from virtual particle–anti-particle pairs produced in the curved space–time near the black hole's event horizon. The particles are produced within the constraints of Heisenberg's energy–time uncertainty relation, with zero net energy. This means that one particle in the pair will possess positive energy and the other negative energy. Under normal circumstances, the pair would quickly annihilate. But if the negative-energy particle is drawn into the black hole before it can be annihilated, it can acquire energy and become a real particle. Another way of looking at this is to think of the negative-energy particle as having negative mass. As the negative-mass particle falls into the back hole, it gains mass and becomes a real particle.

The positive-energy particle created in the erstwhile pair may escape, to all intents and purposes appearing as though it had been emitted by the black hole.

The particles or anti-particles so emitted can be of any and all types. As negative-energy particles spill through the event horizon, the black hole loses mass and its area decreases. This apparent reduction in entropy is more than compensated for by the entropy of the emitted radiation, so there is no violation of the second law of thermodynamics. As the area decreases, the 'temperature' of the black hole increases, as does the rate of emission. The black hole 'evaporates', eventually disappearing altogether in an explosion small by astronomical standards but, as Hawking estimated, still equivalent to a million 1 megaton hydrogen bombs.

For a black hole with a mass close to that of the sun the evaporation process would take longer than the present age of the universe, but there would have been plenty of time for much smaller black holes formed in the big bang to have evaporated by now.

Hawking fretted over his calculations through Christmas 1973. He submitted a short paper to the journal *Nature* in January. He wrote:

> ...Bardeen, Carter and I considered that the thermodynamical similarity between [the surface gravity] and temperature was only an analogy. The present result seems to indicate, however, that there may be more to it than this.

Hawking presented his results officially at the second conference on quantum gravity held at the Rutherford–Appleton Laboratory near Oxford in February 1974 (the *Nature* paper was published in March). Whilst his conversion was complete, the rest of the physics community was still very sceptical. The chairman of the session at which Hawking presented his paper, British physicist John Taylor from King's College, London, declared that it was all nonsense. Curiously, Zeldovich was also reluctant to accept the idea. Having set Hawking on his journey, he could not accept that a black hole would radiate after it had stopped spinning.

It took several years for the rest of the community to catch up. Hawking's welding of quantum field theory and general relativity was novel but it was not arbitrary. Others began to understand how to apply the same methods, and what they found confirmed Hawking's discovery. Hawking, in the meantime, had learned to throw caution to the winds. Having satisfied himself that Bekenstein's instincts were right all along, he declared without further proof that the four laws of black hole physics he had worked out with Carter and Bardeen were indeed the laws of thermodynamics. 'I would rather be right than rigorous,' he told Thorne.

Zeldovich also capitulated. Thorne handed over a bottle of White Horse scotch whisky in September 1975.

Although Hawking's discovery had not been based on a fully fledged quantum theory of gravity, his successful application of quantum field theory in curved space–time—and the revelations that followed—sparked renewed interest in the field. It was a beacon of hope in a sea of frustration. A few years later it would lead Hawking to resurrect the covariant approach based on Feynman quantization of general relativity, this time in a four-dimensional space, or four-space.

Hawking's achievements were recognized by his election to a Royal Society Fellowship later in 1974. At 32, he was one of the youngest Fellows in the Society's history. His condition continued to deteriorate, however. His speech had now become so slurred that few could understand him. Routine physical activities required major effort. Desperate for help, Jane persuaded him to allow one or more of his graduate students into his close circle of carers.

He was inducted into the Royal Society in May 1974. He had to be carried up the wheelchair-unfriendly steps into the building. Unable to walk to the podium to sign his name in the Society's roll book, he was handed the book, which he signed with great difficulty.

The astronomer Carl Sagan was attending a meeting on the search for extraterrestrial life in another part of the building. He was drawn to the ceremony during a coffee break and, standing in the doorway, witnessed the signing. 'In the front row a young man in a wheelchair was, very slowly, signing his name in a book that bore on its earliest pages the signature of Isaac Newton,' Sagan wrote. 'When at last he finished, there was a stirring ovation. Stephen Hawking was a legend even then.'

38

The First Superstring Revolution

Aspen, August 1984

The success of the standard model in the mid-1970s led inevitably to questions about its SU(3) × SU(2) × U(1) symmetry properties. Why this particular combination? Was there a higher symmetry group that would properly unify both the strong force and the electro-weak force? A number of different approaches had been tried, but in 1974 Sheldon Glashow was convinced he had a solution. With Harvard postdoctoral fellow Howard Georgi, he developed a so-called grand unified theory based on the SU(5) symmetry group. They declared it the 'gauge group of the world'.

The theory proliferated more Higgs particles. It also allowed transformations between quarks and leptons. This meant that a quark inside a proton could in principle transform into a lepton. 'And then I realized that this made the proton, the basic building block of the atom, unstable,' Georgi said. 'At that point I became very depressed and went to bed.'

Admittedly, the time required for proton decay was longer than the age of the universe, but it was a prediction that nevertheless could be checked experimentally. By the early 1980s, the experimental results appeared unequivocal: the proton is more stable than Georgi and Glashow's SU(5) quantum field theory suggested.[1]

[1] These experiments involved searching for a single proton decay event from a large volume of protons shielded from cosmic rays. As Carlo Rubbia explained: '...just put half a dozen graduate students a couple of miles underground to watch a large pool of water for five years.' Quoted in Woit, p. 104.

Another approach emerged in the early 1970s from theorists in the Soviet Union and was independently rediscovered in 1973 by CERN physicists Julius Wess and Bruno Zumino. This was called supersymmetry. It featured 'super-multiplets' of particles which, for the first time, connected the matter particles—fermions—with the bosons that carried forces between them. It also proliferated more particles. For every fermion, the theory predicted a corresponding boson. This meant that for every particle in the standard model, the theory required a massive supersymmetric partner with a spin different by ½. The partner for the electron was called the selectron (a shortening of supersymmetric-electron). Quarks are partnered by squarks. Supersymmetric partners of the photon, W, and Z particles are the photino, wino, and zino. These particles may exist at the TeV energy scale, but to date they have not been found.

In supersymmetry, a 'super-gauge' particle acting on a target fermion or boson changes the spin of the target particle by ½. This kind of change affects the space–time properties of the target particle, such that it is slightly displaced.[2] *Forces such as electromagnetism cannot do this. They can change the direction of motion, momentum, and energy of the particle but they cannot displace it in space–time in this way. This displacement is equivalent to a gauge transformation characteristic of the gravitational force. A super-gauge theory is also therefore a theory of gravity. Such theories are collectively called supergravity, and the super-gauge particle is the gravitino, with spin 3⁄2, the super-partner of the graviton.*

There are many different types of supersymmetric transformation, and theories with more than one transformation are called extended supersymmetric theories. These theories are classified according to the number of supersymmetry 'generators' they posses. Theories with 32 generators, called N = 8 supersymmetry, automatically produce gravitons. They are also theories of supergravity.[3]

For a time, theories of N = 8 supergravity were very popular. In April 1980, Stephen Hawking used his inaugural lecture as Lucasian Professor of Mathematics at Cambridge, a position once held by Newton and Dirac, to declare that the end of physics was in sight. He believed that there was a 50:50 chance that N = 8 supergravity would prove to be the ultimate unifying theory for all physical forces.

But there was yet another approach. It had lurked in the background, largely ignored or dismissed. Yet it would soon overtake all these efforts, to become the dominant candidate as the theory of everything.

[2] Actually, a supersymmetry transformation is equivalent to the square-root of an infinitesimal translation in space.

[3] Theories with more than 32 generators are possible but predict massless particles with spin greater than 2, which are thought not to exist in nature.

In 1968, whilst puzzling over experimental data from high-energy strong-force particle collisions, young Italian postdoctoral physicist Gabriele Veneziano at CERN had noticed a pattern. The probabilities for two particles to scatter from each other at different collision angles—called scattering amplitudes—appeared to conform to a formula first devised by the eighteenth-century Swiss mathematician Leonhard Euler. This is Euler's beta function, or the Euler integral of the first kind.

Veneziano had no idea why the data conformed to this function and could therefore say nothing about the underlying physics. But identifying elegant, if empirical, mathematical connections between data is usually an important first step to a subsequent revelation.

So it was to prove.

Told about Veneziano's discovery by an enthusiastic colleague, young Yeshiva University professor Leonard Susskind was struck by the simplicity of the mathematical relationship describing the scattering amplitudes. 'I worked on it for a long time,' he later explained, 'fiddled around with it, and began to realize that it was describing what happens when two little loops of string come together, join, oscillate a little bit, and then go flying off. That's a physics problem that you can solve. You can solve exactly for the probabilities for different things to happen, and they exactly match what Veneziano had written down. This was incredibly exciting.'

Similar discoveries were made independently by Danish physicist Holger Nielsen at the Niels Bohr Institute in Copenhagen and by Yoichiro Nambu in Chicago. Instead of treating fundamental particles as point-particles, they were imagined instead as strings—small, one-dimensional filaments of energy. In such a theory the different particles would not be translated into different strings, but into different *vibrational patterns* in one common string type. The mass of a particle is just the energy of its string vibration, with charge and spin being more subtle manifestations. It was a breathtaking idea.

Susskind drafted a paper outlining a quantum theory of strings and submitted it to the journal *Physical Review Letters*. It was promptly rejected. Susskind was deeply perturbed. He returned home nervous and upset. He took a tranquiliser and slept for a while, waking to join some friends who were visiting that evening for a few drinks. The combination of tranquiliser and alcohol didn't agree with him, and he passed out.

Having tried—and failed—to get the editor of *Physical Review Letters* to reconsider, Susskind resubmitted his paper to the journal *Physical Review D* in July 1969. It was published in 1970, to no acclaim. The reaction of the physics community to the idea of particles as strings was typified by an encounter with Murray Gell-Mann. After lecturing at an international conference on fundamental high-energy interactions held at Coral Gables in Florida, Gell-Mann headed back to his motel. He got stuck in the elevator with Susskind.

'What do you do?' Gell-Mann asked.

'I'm working on this theory that hadrons are like rubber bands, these one-dimensional stringy things,' Susskind explained.

Gell-Mann just laughed.

But, in fairness, string theory in this early form did not seem to have an awful lot going for it. For one, it was a theory of bosonic strings only—it could not accommodate 'particles'—string vibrational patterns—with half-integral spin. This was a significant stumbling block for a theory that was supposed to be about the strong nuclear force. It was also a theory in a twenty-six-dimensional space–time, rather than the more familiar four. Worse still, it predicted the existence of tachyons, hypothetical particles that travel only at speeds faster than light and wreak untold damage on the principles of cause-and-effect.

But within a matter of a few months one of the problems of early string theory had been solved. In the spring of 1971 French physicist Pierre Ramond, working at the National Accelerator Laboratory in Chicago, figured out how to make the transformations between bosons and fermions that formed the basis for what was later to be called supersymmetry. It meant that the string (or, more correctly, supersymmetric string or *superstring*) vibrational patterns for fermions had been found.

Within a few years some of the other problems had been resolved, too. The number of dimensions required for the superstrings to vibrate in had reduced from twenty-six to ten: nine spatial dimensions and one time dimension. Though this was still a lot more than the four dimensions of our everyday world, it was at least a step in the right direction. And the tachyons were now gone.

In Princeton, American theoretical physicist John Schwarz watched the development of early string theory with great interest. He had also been struck by the implications of Veneziano's work and in 1969 began collaborating on string theory with David Gross. They were joined in Princeton by two French theorists, André Neveu and Joël Scherk who, by American standards, had the equivalent of PhDs but had yet to secure their *doctorat* in the French system. They were assigned to Schwarz as graduate students, but when he suggested they take a course in quantum mechanics they explained that they didn't need one. Schwarz sent them away.

They came back a short time later to show him some results they had derived for the scattering of strings. It quickly became clear to Schwarz that they really didn't need a graduate course on quantum mechanics.

Schwarz was among those physicists who identified in 1972 that superstrings would need to vibrate in just ten space–time dimensions, but his work on string theory was not enough to earn him a tenured position at Princeton. Gross, however, was promoted. It was about this time that he resolved to finish off quantum field theory once and for all, by showing that there was no class of renormalizable theory that was asymptotically free.[4]

The early 1970s was a bad time in the American market for academic physics. A few months later Schwarz was offered a research associate position at Caltech.

The summer of 1973 saw asymptotic freedom and the birth of what was to become QCD, which quickly became established as the theory of the strong force. There seemed little need for any further work on strings. But Schwarz and Scherk were not ready to let go. Scherk came to Caltech in January 1974 and they resolved to continue their collaboration:

> I think we were kind of struck by the mathematical beauty; we found the thing a very compelling structure. I don't know that we said it explicitly, but we must have both felt that it had to be good for something, since it was just such a beautiful, tight structure. So, one of the problems that we had had with the string theory was that in the spectrum of particles that it gave, there was one that had no mass and two units of spin. And this was just one of the things that was wrong for describing strong nuclear forces, because there isn't a particle like that. However, these are exactly the properties one should expect for the quantum of gravity.

[4] See Chapter 26, p. 261–262.

If the problem of providing a theoretical description of the strong force had been solved, then the problem of providing a quantum theory of gravity had not. Superstring theory not only promised to accommodate all the particles of the Standard Model then known, it also predicted a particle with the properties of the graviton. Superstring theory was not a theory of the strong force. It was potentially a theory of *everything*.

It was a revelation. Superstrings come in two forms: open and closed. Open strings have loose ends that can be thought of as representing charged particles and their anti-particles, one at either end, with the string vibration representing the particle carrying the force between them. Open strings therefore predict both matter particles and their forces. But the theory also *demands* closed strings. When a particle and anti-particle annihilate, the two ends of the string join up and the result is a closed string.

But if there are closed strings then there are also gravitons and the force of gravity. Rather than try to force fit quantum theory and general relativity together, superstring theory appeared to be saying that all nature's forces are just different vibrational patterns in open and closed strings. In superstring theory these forces are automatically unified.

It offered an enticing possibility. All the 'fundamental' particles then known—their masses, charges, spins—the forces between them and all the Standard Model parameters that could not be derived from first principles, could be subsumed into a single theory with just two fundamental constants. These were the constants that determine the tension of the string and the coupling between strings.

Nobody was interested.

In the meantime, Hawking declared that black holes could radiate and momentum began to build for supersymmetry and supergravity. Schwarz continued his collaboration with both Neveu and Scherk and published papers on supersymmetric Yang–Mills field theories. He also collaborated with British physicist Michael Green at Queen Mary College, London. Together they resolved various murky aspects of the different types of superstring theory that had by then emerged, which they called Type I, Type IIA, and Type IIB.

Although their work continued to be largely ignored, the theory was winning a few advocates. Among them was Gell-Mann, who had by now been won round, and Princeton mathematical physicist Edward Witten.

By the early 1980s, Witten was already a phenomenon. He had studied history and linguistics at Brandeis University, graduating in 1971. He went on to study economics at the University of Wisconsin and contributed to George McGovern's presidential campaign. After McGovern's overwhelming defeat by Richard Nixon in 1972, Witten abandoned politics and moved to Princeton to study mathematics. He transferred to physics shortly afterwards. He became one of Gross's graduate students, securing his PhD in 1976. Four years later, he was a tenured professor at Princeton.

Schwarz and Green would visit Princeton and discuss their progress with Witten. As Schwarz later explained: '... in our discussions with him it became very clear that one of the really important problems that we had to think about was something called anomalies. We had been aware of this ourselves to some extent, but I think in our discussions with Edward it became even clearer to us how important this was.'

These 'gauge anomalies' relate to the mathematical consistency of the superstring theory after making certain quantum corrections. If the corrections destroy certain symmetries, then the theory might no longer make mathematically consistent predictions. Different versions of the theory have different symmetry properties. Type IIA superstring theory is mirror-symmetric, and the physicists could be confident that this theory would exhibit no anomalies. But look in a mirror. The world we inhabit is not mirror-symmetric.

In 1984, Witten and Spanish physicist Luis Alvarez-Gaumé pointed out that there could be further anomalies due to the gauge field of gravitation, but also showed that in a low-energy approximation Type IIB theory these anomalies cancelled. This seemed encouraging, but the Type IIB theory cannot accommodate Yang–Mills fields of the kind required by the Standard Model.

This left all hopes pinned on the Type I superstring theory. In the summer of 1984, Green and Schwarz retreated to Aspen to think on the problem some more.

The Type I superstring theory in principle accommodates an infinite number of different symmetry groups. Green and Schwarz had to find one symmetry group in which the gauge anomalies would cancel. The question was: which one?

A clue came from studying a number of Feynman diagrams derived from string interactions. A couple of these diagrams gave formulas for the anomalies that looked very similar. Schwarz wondered if these contributions might cancel for a specific choice of symmetry group. They then went into a seminar. At the end of the seminar, Green said: 'SO(32)'.

It checked out. In a low-energy approximation, both Yang–Mills and gravitational anomalies canceled in a Type I superstring theory based on the symmetry group SO(32), a group of rotations in a 32-dimensional space.

Before Green and Schwarz could tell anyone about what they had discovered, Schwarz was obliged to act the part of Gell-Mann in a cabaret that had been organized at the Aspen Center. The same cabaret had been put on ten years before. Gell-Mann himself had jumped up from the audience and declared that he had found the theory of everything, at which point he was carried away by men in white coats.

Schwarz now reprised the role, but with a twist. He ran up on stage and instead declared: 'I figured out how to do everything. Based on string theory with a gauge group SO(32), the anomalies cancel! It's all consistent! It's a finite quantum theory of gravity! It explains all the forces!' The audience, not realizing this was the first public announcement of their discovery, laughed as Schwarz was carried off.

Green and Schwarz returned to Caltech and started to write up their results. Witten had heard rumours from Aspen, and called to find out more. They sent him a draft manuscript by Federal Express.

Things then happened very quickly.

The Green–Schwarz paper was submitted to *Physics Letters* in September 1984. Witten submitted his first paper on superstrings to the same journal later that same month. At Princeton, David Gross, Jeffrey Harvey, Emil Martinec, and Ryan Rohm found another version of the theory, called heterotic (or hybrid) superstring theory, in which the anomalies cancelled. They submitted their paper to the journal *Physical Review Letters* in November 1984.

The first superstring revolution had begun.

There remained the puzzle of strings vibrating in a ten-dimensional space–time. To a certain extent, the idea of spatial dimensions additional to the familiar right-left, backwards-forwards, up-down dimensions had been anticipated by German mathematician Theodor Kaluza in 1919. He sent Einstein a draft paper showing how the two forces then known—electromagnetism and gravity—could be unified in a single theoretical framework which required four spatial dimensions. In 1926, Oscar Klein had suggested that the extra dimension might be rolled up so small and tight, with a radius equal to the Planck length, that it would remain forever inscrutable. The resulting framework became known as Kaluza–Klein theory.

But why stop at one hidden dimension? If there could be one such dimension, why couldn't there be two, or three, or nine? The point about superstring theory was that it demanded precisely nine spatial dimensions for the strings to vibrate in, no more and no less. It was necessary to assume that the six additional spaces are curled up in a small bundle so that we can't experience them. Superstring theory had something to say about the shape of these extra dimensions, but it had nothing to say about their size.

In 1985, Princeton physicists Philip Candelas, Gary Horowitz, Andrew Strominger, and Witten published a paper explaining that the topology of the six extra dimensions had already been found as the solution to an abstract problem in mathematics by Eugenio Calabi in 1957 and Shing-tung Yau in 1978. According to this theory, at every point in our familiar three-space, there lies a six-dimensional Calabi–Yau shape, so small that it is beyond the reach of experimental instruments.

Each 'hole' in the Calabi–Yau shape gives rise to a family of low-energy string vibrations. A Calabi–Yau shape would therefore produce three families of vibrational patterns, corresponding to the three generations of particles in the Standard Model.

But this was not plain sailing. Although the theory demands that the extra dimensions curl up into a Calabi–Yau shape, there are potentially hundreds of thousands of such shapes that would satisfy this requirement. Each candidate comes with its own set of free constants which determine its size and shape. Each produces a different version of particle physics.

These were not the only problems. The low-energy vibrational patterns correspond to massless particles, yet the three generations of particles of the Standard Model all have mass. The theory also predicts many more vibrational patterns than there are particles.

Despite the great rush of enthusiasm, it was clear that superstring theory was quickly losing any sense of uniqueness and hence a capacity to make unique predictions. It would have some way to go before it could fulfil its early promise as *the* theory of everything.

39

Quanta of Space and Time

Santa Barbara, February 1986

Superstring theory promised much, and momentum built through the mid-1980s as more theoreticians began to grapple with its structure. As this was a quantum field theory of strings, it seemed that the covariant approach to quantum gravity, favoured largely by the particle physics community, was winning out.

But, for those physicists grounded in general relativity, there was still (at least) one significant objection to superstring theory. In order for the strings to vibrate, superstring theory had to assume a background space–time for them to vibrate in. The theory was not background-independent, and there appeared to be no easy way to free it from this dependence. Many superstring theorists did not even appear to acknowledge that this was a problem.

The canonical approach to quantum gravity had stalled with the Wheeler–DeWitt equation. The dying embers had been stirred in 1983 by Hawking and American physicist James Hartle at the Enrico Fermi Institute in Chicago. They had derived a wavefunction of the universe from a specific solution of the Wheeler–DeWitt equation and had used Feynman's path-integral approach to explore aspects of big bang cosmology.

Their solution was based on a 'no boundary' assumption, applied to the finite space–time of a simple model universe.[1] *No boundary to space–time implies that there was no beginning to the universe, as we commonly understand the meaning of the word 'beginning'.*

[1] To begin to understand what it means for space–time to have no boundary, think how you might answer the question: 'What lies north of the north pole?'

When at a conference at the Vatican in 1981 Pope John Paul II suggested that cosmologists confine their speculations to moments after the Creation, Hawking did not have the courage to explain that his no-boundary assumption meant that there could be no moment of Creation. 'I had no desire to share the fate of Galileo,' he explained, 'with whom I feel a strong sense of identity, having been born exactly 300 years after his death!'

The ground-state wavefunction was found to correspond to an expanding universe. Excited states were also found which expanded and then collapsed, but also had a finite probability of tunnelling to states describing continually expanding universes. But the approach was to prove to be another blind alley, and enthusiasm quickly waned. More gloom gathered around the canonical approach.

As superstring theorists wrestled with string vibrations in ten dimensions, curled-up Calabi–Yau spaces, and a general loss of uniqueness, a series of discoveries was being made that would resurrect the canonical approach and provide a genuine rival to superstrings as a theory of quantum gravity. A rival that was, moreover, completely independent of any assumed space–time background.

It would become known as loop quantum gravity.

In August 1982 Amitabha Sen at the University of Maryland submitted a paper to the journal *Physics Letters* in which he proposed to recast general relativity in terms of a three-dimensional 'spin system'. This is a system based on 'spinors', or spin-vectors first developed by the French mathematician Élie Cartan in 1913. Spinors had first been deployed in quantum theory by Pauli in May 1927, in the form of his spin matrices. This basic spinor structure had subsequently dropped out of Dirac's relativistic wave equation, for both the electron and positron.

In Sen's development, the spinors had no physical meaning. They were a mathematical device which allowed him to recast general relativity in a space of complex vectors more comprehensive in terms of its ability to accommodate geometrical information. What he found was a new way to express the constraints of the ADM Hamiltonian formalism of general relativity.

The formalism was *much* simpler.

Sen's work was picked up a few years later by Indian physicist Abhay Ashtekar at Syracuse University in New York. Ashtekar had studied gravitational physics and general relativity in Chicago, securing his PhD in 1974. He recognized that Sen's use of spinors could be the basis for a complete reformulation of the constrained Hamiltonian form of general relativity, leading to a much simpler version of the Wheeler–DeWitt equation. By changing from a space–time metric to a space of spin-connections, on which the spinors are propagated, the theory begins to look remarkably like a gauge field theory. 'One could now import into general relativity techniques that have been highly successful in the quantization of gauge theories,' he later wrote.

Lee Smolin, recently appointed as an assistant professor at Yale University, heard about Ashtekar's work and invited him to Yale to give a seminar on the subject. Smolin had graduated in physics and philosophy and gained his PhD at Harvard in 1979. He had held postdoctoral positions at the Institute for Advanced Study, the Institute for Theoretical Physics in Santa Barbara, and the Enrico Fermi Institute in Chicago before joining the faculty at Yale. He had seen Sen's papers and had pondered their significance, but he had not followed them up. Now he learned first-hand from Ashtekar what these ideas promised.

In December 1985 Ashtekar submitted a paper to *Physical Review Letters*, outlining his approach and suggesting '...new ways of attacking a number of problems in both classical and quantum gravity.' In January of the following year he relocated to Santa Barbara to act as the coordinator for a six-month workshop on quantum gravity held at the National Science Foundation Institute for Theoretical Physics at the University of California.

Trading an east coast for a west coast winter might be regarded by many as a smart move. Despite having only just joined the faculty at Yale, Smolin had managed to persuade his department to allow him to attend the Santa Barbara workshop. On arrival, he recruited two colleagues, Paul Renteln from Harvard University and Theodore Jacobson from the University of Maryland, to help him push forward the programme that Sen and Ashtekar had started.

Just a month later, in February 1986, Smolin and Jacobson made a breakthrough. They had sat in a small classroom scrawling up on a blackboard possible solutions to the simpler version of the Wheeler–DeWitt equation that had resulted from Sen and Ashtekar's reformulation. Something quite remarkable had happened:

> All of a sudden we realized that our second or third guess, which we had written on the blackboard in front of us, solved the equations exactly. We

> tried to compute a term that would measure how much our results were in error, but there was no error term. At first we looked for our mistake, then all of a sudden we saw that the expression we had written on the blackboard was spot on: an exact solution of the full equations of quantum gravity.

The moment was etched on Smolin's memory. He remembered that it was a sunny day, and that Jacobson was wearing a T-shirt. '... then again, it is always sunny in Santa Barbara and Ted always wears a T-shirt.'

They had found solutions to the Wheeler–DeWitt equation based on 'Wilson loops', named for American theoretician Kenneth Wilson. Wilson loops had been introduced into quantum theory in the 1970s in attempts to find an exactly solvable (i.e. non-perturbative) formulation of QCD. They can be thought of as closed loops of force. We tend to think of the magnetic field around a bar magnet in terms of the 'lines of force' drawn between its poles. These can be revealed in the laboratory by sprinkling iron filings on a piece of paper held above the magnet. In the absence of charged particles the field excitations become closed loops.

Wilson had constructed quantized loops of force on a space–time that was not continuous, but composed of a discrete lattice of extremely small dimensions. Particles could exist only at the nodes (intersections) of the lattice and field lines only along the lattice edges. In essence, according to Wilson's analysis, space–time *is* the lattice: there is no space–time between the lattice nodes and edges.

Smolin and Jacobson had dispensed with the lattice and had worked only with the loops as the fundamental 'observables' of the theory. What they discovered was that, provided they did not intersect or possess sharp kinks, any one of an infinite number of these quantized loops could serve as solutions to the Wheeler–DeWitt equation. There was no need for a background space–time framework. This was not a theory of loops existing in space and time. Rather, the relationships between the loops *define* space. Smolin wrote:

> ...even at the start we knew that we had in our grasp a quantum theory of gravity that could do what no theory before it had done—it gave us an exact description of the physics of the Planck scale in which space is constructed from nothing but the relationships among a set of discrete elementary objects.

There was just one more hurdle to clear.

In order to qualify as a fully background-independent theory, the solutions that Smolin and Jacobson had obtained had to be shown to be solutions of a second set of equations, called diffeomorphism constraints. These are the equations that could demonstrate that the solutions conformed to the general covariance requirements of general relativity. They would demonstrate that the solutions were completely independent of any choice of coordinate system. In general relativity, all coordinate systems must be equivalent.

This was meant to be the easy part, but it proved stubbornly resistant. Working back at Yale, Smolin and his colleagues grappled with the diffeomorphism constraints and drew a blank. The theory depended not only on the loops themselves but also on the field flowing around the loops, and this made for an intractable problem.

They were getting nowhere fast. Then, in October 1987, a young Italian theorist called Carlo Rovelli arrived at Yale. Rovelli had gained his PhD at the University of Padua a year earlier. He and Smolin had met towards the end of the Santa Barbara workshop, and Rovelli had subsequently written asking if he could come to Yale and spend some time collaborating with Smolin on quantum gravity. On his arrival, Smolin had to explain to his visitor that there was nothing to be done, as they had become completely stuck.

There was an awkward silence. Then Smolin suggested that they go sailing instead. Happily, Rovelli was a keen sailor.

Smolin saw nothing of Rovelli the next day. On the third day, Rovelli appeared at the door to Smolin's office and declared: 'I've found the answer to all the problems.'

He had reformulated the theory once more, using the loops themselves as 'basis' states. This removed the dependence on the field, leaving only the loops. The result was a theory that satisfied the diffeomorphism constraints and was genuinely background-independent.[2]

[2] After making this second breakthrough, Smolin discovered to his dismay that the loop representation they had discovered had already been applied to QCD (but not quantum gravity) in 1980 by Uruguayan physicist Rodolfo Gambini and his collaborator Antoni Trias.

In this formulation of quantum gravity, the loops are the quanta of space. As Rovelli later wrote:

> The loops *are* space because they are the quantum excitations of the gravitational field, which is the physical space. It therefore makes no sense to think of a loop being displaced by a small amount in space. There is only sense in the relative location of a loop with respect to other loops, and the location of a loop with respect to the surrounding space is only determined by the other loops it intersects. A state of space is therefore described by a net of intersecting loops. There is no location *of* the net, but only location *on* the net itself; there are no loops on space, only loops on loops.

Space is a result of the way loops knot and link together, forming a 'weave', stitching together the very fabric of the universe. No longer was the play of nature just about the actors strutting and fretting their hour upon the stage. It was also as much about the stage, and the set; the background against which the play is acted out.

Rovelli gave the first seminar on loop quantum gravity at Syracuse University. Ashtekar was very appreciative. The first public announcement was made at an international conference on gravitation and cosmology held in Goa, India, in December 1987.

As they continued to work on the theory, Smolin and Rovelli realized that they had stumbled across a structure already familiar. They discovered parallels between the networks of loops and a mathematical structure developed by Roger Penrose in 1971 called *spin networks*. At its simplest, a spin network is a diagram representing the states of, and interactions between, particles and fields.

In Penrose's original formulation, each vertex or 'node' of the network represents an event, such as two particles colliding or a single particle breaking up into two or more fragments. Each line joining one event to another is labelled with the spin number, a measure of the angular momentum of the particle (in units of $\hbar/2$), with even spin numbers for bosons and odd numbers for fermions. The particles then interact in ways in which the total angular momentum is conserved. Penrose's motivation in developing spin networks was to find a way to eliminate the space–time background and deal only with relationships between the particle events.

Initially rather nervous about using spin networks, Smolin visited Penrose in 1994 and learned directly how to use them from their inventor. He took what he had learned back to Rovelli in Verona, Italy, and together they spent the summer of 1994 working out how to apply spin networks in loop quantum gravity. What they found was that each spin network, formed from a specific combination of loops, represents a quantum state of geometry. For each of these states, the nodes are volume elements, characterized by the number of quanta of volume, and the links between them are elements of area, characterized by the number of quanta of area.

By solving the equation for the mathematical 'operator' for area, they were able to determine the 'eigenstates' of area and so deduce the dimensions of a single quantum. What they found is that every area is composed of an integral number of fundamental quanta with dimensions about the square of the Planck length. Similarly, by solving the equation for the 'operator' for volume, they were also able deduce the dimensions of a single quantum of volume. They found that every volume is composed of an integral number of fundamental quanta given by the cube of the Planck length.

These are unimaginably small dimensions. About 10^{65} quanta of volume will fit inside a single proton.[3] At these dimensions space–time is no longer continuous. There is no area smaller than a quantum of area, no volume smaller than a quantum of volume.

And, of course, the interconnectedness of space and time means that time, viewed as the rearrangement of links between the loops or evolution of the spin network, must be similarly quantized, with the dimensions of the quanta determined by the Planck time. It had been believed for some time that the Planck scale should somehow represent an ultimate limit, beyond which it is no longer possible to ask sensible questions about the nature of space and time. Now loop quantum gravity had furnished the reason.

There is no restriction on the size of a spin network. As Smolin explained: 'If we could draw a detailed picture of the quantum state of

[3] That's 10 raised to the power 65, or 1 followed by 65 zeros.

our universe—the geometry of its space, as curved and warped by the gravitation of galaxies and black holes and everything else—it would be a gargantuan spin network of unimaginable complexity, with approximately 10^{184} nodes.'

There was still plenty to do. Loop quantum gravity was recognized as a strong candidate for a quantum theory of gravity. At this time, it was not considered as a candidate theory of everything, like superstring theory. Although it had provoked little interest at the time, the fact that superstring theory accommodated a particle with the expected properties of the graviton held much appeal for particle physicists. They liked particles. In loop quantum gravity, however, the graviton appeared to have gone missing.

And loop quantum gravity was a theory of space and time. The familiar particles of the Standard Model had yet to be introduced to it.

40

Crisis? What Crisis?

Durham, Summer 1994

The Standard Model could not be advanced beyond the position it had reached in the late 1970s. The Higgs boson remained out of reach of CERN and Fermilab physicists. Grand unified theories had come, and gone. Supersymmetry looked promising, but proliferated supersymmetric partners for every particle in the Standard Model; particles for which there was no experimental evidence. For a time, Hawking had declared N = 8 supergravity to be the real deal, the physicists' holy grail. But it was not.

Superstring theory, meanwhile, had suffered an embarrassment of riches. In addition to Type I string theory, based on the symmetry group SO(32), Type IIA, Type IIB, and heterotic string theories, a fifth version was formulated. This was a variant of heterotic string theory, also based on SO(32). As theories proliferated, interest waned.

Witten then made a bold conjecture. A number of interesting dualities between string theory and quantum field theories had emerged in the early-mid 1990s.[1] *At a superstring theory conference at the University of Southern California in March 1995, Witten speculated that the duality relations between the five string theories meant that they might all be subsumed into one, overarching structure. He called it M-theory. He was not specific on the meaning or significance of 'M'.*

[1] For example, S-duality, or strong–weak duality, allows states with a coupling constant g in one type of theory to be mapped to states with coupling $1/g$ in its dual theory. This means that the techniques of perturbation theory, normally applicable only in quantum field theories for systems of low energy and weak coupling, can be applied to the high-energy, strong-coupling regimes of string theory.

Witten could not formulate M-theory, he could only speculate that the theory must exist. If it does exist it will require seven extra spatial dimensions in addition to our familiar space–time. The seventh dimension had been missed in the earlier versions of string theory because it is so much smaller than all the others. A further revelation from M-theory is that it is not just a theory of one-dimensional strings. It accommodates higher-dimensional objects, called membranes. Two-dimensional membranes are referred to as two-branes, three-dimensional membranes as three-branes, and so on.

The M-theory conjecture sparked a second superstring revolution. Superstrings were 'hot' once again, and demand from prestigious academic institutions for superstring theorists exploded. As a new century dawned, superstring theory was fast becoming an industry.

Advances were made in loop quantum gravity also during this time, but this was the territory of the relativists, not particle physicists, and was therefore much less fashionable.

Loop quantum gravity was placed on a firmer mathematical foundation. The theorists began to understand how changes in the spin networks reflect the curvature of space–time in the presence of mass. Masses were predicted to attract each other according to the laws first devised by Newton. The graviton appeared in low-energy approximations to the theory as a collective excitation of the spin network, much like phonons (the quantum particle equivalent of sound waves) are collective excitations in the lattice of atoms or ions that make up a solid.

In 1996, the theory was used to derive the Bekenstein–Hawking formula for the entropy of a black hole, an accomplishment paralleled in superstring theory at around the same time. The theory was also found to make the heretical prediction that the speed of light may not after all be absolute, but may depend on the frequency of its photons. This means that it may actually be possible to put the theory to the test. The effect is small, but when accumulated over several billion light-years this difference may be measureable. Observations of gamma-ray photons from bursts that signal distant cosmic explosions are thought to provide a fertile testing ground.[2]

Superstring theory may have dominated the landscape, but there were also other programmes, such as topos theory, twistor theory, technicolour, topological quantum field theory, to name a few. There was progress, of a kind, although the basic problems remained stubbornly unyielding.

[2] NASA's Fermi Gamma-ray Telescope, launched on 11 June 2008, is believed to offer the required sensitivity to measure these small differences.

There was, however, yet another problem to be addressed. A quantum theory of gravity would first require the unification of general relativity and quantum theory at the fundamental level of interpretation.

This was not good news.

Bryce DeWitt had identified the problem of interpretation in his early work on the Wheeler–DeWitt equation. He had reached for Everett's 'many worlds' interpretation as a way of resolving the quantum measurement problem, avoiding the need to invoke an external observer, sitting outside the universe, to make a 'measurement' and collapse the wavefunction of the universe. To bridge the gulf between the quantum and classical realms, quantum theory demands a privileged frame of reference within which classical outcomes can be obtained from quantum measurements. General relativity removes all such privileged frames.

It was a curious juxtaposition. Just as experiments designed to explore the nature of physical reality at the quantum level were confirming the supremacy of Bohr's complementarity and the Copenhagen interpretation—and so deepening the quantum measurement problem—so quantum cosmologists were concluding that the original Copenhagen interpretation just couldn't be right.

But neither was many worlds. When Everett had first presented his theory in 1957, he had written of the observer state 'branching' into different states and drew an analogy with the branching of a tree. However, variants of Everett's original interpretation assumed that the world with which we are familiar is but one of a very large number (possibly an infinite number) of *parallel* worlds. Thus, instead of the world splitting into separate branches as a result of a quantum event, the different terms of the superposition are partitioned between a number of already existing parallel worlds.

We can imagine that, in one of these worlds, a version of you records the outcome of a measurement of the spin of an electron. You observe the result spin-up. In another world, a parallel version of you records the result spin-down. DeWitt called it 'schizophrenia with a vengeance'.

The parallel worlds may interact and merge. Indeed, it has been argued that we obtain indirect evidence for such a merging of worlds every time we perform an interference experiment. Having initially embraced his student's work, Wheeler came to reject it, arguing that it carried too much 'metaphysical baggage'. Many physicists dismissed it as cheap on assumptions, but expensive with universes. It led Gell-Mann to complain:

> One distinguished physicist, well versed in quantum mechanics, inferred from certain commentaries on Everett's interpretation that anyone who accepts it should want to play Russian roulette for high stakes, because in some of the 'equally real' worlds the player would survive and be rich.

Clearly some kind of alternative interpretation was needed.

In 1984, American physicist Robert Griffiths embarked on a research programme designed to construct a new, alternative interpretation of quantum theory, based on the notion of 'consistent histories'. Griffiths used the word history much as we use it in common language, as a summary of a series of connected events that unfolded in some recent past.

In this respect, Griffiths' histories are similar to Feynman's histories, except that Feynman's approach was focused on the specific dynamics associated with the motion of quantum particles from one place to another, whereas Griffiths' approach is more concerned with the logical framework and interpretation of quantum theory. Alternative Feynman histories refer to different paths a quantum particle might take from here to there, whereas Griffiths' histories refer to different descriptions of events unfolding in time that, individually, serve as different but internally consistent ways of looking at the process.

In this interpretation, there is no 'correct' history, no right way to look at a quantum event. There are, instead, many possible consistent histories—all equally valid. We choose which histories to apply depending on the type of experiment we wish to perform.

If we recognize some of the possible alternative histories as 'particle histories' (with which-way trajectories) and some as 'wave histories' (resulting in interference effects), then the consistent histories interpretation is a restatement of Bohr's principle of complementarity in the language of probability. In fact, Griffiths claimed that the consistent histories interpretation is 'Copenhagen done right'. He wrote:

> ...just as there is no single 'correct' choice of consistent [history] for describing the system, there is no single arrangement of apparatus which can be used to verify the predictions obtained using different [histories].

In 1991 Gell-Mann and Hartle extended Griffiths' consistency conditions and introduced the concepts and principles of decoherence. The consistent histories interpretation is now often referred to as 'decoherent histories'. The ambition of this interpretation was subsequently described by Gell-Mann:

> We believe Everett's work to be useful and important, but we believe that there is much more to be done. In some cases too, his choice of vocabulary and that of subsequent commentators on his work have created confusion. For example, his interpretation is often described in terms of 'many worlds', whereas we believe that 'many alternative histories of the universe' is what is really meant. Furthermore, the many worlds are described as being 'all equally real', whereas we believe it is less confusing to speak of 'many histories, all treated alike by the theory except for their different probabilities'.

That seemed to do the trick. It was no longer necessary to invoke an external observer: wave–particle duality and other curious quantum phenomena are simply manifestations of the different, but consistent, histories that evolve in the quantum domain in our one, and only, universe. And the quantum domain—with all its superpositions—evolves smoothly into our more familiar classical domain through the mechanism of decoherence.[3]

But this was not quite the end of the matter. Sensing that this was not so straightforward, British theorist Fay Dowker, a former student of Hawking, set to work with her colleague Adrian Kent. At a conference on quantum gravity held at the University of Durham in the summer of 1994, Dowker described what they had found. Her lecture left a lasting impression on Smolin, who was in the audience. It was one of the most dramatic experiences in his scientific career, as he explained six years later:

> While the 'classical' world we observe, in which particles have definite positions, may be one of the consistent worlds described by a solution to the theory, Dowker and Kent's results showed that there had to be an infinite number of other worlds, too. Moreover, there were an infinite number

[3] We'll set aside for the moment our concern that decoherence explains the absence of classical superpositions but does not solve the measurement problem. See Chapter 34.

> of consistent worlds that have been classical up to this point but will not be anything like our world in five minutes' time. Even more disturbing, there were worlds that were classical now that were arbitrarily mixed up superpositions of classical at any point in the past. Dowker concluded that, if the consistent histories interpretation is correct, we have no right to deduce from the existence of fossils now that dinosaurs roamed the planet a hundred million years ago.

There is no 'correct' family of histories that emerges as a result of some law of nature. The theory regards all possible histories to be equally valid and our choice of history therefore depends on the kinds of questions we ask. This leaves us with a significant context-dependence, in which our ability to make sense of the theory depends on our ability to ask the 'right' questions.

If we want a history in which dinosaurs roamed the earth, then we must ask a question for which such a history is appropriate. Of course, this is just the same as observing that if we want to prove that the electron is a wave, then we must set up an experiment designed to reveal electron diffraction or interference.

The multiplicity of worlds or parallel universes implied by Everett's original interpretation are simply traded in the decoherent histories interpretation for many histories, all equally 'real'. 'So conventional quantum cosmology,' Smolin concluded, 'seems to be a theory in which we can formulate the answers, but not the questions.'

After a century of great achievement, we still don't have a definitive, unified quantum theory which accommodates all the known fundamental particles and all the known forces between them. The twenty parameters of the Standard Model must still be determined by experiment. We still do not know for sure how particles acquire mass. And, whilst we must acknowledge that physical reality at the quantum level is far stranger than we could have possibly imagined, we're still unsure how it should be interpreted. And the situation has changed little in the last thirty years or so.

Crisis.

Perhaps we've been here before.

In 1934 Robert Oppenheimer had complained bitterly about the state of theoretical physics in a letter to his brother Frank:

> As you undoubtedly know, theoretical physics—what with the haunting ghosts of neutrinos, the Copenhagen conviction, against all evidence, that cosmic rays are protons, Born's absolutely unquantizable field theory, the divergence difficulties with the positron, and the utter impossibility of making a rigorous calculation of anything at all—is in a hell of a way.

Just thirteen years later Isidor Rabi was declaring that the previous eighteen years had been the most sterile of the century.

Physics, like all the sciences, has a tendency to stumble from one crisis to another. Great moments of inspired illumination may be followed by long periods of confused fumbling in the dark. It might be argued that crisis is a natural and preferred state. For only a crisis is likely to sponsor the kinds of outrageous ideas needed to take our understanding of the world forward to the next stage. Only a crisis could push Heisenberg to wander aimlessly around Fælled Park late at night, wondering if nature could possibly be as absurd as it seemed. Only in moments of desperation can we find inspiration.

But this crisis is rather different.

The problem with superstring theory is that although it is the dominant physical theory of modern times, it has not so far been used to make a single prediction that could be confirmed or otherwise in a laboratory, or in a particle accelerator. This is a unique situation. Richard Feynman was already expressing his concerns in the late 1980s:

> I don't like that they're not calculating anything. I don't like that they don't check their ideas. I don't like that for anything that disagrees with an experiment, they cook up an explanation—a fix up to say 'Well, it still might be true.'

In his book *Facts and Mysteries in Elementary Particle Physics*, published in 2003, Martinus Veltman did not even wish to acknowledge supersymmetry and superstrings:

> The fact is that this is a book about physics, and this implies that the theoretical ideas discussed must be supported by experimental facts. Neither supersymmetry nor string theory satisfy this criterion. They are figments of the theoretical mind. To quote Pauli: they are not even wrong. They have no place here.

If this were a few theoreticians plugging away in an unfashionable area of physics, as Yang, Mills, Glashow, Salam, and Weinberg had plugged away with quantum field theory in the 1950s and 1960s, then we might agree that no real harm is done. But such is the fashion for superstring theory that it has come to dominate modern theoretical physics, particularly in America. Mathematician Peter Woit used Pauli's acerbic quote as the title for his 2006 book denouncing the failure of superstring theory:

> The failure of the superstring theory programme must be recognized and lessons learned from this failure before there can be much hope of moving forward. As long as the leadership of the particle theory community refuses to face up to what has happened and continues to train young theorists to work on a failed project, there is little likelihood of new ideas finding fertile ground in which to grow.

Smolin added his voice to the growing chorus in 2006, with his book *The Trouble With Physics*.

In July 2009, Weinberg delivered a lecture at CERN on the varying fortunes of quantum field theory. He had this to say:

> I don't want to discourage string theorists, but there's just the possibility that maybe that isn't the way the world is, that the world is much more like we've always known, that is, the Standard Model and general relativity.

EPILOGUE

A Quantum of Solace?

Geneva, March 2010

'It's a fantastic moment,' declared Lyn Evans, LHC project manager at CERN in Geneva. 'We can now look forward to a new era of understanding about the origins and evolution of the universe.'

Sadly, Evans' delight was to be rather short-lived. The LHC was switched on at 10:28 am local time on 10 September 2008. Physicists crammed into the small control room cheered as a single flash of light appeared on a monitor, signifying that high-speed protons had been steered all the way around the machine's 27-kilometre ring at the operating temperature of −271 °C, just two degrees above absolute zero. Though somewhat unspectacular (and something of an anti-climax for the estimated one billion people thought to have watched the moment on television), it represented the culmination of two decades of unstinting effort by armies of physicists, designers, engineers, and construction workers.

Another proton beam was sent around the second, anticlockwise, ring, at 3 pm later that day. Trouble began shortly afterwards. Just nine days later an electrical bus connection between two of the superconducting magnets short-circuited. Electricity arced, punching a hole in the magnets' helium enclosure. Helium gas leaked into sector 3–4 of the LHC tunnel, and in the subsequent explosion 53 magnets were damaged and the proton tubes were contaminated with soot. There was no hope of repair before the scheduled winter shut-down, and a restart was tentatively fixed for spring 2009. But there were more problems, and at a meeting in Chamonix in February 2009 CERN managers took the decision to commission further work. The restart date was pushed back.

At the beginning of September 2009, almost a year after it had first been switched on, the last of the LHC's eight sectors began its cool-down procedure. All eight sectors were back at their operating temperature by the end of October, with a restart scheduled for November. Despite the increased cost of electricity during the winter months, the LHC was operated through the winter of 2009–10, primarily so that CERN physicists could stay ahead of their rivals at Fermilab's Tevatron, who were snapping at their heels.

Through the first few months of 2010, protons around the two rings were accelerated to 3.5 TeV before being brought into head-on collision. The first 7-TeV proton–proton collisions—the highest energy collisions engineered on earth—were recorded on 30 March, representing the beginning of the LHC research programme. Candidate events involving W bosons were already being reported in early April; candidate Z boson events were reported in May. This collision energy will be maintained for between 18 and 24 months as data are gathered. Researchers will first look for particles already known to the Standard Model, before pushing on in search of new physics, with the first preliminary scientific reports anticipated in the summer of 2010. There will follow a year-long shut-down, after which the proton energies will be increased to achieve the LHC's target collision energy of 14 TeV.

There is a lot at stake. For the politicians, bureaucrats, and science managers, getting some results (any results) from an experiment that has so far cost £3.5 billion will no doubt help to provide some justification for the endeavour.

But the greatest source of value to be derived from the LHC is what it promises to do for the future of physics.

The current crisis in physics is the direct result of a lack of experimental data. Theory must have the discipline of experiment if it is to remain focused on the things that really matter, the things that manifestly happen in the real world. Without experiment, theory risks a retreat into metaphysics, into relatively idle speculation about how many angels can fit on the head of a pin, or how many unseen (and unseeable) spatial dimensions are required for superstrings to vibrate in.

Science is not philosophy. Yet, without recourse to experiment, it is perhaps inevitable that science becomes more speculative and metaphysical. This drift into metaphysics is not always apparent. It can sometimes be obscured by the language of the abstruse mathematics that is deployed.

Modern theoretical physics is filled with dense, impenetrable, complex mathematical structures. These are not for the uninitiated. Those on the outside looking in may be deluded into thinking that such high mathematics must bring with it rigour, a sense of the absolute, the clarity of the difference between right and wrong.

But this is not so. Yes, there is rigour, there is right and wrong within the rules of the esoteric language, but the answers you can hope to get will still depend on the questions you choose to ask, and the way you ask them. And while the structures may be internally rigorous, they can still be inappropriately applied. The history of science is littered with failed theories, all of which no doubt were mathematically rigorous and looked good on paper at the time they were written down.

The LHC has been called a 'big bang' machine, capable of reproducing conditions not seen since the birth of our universe. For the Higgs boson to reveal itself, it will be necessary not just for protons to collide at high energy, but for their quark constituents to collide head-on. In theory, the conditions created in the LHC should generate a Higgs boson every couple of hours. The mass of the Higgs boson is hard to pin down theoretically, but the LHC's ATLAS[1] and Compact Muon Solenoid (CMS) experiments are set up to search for tell-tale decay signatures (for example, into bottom–anti-bottom pairs, and into combinations of two high-energy photons, Z particles, W particles, and tau particles).

Detection is no simple matter. ATLAS, for example, is about half as big as Notre Dame Cathedral in Paris and weighs as much as the Eiffel Tower. The CMS detector is 21 metres long, 15 metres in diameter, and weighs as much as 30 jumbo jets. The ATLAS and CMS collaborations each employ about 3000 scientists and engineers.

If the Higgs weighs between 150 and 400 GeV, then it should be found quite quickly and its mass could be measured with a precision of the order of one per cent. Its discovery will indeed be a famous result.[2] The Higgs will formally enter the lexicon of the Standard Model, transformed

[1] ATLAS stands for A Toroidal LHC ApparatuS. Aside from ATLAS and CMS, the LHC is also home to four other detector facilities. The purpose of ALICE (A Large Ion Collider Experiment) is to look for quark–gluon plasmas. LHCb (LHC-beauty) is designed to study CP-violation in bottom-quark decays. LHCf (LHC-forward) will be used to test devices designed to detect cosmic rays. Finally, TOTEM (Total Elastic and diffractive cross-section Measurement) is designed to carry out high-precision measurements on protons.

[2] Tantalizing glimpses of the Higgs boson were observed with a mass of 115 GeV in the final days of CERN's Large Electron–Positron collider. However, the evidence was not deemed to be sufficiently substantial to warrant a continuation of the LEP programme, beyond the point at which the development of the LHC would have been jeopardized. After much agonizing discussion, the decision was taken to shut the LEP down. See Sample, pp. 202–23.

from something that *ought* to be there to something that really *is* there, and no doubt prompting the Swedish Academy to award a Nobel Prize for the Higgs mechanism.

It is also hoped that the LHC may have some light to shed on the existence, or otherwise, of supersymmetric partners for the particles of the Standard Model. It may also have something to say about dark matter and dark energy, thought to account for the mass 'missing' from the observable universe, but required to explain its large-scale structure and behaviour.

But the real hope is that the LHC will reveal something unexpected, something that doesn't fit quite so neatly into what physicists already know about the quantum world. Even a negative result might provide a rallying cry, a call to arms to theoretical physicists for too long numbed by the comfort of the Standard Model.

'I think it will be much more exciting if we don't find the Higgs,' said Hawking. 'That will show something is wrong, and we need to think again.

'I have a bet of $100 that we won't find the Higgs.'

Hawking's confidence stems from his belief that evidence for the Higgs boson will be obscured by the creation of virtual black holes. Higgs is not convinced.

'It was a bit of a cheek,' he said. 'I am very doubtful about his calculations.'

Following an exceptional performance by the LHC at collision energies of 7 and 8 TeV through 2011 and the first half of 2012, on 4 July 2012 both the ATLAS and CMS collaborations declared that they had discovered a new Higgs-like boson with a mass around 125-126 GeV. Further work is needed to identify the particle's properties fully, but for now it looks very much like a Standard Model Higgs. Hawking told the BBC News: 'It seems I have just lost $100.' Readers interested to learn more about the final stages of the search for the Higgs might want to consult *Higgs: The Invention and Discovery of the 'God Particle'*, published by Oxford University Press.

NOTES AND SOURCES

Prologue: Stormclouds

2 'Hitherto we have explain'd...' Isaac Newton, *The Mathematical Principles of Natural Philosophy*, Book II (1729 English translation), p. 392. The Latin 'hypotheses non fingo' can also be translated as 'I feign no hypotheses'.

2 'Destitute of every species of merit,' attributed to Lord Brougham, *Edinburgh Review*. Quoted in Hecht and Zajac, p. 5.

4 'There is nothing new to be discovered in physics now,' attributed to Lord Kelvin. Quoted in Isaacson, p. 90, but see also the footnote on p. 575.

1: The Most Strenuous Work of My Life

7 'Dangerous enemy of progress', Max Planck, letter to Wilhelm Ostwald, 1 July 1893. Quoted in Heilbron, p. 15.

7 'Have to be abandoned in favour of the assumption of continuous matter', Max Planck, *Physikalische Abhandlungen und Vorträge*, Volume 1, Vieweg, Braunschweig, 1958, p. 163. Quoted in Heilbron, p. 14.

12 'The possibility of establishing units of length, mass, time and temperature...', Max Planck, *Physikalische Abhandlungen und Vorträge*, Volume 1, Vieweg, Braunschweig, 1958, p. 666. Quoted in Hermann, p. 11.

13 'I therefore feel justified in directing attention...', Max Planck, *Verhandl. Der Deutsche Physikalische Gesellschaft*, **2**, 1900, p. 202. Quoted in Kuhn, p. 97.

13 'Completely satisfactory agreement in all cases', Max Planck, *Physikalische Abhandlungen und Vorträge*, Volume 3, Vieweg, Braunschweig, 1958, p. 263. Quoted in Hermann, p. 15.

13 'Some weeks of the most strenuous work of my life'. Max Planck, Nobel Lecture, 2 June 1920, in *Nobel Lectures: Physics 1901–1921*, Elsevier, Amsterdam 1967. See http://nobelprize.org/nobel_prizes/physics/laureates/1918/planck-lecture.html

14 'I busied myself, from then on, that is, from the day of its establishment...', Max Planck, Nobel Lecture, 2 June 1920, in *Nobel Lectures: Physics* 1901–1921, Elsevier, Amsterdam 1967. See http://nobelprize.org/nobel_prizes/physics/laureates/1918/ planck-lecture.html

15 'Briefly summarised, what I did can be described as simply an act of desperation', Max Planck, letter to Robert Williams Wood, 7 October 1931. Quoted in Hermann, p. 21.

15 'We therefore regard...', Max Planck, lecture to the Berlin Physical Society, 14 December 1900. Quoted in Isaacson, p. 96.

15 '...since it has the dimensions of a product of energy and time', Max Planck, *Physikalische Abhandlungen und Vorträge*, Volume 3, Vieweg, Braunschweig, 1958, p. 266. Quoted in Hermann, p. 19.

16 'Felt that he had possibly made a discovery of the first rank', Heisenberg, *Physics and Philosophy*, p. 19.

2: Annus Mirabilis

19 'I promise you four papers in return...', Albert Einstein, letter to Conrad Habicht, 18 or 25 May 1905. Quoted in Isaacson, p. 93.

21 'A profound formal difference exists...', Albert Einstein, *Annalen der Physik*, 17, 1905, p. 132. This quote is from the English translation published in Stachel, p. 177.

22 'If monochromatic radiation...', Albert Einstein, *Annalen der Physik*, **17**, 1905, pp. 143–144. English translation quoted in Stachel, p. 191.

22 '...in the propagation of a light ray...', Albert Einstein, *Annalen der Physik*. 17, 1905, p. 133. English translation quoted in Stachel, p. 178.

24 'That he may sometimes have missed the target in his speculations...', Pais, *Subtle is the Lord*, p. 382.

3: A Little Bit of Reality

25 'This Einstein must be a clever fellow...', attributed to George von Hevesy. Quoted in Kuhn, p. 215.

27 'This is wrong', Pais, *Niels Bohr's Times*, p. 120.

28 'The whole thing was very interesting in Cambridge...', Pais, *Niels Bohr's Times*, p. 121.

29 'This seems to be nothing else than what was to be expected...', Pais, *Niels Bohr's Times*, p. 137.

29 'Perhaps I have found out a little about the structure of atoms...', Niels Bohr, letter to Harald Bohr, 19 June 1912, quoted in French and Kennedy, p. 76.

32 '...the dynamical equilibrium of the systems in the [stable orbits] can be discussed...', Niels Bohr, *Philosophical Magazine*, **26**, 1913, p 7. This paper is reproduced in French and Kennedy, p. 83.

32 'I do not know if you appreciate the fact that long papers have a way of frightening readers', quoted in French and Kennedy, p. 77.

4: la Comédie Française

36 'All movements of bodies were supposed to be relative to the light-carrying ether...', Albert Einstein, letter to Maurice Solovine (undated), quoted in Isaacson, p. 131.

36 '...its mass decreases by [E/c^2]...', Albert Einstein, *Annalen der Physik*, **18**, 1905, p. 641. From the English translation in Stachel, p. 164.

37 'The notion of a quantum makes little sense, seemingly, if energy is to be continuously distributed through space...', de Broglie, p. 8.

37 'After long reflection in solitude and meditation...', Louis de Broglie, from the 1963 re-edited version of his PhD thesis. Quoted in Pais, *Subtle is the Lord*, p. 436.

38 'An electron is for us the archetype of [an] isolated parcel of energy...', de Broglie, p. 8.

38 'Represents a spatial distribution of *phase*, that is to say, it is a "Phase wave"', de Broglie, p. 10.

40 'Thus, the resonance condition can be identified with the stability condition from quantum theory', de Broglie, p. 29.

41 'He [de Broglie] has lifted a corner of the great veil', Albert Einstein, letter to Paul Langevin, 16 December 1924. Quoted in Moore, p. 187.

41 '...de Broglie has undertaken a very interesting attempt...', Albert Einstein, letter to Hendrik Lorentz, 16 December 1924. Quoted in Pais, *Subtle is the Lord*, p. 436.

41 'Some wit, in fact, dubbed de Broglie's theory "la Comédie Française"', Gamow, *Thirty Years that Shook Physics*, p. 81.

42 'From this we see why certain orbits are stable; but, we have ignored passage from one to another stable orbit...', de Broglie, p. 29.

42 'Would be very unhappy to renounce complete causality', Albert Einstein, letter to Max Born, 27 January 1920. Quoted in Pais, *Subtle is the Lord*, p. 412.

5: A Strangely Beautiful Interior

44 'Honourable funeral', Pais, *Niels Bohr's Times*, p. 238.

44 '...as soon as it became clear that I was about to work in Professor Bohr's Institute...', Heisenberg, *Physics and Beyond*, p. 45.

45 'He looked like a simple peasant boy...', Max Born, quoted in van der Waerden, p. 19.

45 'A large number of brilliant young men from every part of the world...', Heisenberg, *Physics and Beyond*, p. 45.

46 'But now and then our papers also tell us about more ominous, anti-Semitic, trends in Germany...', Heisenberg, *Physics and Beyond*, p. 55.

47 'These models have been deduced, or if your prefer guessed, from experiments...', Heisenberg, *Physics and Beyond*, p. 41.

48 'When the first terms seemed to accord with the energy principle, I became rather excited...', Heisenberg, *Physics and Beyond*, p. 61.

49 'He had written a crazy paper and did not dare send it in for publication...', Max Born, quoted in Pais, *Niels Bohr's Times*, p. 278.

50 'Only much later did I learn from Born that it was simply a matter here of multiplying matrices...', Werner Heisenberg, manuscript of a lecture intended for delivery in Göttingen, May 1975. Subsequently published in Heisenberg, *Encounters with Einstein*, p. 45.

50 '...on the one hand, your results, especially concerning the general definition of the differential quotient...', Werner Heisenberg, letter to Paul Dirac, 20 November 1925. Quoted in Kragh, *Dirac: A Scientific Biography*, p. 20.

6: The Self-rotating Electron

51 'How can one look happy when he is thinking about the anomalous Zeeman effect?', quoted in Enz, p. 92.

52 'Whoever studies this mature and grandly conceived work...', Albert Einstein, quoted in Pais, *Niels Bohr's Times*, p. 200.

53 'In the classification adopted, the remarkable feature emerges that the number of electrons in each completed level...', Edmund Stoner, quoted in Enz, p. 122.

54 'There can never be two or more equivalent electrons in the atom...', Wolfgang Pauli, quoted in Enz, p. 122.

54 'Offer itself automatically in a natural way', Wolfgang Pauli, quoted in Enz, p. 122.

55 'I am the one who *rejoices most* about it...', Werner Heisenberg, postcard to Wolfgang Pauli, 15 December 1924. Quoted in Enz, p. 124.

57 'For some reason I had imagined him [Pauli] as being much older and as having a beard...', Ralph de Laer Kronig, quoted by Enz, p. 111.

58 'A new Copenhagen heresy', Wolfgang Pauli, quoted in Pais, *Niels Bohr's Times*, p. 243.

58 'I should not have mentioned the matter [to Kramers] at all...', Ralph de Laer Kronig, letter to Niels Bohr, 8 April 1926. Quoted in Pais, *Niels Bohr's Times*, p. 244.

7: A Late Erotic Outburst

60 'Heisenberg's form of quantum theory completely avoids a mechanical-kinematic visualization of the motions of electrons...', Wolfgang Pauli, *Zeitschrift für Physik*, **36**, 1926, p. 336. An English translation is available in van der Waerden, p. 387.

61 'You know it would be easier to live with a canary than a racehorse...', Arthur I. Miller, in Farmelo, p. 117.

62 'A very notable contribution', Albert Einstein, quoted in Pais, *Niels Bohr's Times*, p. 241.

63 'Naturally de Broglie's consideration in the framework of his comprehensive theory is altogether of far greater value...', Erwin Schrödinger, letter to Albert Einstein, 3 November 1925. Quoted in Moore, p. 192.

63 'So in one of the next colloquia, Schrödinger gave a beautifully clear account...', Felix Bloch, *Physics Today*, **29**, 1976, p. 23.

65 'At the moment I am struggling with a new atomic theory...', Erwin Schrödinger, letter to Wilhelm Wien, 27 December 1925. Quoted in Moore, p. 196.

66 'In the same natural way as the integers specifying the number of nodes in a vibrating string', Erwin Schrödinger, *Annalen der Physik*. 79, 1926, p. 361. Quoted in Moore, p. 202.

66 'Late, erotic outburst in his life', Hermann Weyl, quoted in Moore, p. 191.

67 'I was absolutely unaware of any generic relationship with Heisenberg...', Erwin Schrödinger, *Annalen der Physik*. **79**, 1926, p. 735. English translation quoted in Moore, p. 211.

67 'The more I think about the physical portion of the Schrödinger theory...', Werner Heisenberg, letter to Wolfgang Pauli, 8 June 1926. Quoted in Cassidy, p. 215.

8: Ghost Field

73 'Only a further extension of the theory; which in all likelihood will be very laborious...', Max Born, *Problems of Atomic Dynamics*, MIT Press, Cambridge, Massachusetts, 1926. Quoted in Beller, p. 31. Born's book was the first to present the new quantum mechanics.

74 '...only one interpretation is possible, [the wavefunction] gives the probability for the electron...', Max Born, *Zeitschrift für Physik*. **37**, 1926, p. 863–867. An English translation is available in Wheeler and Zurek, pp. 52–55. The quotations (and associated footnote) appear on p. 54.

75 'Schrödinger's quantum mechanics therefore gives quite a definite answer to the question of the effect of the collision...', Max Born, *Zeitschrift für Physik*. **37**, 1926, p. 863–867. See Wheeler and Zurek, p. 54.

77 'In classical dynamics the knowledge of the state of a closed system...', Max Born, *Nature*, **119**, 1927, pp. 354–357. Reproduced in Born, pp. 6–12. This quotation appears on p. 6.

77 'The classical theory introduced the microscopic coordinates which determine the individual process...', Born, p. 10.

78 'From an offprint of Born's last work in the *Zeitsch. f. Phys.* I know more or less how he thinks of things...', Erwin Schrödinger, letter to Wilhelm Wien, 26 August 1926. Quoted in Moore, p. 225.

78 'Quantum mechanics is very impressive. But an inner voice tells me that it is not yet the real thing...', Albert Einstein, letter to Max Born, 4 December 1926. Quoted in Pais, *Subtle is the Lord*, p. 443.

9: All This Damned Quantum Jumping

79 'Deepest formulation of the quantum laws', Max Born, *Zeitschrift für Physik*. **37**, 1926, p. 863–867. See Wheeler and Zurek, p. 52.

80 '...told me rather sharply that while he understood my regrets...', Heisenberg, *Physics and Beyond*, p. 73.

81 'Surely you realise that the whole idea of quantum jumps is bound to end in nonsense...', This, and the subsequent dialogue, is adapted from Heisenberg's reconstruction of the debate, which appears in Heisenberg, *Physics and Beyond*, pp. 73–76.

84 'Bohr's...approach to atomic problems...is really remarkable...', Erwin Schrödinger, letter to Wilhelm Wien, 21 October 1926. Quoted in Moore, p. 228.

86 'I remember discussions with Bohr which went through many hours till very late at night...', Heisenberg, *Physics and Philosophy*, p. 30.

86 '...where I could think about these hopelessly complicated problems undisturbed', Heisenberg, *Physics and Beyond*, p. 77.

10: The Uncertainty Principle

87 'It would have been beautiful if you were right...', Max Born, letter to Erwin Schrödinger, 6 November 1926. Quoted in Beller, p. 36.

87 'Schrödinger's achievement reduces itself to something purely mathematical; his physics is quite wretched', Max Born, letter to Albert Einstein, 30 November 1926. Quoted in Pais, *Niels Bohr's Times*, p. 288.

88 '...was witnessing the fertility of the particle concept every day...', Max Born, *My Life and Views*, Scribner's, New York, 1968, p. 55. Quoted in Pais, *Niels Bohr's Times*, p. 288.

89 'I am now however, convinced with all the fervour of my heart...', Wolfgang Pauli, letter to Werner Heisenberg, 19 October 1926. Quoted in Cassidy, p. 232. The emphasis is Pauli's.

89 'One may view the world with the p-eye and one may view it with the q-eye...', Wolfgang Pauli, letter to Werner Heisenberg, 19 October 1926. Quoted in Enz, p. 141.

90 '...when I realised fairly soon that the obstacles before me were insurmountable...', Heisenberg, *Physics and Beyond*, p. 77.

90 'But you don't seriously believe that none but observable magnitudes must go into a physical theory?', This, and subsequent dialogue, is taken from Heisenberg's reconstruction: Heisenberg, *Physics and Beyond*, p. 63.

91 'Can quantum mechanics represent the fact that an electron finds itself approximately in a given place...', Heisenberg, *Physics and Beyond*, p. 78.

92 'Thus, the more precisely the position is determined, the less precisely the momentum is known, and conversely...', Werner Heisenberg, *Zeitschrift für Physik*, **43**, 1927, pp. 172–198. An English translation is provided in Wheeler and Zurek, pp. 62–84. This and the subsequent quote appear on pp. 64–65.

93 '"When we know the present precisely, we can predict the future," is not the conclusion but the assumption...', Werner Heisenberg, *Zeitschrift für Physik*, **43**, 1927, pp. 172–198. See Wheeler and Zurek, p. 83. According to Beller (see pp. 99–101), in reaching this conclusion Heisenberg was influenced by a paper published by physicist H.A. Sentfleben in 1923.

94 'I wanted to get Pauli's reactions before Bohr was back...', Werner Heisenberg, interview by Thomas Kuhn, 19 February 1963, Niels Bohr Archive. Quoted in Pais, *Niels Bohr's Times*, p. 304.

11: The 'Kopenhagener Geist'

97 'Then Niels Bohr returned from his skiing holiday and we had a fresh round of difficult discussions', Heisenberg, *Physics and Beyond*, p. 79.

100 'I remember that it ended with my breaking out in tears because I just couldn't stand this pressure from Bohr', Werner Heisenberg, interview by Thomas Kuhn, 25 February 1963, *Archive for the History of Quantum Physics*. Quoted in Jammer, p. 65.

101 'Well, we have a consistent mathematical scheme and this consistent mathematical scheme tells us everything which can be observed...', This, and subsequent quotes, are taken from Werner Heisenberg, interview by Thomas Kuhn, 25 February 1963, *Archive for the History of Quantum Physics*. Quoted in Pais, *Niels Bohr's Times*, p. 310.

101 'Naturally, if one starts with [wave–particle duality], one can do everything without fear of contradiction...', Werner Heisenberg, letter to Wolfgang Pauli, 16 May 1927. Quoted in Cassidy, p. 238.

102 'In this connection Bohr has brought to my attention that I have overlooked essential points in the course of several discussions in this paper...', This, and the subsequent quote, are taken from Werner Heisenberg, *Zeitschrift für Physik*, **43**, 1927, pp. 172–198. See Wheeler and Zurek, pp. 83–84.

102 'Kopenhagener Geist der Quantentheorie', Heisenberg, *The Physical Principles of the Quantum Theory*, preface.

12: There is No Quantum World

104 'Bohr dictated and the next day all he had dictated was discarded and we began anew', Oskar Klein, interview with Léon Rosenfeld and J. Kalckar, 7 November 1968, *Niels Bohr Archive*. Quoted in Pais, *Niels Bohr's Times*, p. 311.

104 '...I shall try, by making use only of simple considerations and without going into any details of technical mathematical character...' Niels Bohr, *Nature*, **121**, 1928, pp. 580–590. This article was reprinted in Niels Bohr, *Atomic Theory and the Description of Nature*, Cambridge University Press, 1934, pp. 52–91. This was reproduced again in Wheeler and Zurek, pp. 87–126. This quote appears in Wheeler and Zurek, p. 87.

105 'On one hand, the definition of the state of a physical system, as ordinarily understood, claims the elimination of all external disturbances...', Niels Bohr, *Nature*, **121**, 1928, pp. 580–590. This quote appears in Wheeler and Zurek, pp. 89–90.

105 'This lecture will not induce any one of us to change his own [opinion] about quantum mechanics', Léon Rosenfeld, interview by Thomas Kuhn

and John Heilbron, 1 July 1963, *Niels Bohr Archive*. Quoted in Pais, *Niels Bohr's Times*, p. 315.

106 'The quantum theory is characterised by the acknowledgement of a fundamental limitation in the classical physical ideas when applied to atomic phenomena...', Niels Bohr, *Nature*, **121**, 1928, pp. 580–590. This quote appears in Wheeler and Zurek, p. 90.

106 'The Copenhagen interpretation of quantum theory starts from a paradox...', Heisenberg, *Physics and Philosophy*, p. 32.

109 'Our actual situation in research work in atomic physics is usually this...', Heisenberg, *Physics and Philosophy*, pp. 45–46.

110 'There is no quantum world. There is only an abstract quantum physical description...', Aage Petersen, in French and Kennedy, p. 305.

110 'The originality of the logical positivists lay in their making the impossibility of metaphysics depend not upon the nature of what could be known but upon the nature of what could be said', A.J. Ayer (ed.), *Logical Positivism*. The Library of Philosophical Movements. The Free Press of Glencoe, 1959, p. 11.

111 'Accordingly, an independent reality in the ordinary physical sense can neither be ascribed to the phenomena nor to the agencies of observation', Niels Bohr, *Nature*, **121**, 1928, pp. 580–590. This quote appears in Wheeler and Zurek, p. 89.

13: The Debate Commences

115 '...several of us came to the conference with great anticipations to learn his [Einstein's] reaction...', Niels Bohr, in Schilpp, pp. 211–212.

117 'The wave ψ then appears as both a pilot wave (Führungsfeld of Mr Born) and a probability wave...', Louis de Broglie, 'The New Dynamics of Quanta', *Proceedings of the Fifth Solvay Congress*, 1928. This quote is from an English translation published in Bacciagaluppi and Valentini, p. 383.

117 'After much reflection back and forth, I come to the conclusion that I am not competent [to give] such a report...', Albert Einstein, letter to Hendrik Lorentz, 17 June 1927. Quoted in Pais, *Subtle is the Lord*, pp. 431–432.

117 '...we consider *quantum mechanics* to be a closed theory...', Max Born and Werner Heisenberg, 'Quantum Mechanics', *Proceedings of the Fifth Solvay Congress*, 1928. English translation from Bacciagaluppi and Valentini, p. 437.

117 'I see nothing in Mr Schrödinger's calculations that would justify this hope', Werner Heisenberg, *Proceedings of the Fifth Solvay Congress*, 1928. English translation from Bacciagaluppi and Valentini, p. 472.

118 'Despite being conscious of the fact that I have not entered deeply enough into the essence of quantum mechanics...', Albert Einstein, *Proceedings of the Fifth Solvay Congress,* 1928. English translation from Bacciagaluppi and Valentini, p. 486.

119 'In my opinion, one can remove this objection only in the following way...', Albert Einstein, *Proceedings of the Fifth Solvay Congress,* 1928. English translation from Bacciagaluppi and Valentini, p. 488.

120 'I feel myself in a very difficult position because I don't understand what precisely is the point which Einstein wants to [make]...', This and subsequent comments by Bohr do not appear in the published proceedings of the conference. They are reconstructed from notes of the meeting available in the Bohr archive. Quoted in Bacciagaluppi and Valentini, p. 489.

121 'Einstein came down to breakfast and expressed his misgivings about the new quantum theory...', Otto Stern, interview with Res Jost, 2 December 1961. Quoted in Pais, *Subtle is the Lord,* p. 445.

124 'Since, however, any reading of the scale, in whatever way performed, will involve an uncontrollable change in the momentum of the [screen]...', Niels Bohr, in Schilpp, p. 220.

124 '...we are presented with a choice of *either* tracing the path of a particle *or* observing interference effects...', Niels Bohr, in Schilpp, p. 217. The emphasis is Bohr's.

125 'On his side, Einstein mockingly asked us whether we could really believe that the providential authorities took recourse to dice-playing...', Niels Bohr, in Schilpp, p. 218.

14: An Absolute Wonder

126 'Aesthetic failure', Wolfgang Pauli, *Atti del Congresso Internazionale dei Fisci, 11–20 Settembre 1927, Como, Pavia, Roma,* 1928. Quoted in Enz, p.160.

127 'He thought it would be good for me to learn French in that way...', Paul Dirac, interview with Thomas Kuhn and Eugene Wigner, April 1962, *Archive for the History of Quantum Physics.* Quoted in Kragh, *Dirac,* p. 2.

128 'The atoms were always considered as very hypothetical things by me...', Paul Dirac, 'Recollections of an Exciting Era', in C. Weiner (ed.), *Exploring the History of Nuclear Physics,* American Institute of Physics, New York, pp. 109–146. Quoted in Kragh, *Dirac,* p. 8.

129 'Herr Pauli regards the relativistic wave equation of *second* order with much suspicion', Johann Kudar, letter to Paul Dirac, 21 December 1926. Quoted in Kragh, *Dirac,* p. 54.

130 'I suddenly realised that there was no need to stick to quantities, which can be represented by matrices with only two rows and columns...', Paul Dirac, in C. Weiner (ed.), *Exploring the History of Nuclear Physics,* American Institute of Physics, New York, pp. 109–146. Quoted in Kragh, *Dirac,* p. 59.

131 '[Dirac] has now got a completely new system of equations which does the spin right in all cases...', Charles Darwin, letter to Niels Bohr, 26 December 1927. Quoted in Kragh, *Dirac,* p. 57.

131 'It was immediately seen as *the* solution...', Léon Rosenfeld, interview with Thomas Kuhn and John Heilbron, July 1963, *Archive for the History of Quantum Physics.* Quoted in Kragh, *Dirac,* p. 62.

132 'I am much more unhappy about the question of the relativistic formulation...', Werner Heisenberg, letter to Niels Bohr, 23 July 1928. Quoted in Kragh, *Dirac,* p. 66.

15: The Photon Box

133 'It has always been the dream of philosophers to have all matter built from one fundamental kind of particle...', Paul Dirac, *Nature,* **126**, 1930, pp. 605–606. Quoted in Kragh, *Dirac,* p. 97.

135 'A good joke should not be repeated too often', Philipp Frank, *Einstein—His Life and Times,* Knopf, New York, 1947, p. 216. Quoted in Jammer, p. 131.

135 'At the next meeting with Einstein at the Solvay Conference in 1930...', Niels Bohr, in Schilpp, p. 224.

135 'It was quite a shock for Bohr... he did not see the solution at once...', Léon Rosenfeld, *Proceedings of the Fourteenth Solvay Conference.* Interscience, New York, 1968, p. 232. Quoted in Pais, *Subtle is the Lord,* pp. 446-7.

135 'This argument amounted to a serious challenge and gave rise to a thorough examination of the whole problem', Niels Bohr, in Schilpp, p. 226.

138 'The discussion, so illustrative of the power and consistency of relativistic arguments...', Niels Bohr, in Schilpp, p. 228.

138 'Free of contradictions...', Albert Einstein, quoted by Hendrik Casimir in a letter to Abraham Pais, 31 December 1977. Quoted in Pais, *Subtle is the Lord,* p. 449.

140 'Since, however, according to the quantum-mechanical formalism...', Niels Bohr, in Schilpp, p. 229. I am assuming that this discussion reflects Bohr's views 18 years previously, at the time Einstein made his challenge.

140 'It may also be added that it obviously can make no difference as regards observable effects...', Niels Bohr, in Schilpp, p. 230.

16: A Bolt from the Blue

142 'A new kind of particle, unknown to experimental physics...', Paul Dirac, *Proceedings of the Royal Society*, **A133**, 1931, pp. 60–72. Quoted in Kragh, *Dirac*, p. 103.

143 'If, without in any way disturbing a system, we can predict with certainty...', Albert Einstein, Boris Podolsky, Nathan Rosen, *Physical Review*. **47**, 1935, pp. 777–780. Reproduced in Wheeler and Zurek, pp. 138–141. This quote appears on p. 138.

144 'No reasonable definition of reality could be expected to permit this', Albert Einstein, Boris Podolsky, Nathan Rosen, *Physical Review*. **47**, 1935, pp. 777–780. Reproduced in Wheeler and Zurek, p. 141.

145 'While we have thus shown that the wave function does not provide a complete description of physical reality...', Albert Einstein, Boris Podolsky, Nathan Rosen, *Physical Review*. **47**, 1935, pp. 777–780. Reproduced in Wheeler and Zurek, p. 141.

145 'Is not a complete theory...', and subsequent quotations, *The New York Times*, 4 May 1935, p. 11. Quoted in Jammer, pp. 189-90.

145 'This onslaught came down upon us as a bolt from the blue...', Léon Rosenfeld, in Stefan Rozenthal (ed.), *Niels Bohr: His Life and Work as Seen by his Friends and Colleagues*, North-Holland, Amsterdam, 1967, pp. 114–136. Extract reproduced in Wheeler and Zurek, pp. 137 and 142–143. This quote appears on p. 142.

146 'Now we have to start all over again...', Paul Dirac, interview with Niels Bohr, 17 November 1962, *Archive for the History of Quantum Physics*. Quoted in Beller, p. 145.

146 'They do it "smartly", but what counts is to do it right', Léon Rosenfeld, in Stefan Rozenthal (ed.), *Niels Bohr: His Life and Work as Seen by his Friends and Colleagues*, North-Holland, Amsterdam, 1967, pp. 114–136. Extract reproduced in Wheeler and Zurek, p. 142.

146 'It is shown that a certain 'criterion of physical reality' formulated in a recent article with the above title...', Niels Bohr, *Physical Review*. 48, 1935, 696–702. Reproduced in Wheeler and Zurek, pp. 145–151. This quote appears on p. 145.

146 'From our point of view we now see that the wording of the above-mentioned criterion of physical reality...', Niels Bohr, *Physical Review*. 48, 1935, 696–702. Reproduced in Wheeler and Zurek, p. 145.

17: The Paradox of Schrödinger's Cat

149 '[Schrödinger's departure] is all the more to be regretted...', *Deutsche Zeitung*, 24 October 1933. Quoted in Moore, p. 276–277.

149 '*The Times* had said I would be among that year's prize winners...', Erwin Schrödinger, letter to Max Born, 13 January 1943. Quoted in Moore, p. 280.

150 'Academies of homosexuality', Erwin Schrödinger, quoted by Max Born, *My Life: Recollections of a Nobel Laureate*, Taylor & Francis, London, 1978. Quoted in Moore, p. 298.

150 'I was very pleased that in the work that just appeared in *Physical Review* you have publicly called the dogmatic quantum mechanics to account...', Erwin Schrödinger, letter to Albert Einstein, 7 June 1935. Quoted in Fine, pp. 66–67.

150 '...My interpretation is that we do not have a [quantum mechanics] that is consistent with relativity theory...', Erwin Schrödinger, letter to Albert Einstein, 7 June 1935. Quoted in Moore, p. 304.

151 'If two separated bodies, each by itself known maximally...', Erwin Schrödinger, *Naturwissenschaften.* **23**, 807–812, 823–828, 844–849. This three-part assessment of the current situation in quantum mechanics was translated by John D. Trimmer and published the in *Proceedings of the American Philosophical Society*, **124**, 1980, 323–328. The translation is reproduced in Wheeler and Zurek, pp. 152–167. The quote appears on p. 161.

151 'No doubt, however, you smile at me and think that, after all, many a young whore turns into an old praying sister...', Albert Einstein, letter to Erwin Schrödinger, 17 June 1935. Quoted in Fine, p. 68.

151 '...it did not come out as well as I had originally wanted...', Albert Einstein, letter to Erwin Schrödinger, 19 June 1935. Quoted in Fine, p. 35.

152 'The actual difficulty lies in the fact that physics is a kind of metaphysics...', Albert Einstein, letter to Erwin Schrödinger, 19 June 1935. Quoted in Moore, p. 304.

152 'We face similar alternatives when we want to explain the relation of quantum mechanics to reality...', Albert Einstein, letter to Erwin Schrödinger, 19 June 1935. Quoted in Fine, p. 69.

153 'What I have so far seen by way of published reactions is less witty...', Erwin Schrödinger, letter to Albert Einstein, 13 July 1935. Quoted in Fine, p. 74.

153 'In the beginning the ψ-function characterizes a reasonably well-defined macroscopic state...', Albert Einstein, letter to Erwin Schrödinger, 8 August 1935. Quoted in Fine, p. 78.

153 'I am long past the stage where I thought that one can consider the ψ-function as somehow a direct description of reality', Erwin Schrödinger, letter to Albert Einstein, 19 August 1935. Quoted in Fine, p. 82.

154 'Contained in a steel chamber is a Geigercounter prepared with a tiny amount of uranium...', Erwin Schrödinger, letter to Albert Einstein, 19 August 1935. Quoted in Fine, pp. 82–83.

154 'The ψ-function of the entire system would express this [superposition] by having in it the living and dead cat (pardon the expression) mixed or smeared in equal parts', Erwin Schrödinger, *Naturwissenschaften*. **23**, 807–812, 823–828, 844–849. Reproduced in Wheeler and Zurek, pp. 152–167. The quote appears on p. 156.

157 '...your cat shows that we are in complete agreement concerning our assessment of the character of the current theory...', Albert Einstein, letter to Erwin Schrödinger, 4 September 1935. Quoted in Fine, p. 84.

158 '...there is no practically possible way of observing the neutrino', Hans Bethe and Rudolf Peierls, quoted in Kragh, *Quantum Generations*, p. 181.

158 'Who ordered that?', Isidor Rabi, quoted in Kragh, *Quantum Generations*, p. 204.

Interlude: The First War of Physics

159 'Oh what idiots we have all been! Oh but this is wonderful...', Niels Bohr, quoted in Frisch, p. 116.

160 'Yes, it would be possible to make a bomb...', Niels Bohr, quoted in Wheeler, p. 44.

160 'Extremely powerful bombs of a new type', Albert Einstein, letter to Franklin D. Roosevelt, 2 August 1939. This letter is reproduced in Snow, p 178.

160 'It was from September 1941 that we saw an open road ahead of us, leading to the atomic bomb', Werner Heisenberg, quoted in Irving, p. 114.

161 It might be a good thing if you could discuss the whole subject with Niels in Copenhagen...', Heisenberg, *Physics and Beyond*, p. 181.

161 'Grave consequences in the technique of war', Werner Heisenberg, quoted in Rhodes, p. 384.

161 '[Heisenberg] had agreed to sup with the devil...', Rudolf Peierls, quoted in Rhodes, p. 386.

161 'In Tisvilde, the beautiful vacation home of the Bohrs...', Elisabeth Heisenberg, p. 77.

162 'Here I am once again in the city which is so familiar to me and where a part of my heart has stayed stuck ever since that time fifteen years ago...', Werner Heisenberg, letter to Elisabeth Heisenberg, September 1941. A facsimile, transcription and English translation of this letter can be viewed online at www.werner-heisenberg.unh.edu.

162 'Human concerns and unhappy events of these times...', and subsequent quotations, Werner Heisenberg, letter to Elisabeth Heisenberg, September 1941

163 '[Heisenberg] stressed how important it was that Germany should win the war...', Stefan Rozental, quoted in Pais, *Niels Bohr's Times*, p. 483.

164 '... in vague terms you spoke in a manner that could only give me the firm impression that, under your leadership, everything was being done in Germany to develop atomic weapons...', Niels Bohr, draft letter to Werner Heisenberg. The Bohr drafts are reproduced in Dörries, pp. 101–179. This quote appears on p. 109. This should be read in its proper context. In drafting this, and subsequent letters and notes to Heisenberg, he was reacting—many years after the event—to Heisenberg's version of events that had been incorporated in the 1957 Danish translation of Robert Jungk's popular book, *Brighter than a Thousand Suns.*

164 'It had to make a very strong impression on me that at the very outset...', Niels Bohr, draft letter to Werner Heisenberg. Reproduced in Dörries, p. 163.

165 'And that it was, so to speak, the natural course in this world that the physicists were working in their countries on the production of weapons', Werner Heisenberg, quoted by Helmut Rechenberg in Dörries, p. 69.

165 'You know, I'm afraid it went badly wrong', Werner Heisenberg, quoted by Helmut Rechenberg in Dörries, p. 70.

165 'Our relations with scientific circles in Scandinavia have become very difficult', Werner Heisenberg, quoted by Helmut Rechenberg in Dörries, p. 69.

166 'He made the enterprise seem hopeful...', J. Robert Oppenheimer, quoted in Rhodes, p. 524. This is a quotation from an edited version of a post-war Oppenheimer speech.

167 'In some sort of crude sense which no vulgarity, no humor, no overstatement can quite extinguish...', J. Robert Oppenheimer, 'Physics in the Contemporary World', Second Arthur Dehon Little Memorial Lecture at the Massachusetts Institute of Technology, 25 November 1947. Reproduced in J. Robert Oppenheimer, *The Open Mind*, Simon & Shuster, New York, 1955. This quote appears on p. 88.

18: Shelter Island

171 'The last eighteen years have been the most sterile of the century', Isidor Rabi, quoted in the diary of Karl Darrow, 14 April 1947. Quoted in Gleick, p. 232.

172 'The programs are so crowded that presentation of the papers takes most of the time...', Duncan MacInnes, letter to Frank Jewett, 24 October 1945. Quoted in Schweber, p. 158.

173 'Twenty three of the country's best-known theoretical physicists…', *New York Herald Tribune*, 2 June 1947. Quoted in Schweber, pp. 172–3.

174 'Least improbable explanation', Gregory Breit, Notebook 1946. Breit's appear to be the only extant notes of the conference. Quoted in Schweber, p. 188.

175 'It was the first time people who had all this physics pent up in them for five years could talk to each other…', Julian Schwinger, interview with Robert Crease and Charles Mann, 4 March 1983. Quoted in Crease and Mann, p. 127.

176 'I went on to work out equations of wobbles…', Richard Feynman, *Surely You're Joking, Mr. Feynman!*, p. 174.

177 '…with a clear voice, great rush of words and illustrative gestures sometimes ebullient', Karl Darrow, diary entry 4 June 1947. Quoted in Schweber, p. 190.

179 'At that time no one could follow what he was talking about', Abraham Pais, *J. Robert Oppenheimer: A Life*, p. 115.

179 'There have been many conferences in the world since…', Richard Feynman, interview with Jagdish Mehra, April 1970. Quoted in Mehra, *The Beat of a Different Drum*, p. 217.

180 'I can do that for you. I'll bring it in for you tomorrow', Richard Feynman, *Science*, **153**, 1966, pp. 699–708. This is Feynman's Nobel Prize lecture. The quote appears on p. 705.

19: Pictorial Semi-vision Thing

181 'The magnetic moment of the electron, which came from Dirac's relativistic theory…', Julian Schwinger, interview with Robert Crease and Charles Mann, 4 May 1983. Quoted in Crease and Mann, p. 132.

182 'You should better explain things mathematically and not physically…', Richard Feynman, interview with Jagdish Mehra, April 1970 and January 1988, and interview with Charles Weiner, 1966. Quoted in Mehra, p. 246.

182 'Bohr thought that I didn't know the uncertainty principle…', Richard Feynman, interview with Jagdish Mehra, April 1970 and January 1988, and interview with Charles Weiner, 1966. Quoted in Mehra, p. 248.

182 'So I knew that I wasn't crazy', Richard Feynman, interview with Robert Crease and Charles Mann, 22 February 1985. Quoted in Crease and Mann, p. 139.

183 'The method of calculation was quite new…', Y. Miyamoto, personal recollections. Quoted in Schweber, p. 270.

183 'When I returned from the Pocono Conference, I found a letter from Tomonaga…', Robert Oppenheimer, letter to the Pocono Conference delegates, 5 April 1948. Quoted in Schweber, p. 198.

184 'Tomonaga expressed his [version of QED] in a simple, clear language...', Freeman Dyson, interview with Silvan Schweber, 18–19 November 1984. Quoted in Schweber, p. 502.

185 'The diagram is really, in a certain sense, the picture that comes from trying to clarify visualisation...', Richard Feynman, interview with Silvan Schweber, 13 November 1983. Quoted in Schweber, p. 465.

188 'That's curious, I have decided to switch to theoretical physics for precisely the same reason!', Freeman Dyson. This conversation was recalled by Nicholas Kemmer. Quoted in Schweber, p. 491.

188 'Hans [Bethe] was using the old cookbook quantum mechanics that Dick [Feynman] couldn't understand...', Dyson, p. 54.

189 'He talked of death with an easy familiarity...', Dyson, p. 59.

189 'Marvel of polished elegance', Dyson, p. 66.

190 'On the third day of the journey a remarkable thing happened...', Freeman Dyson, letter to his parents, 18 September 1948. Quoted in Schweber, p. 505.

190 'Nolo contendere. R.O.', Dyson, p. 74.

191 'Well Doc, you're in', Freeman Dyson, quoted in Gleick, p. 270.

191 'To give you a feeling for the accuracy of these numbers, it comes out something like this...', Feynman, *QED*, p. 7.

192 However, one room had been locked, Murray Gell-Mann, interview with James Gleick. Reported in Gleick, p. 277.

20: A Beautiful Idea

193 '..."And now I want to ask you something more: They tell me that you and Einstein are the only two real sure-enough high-brows..."', *Wisconsin State Journal*, 31 April 1929. Quoted in Kragh, *Dirac*, p. 73.

194 The group pestilence, or that pesty group business, Wigner, pp. 116–117.

194 'This may be the first method to derive the root of spectroscopy...', Erwin Schrödinger, quoted in Wigner, p. 117.

194 'Oh, these are old fogeys...', John von Neumann, quoted in Wigner, p. 117.

200 'Occasionally an obsession does finally turn out to be something good', Chen Ning Yang, *Selected Papers with Commentary*, W.H. Freeman, New York, 1983. Quoted by Christine Sutton in Farmelo, p. 241.

200 'There was no other, more immediate motivation...', Robert Mills, telephone interview with Robert Crease and Charles Mann, 7 April 1983. Quoted in Crease and Mann, p. 193.

202 'What is the mass of this B field...' and subsequent quotations, part of a conversation reported by Yang at the International Symposium on the History of Particle Physics, Batavia, Illinois, 2 May 1985. Quoted by Riordan, p. 198.

203 'The idea was *beautiful* and should be published...', Chen Ning Yang, *Selected Papers with Commentary*, W.H. Freeman, New York, 1983. Quoted by Christine Sutton in Farmelo, p. 243.

21: Some Strangeness in the Proportion

206 'There is no excellent beauty that hath not some strangeness in the proportion', Murray Gell-Mann and Edward Rosenbaum, *Scientific American*, July 1957, pp. 72–88. The idea of 'strangeness' was also elaborated around the same time by Japanese physicists Kazuhiko Nishijima and Tadao Nakano (who called it η-charge). Although the term strangeness was retained the theory is sometimes referred to as Gell-Mann–Nishijima theory.

207 'What possible difference does it make?', and subsequent quotations, part of a conversation quoted in Johnson, p. 122.

207 'Temporary kind of solution', J. Robert Oppenheimer, Proceedings of the Sixth Rochester Conference, Section VIII, p. 1, 1956. Quoted in Johnson, p. 133.

207 '...we were all sitting on one bed in a hotelroom, and I could hardly keep still in that position...', Segrè, p. 72.

209 'I do not believe that the Lord is a weak left-hander...', Wolfgang Pauli, letter to Victor Weisskopf, 17 January 1957. Quoted in Crease and Mann, p. 209.

211 'We should care to suggest that a fully acceptable theory of these interactions...', Sheldon Glashow, Harvard University PhD thesis, 1958, p. 75. Quoted in Glashow, Nobel Lecture, 8 December 1979.

212 'What you're doing is good...', Murray Gell-Mann, interview with Robert Crease and Charles Mann, 3 March 1983. Quoted in Crease and Mann, p. 225.

22: Three Quarks for Muster Mark!

214 'Young man, if I could remember the names of these particles, I would have been a botanist', Quoted as 'physics folklore' in Kragh, *Quantum Generations*, p. 321.

216 'It was a big mess. I didn't like it, and I didn't publish it', Murray Gell-Mann, interview with Robert Crease and Charles Mann, Caltech, 21 February 1985. Quoted in Crease and Mann, p. 265.

217 'I worked through the cases of three operators, four operators, five operators, six operators, and seven operators...', Murray Gell-Mann, Caltech Report CALT-68-1214, pp. 22–23. Quoted in Crease and Mann, pp. 264–265.

221 'I pointed out that you could take three pieces and make protons and neutrons...', Robert Serber, telephone interview with Robert Crease and Richard Mann, 4 June 1983. Quoted in Crease and Mann, p. 281.

221 'It was a crazy idea. I grabbed the back of a napkin and did the necessary calculations...', Murray Gell-Mann, interview with Robert Crease and Richard Mann, 3 March 1983. Quoted in Crease and Mann, p. 281.

222 'That's it! Three quarks make a neutron and a proton!', and subsequent quotations. Murray Gell-Mann, interview with Robert Crease and Richard Mann, 3 March 1983. Quoted in Crease and Mann, p. 282.

224 'Oh nonsense, Murray. Don't waste time on a transatlantic call talking about stuff like that', Murray Gell-Mann, interview with Robert Crease and Richard Mann, 3 March 1983. Quoted in Crease and Mann, p. 284.

23: The 'God Particle'

226 'Squalid-state' physics, Murray Gell-Mann, quoted in Johnson, p. 323 and Woit, p. 79. Johnson says that Gell-Mann used this term so frequently that it has entered physics folklore. He also says that Gell-Mann was partly kidding.

227 'Snake in the grass', Abdus Salam, interview with Robert Crease and Charles Mann, 23 February 1984. Quoted in Crease and Mann, p. 241.

228 'I remember being so discouraged by these zero masses that when we wrote our joint paper on the subject...', Steven Weinberg, Nobel lecture, 8 December 1979, p. 545.

229 'It is likely, then, considering the superconducting analogue...', Philip Anderson, *Physical Review*, **130** (1963), p.441, reproduced in E. Farhi and R. Jackiw (eds.), *Dynamical Gauge Symmetry Breaking: A Collection of Reprints*, World Scientific, Singapore (1982), p.50.

232 'I was indignant. I believed that what I had shown could have important consequences in particle physics...', Peter Higgs, in Hoddeson, *et al.*, p. 508.

232 'His amnesia unfortunately persisted through 1966', Peter Higgs, in Hoddeson, *et al.*, p. 510.

233 '...which suddenly seemed to change the role of [Nambu–]Goldstone bosons from that of unwanted intruders to that of welcome friends', Steven Weinberg, Nobel Lecture, 8 December 1979, p. 545.

233 'At some point in the fall of 1967, I think while driving to my office at MIT...', Steven Weinberg, Nobel Lecture, 8 December 1979, p. 548.

233 'My God, this is the answer to the weak interaction!', Steven Weinberg, interview with Robert Crease and Charles Mann, 7 May 1985. Quoted in Crease and Mann, p. 245.

234 'Literally nothing but the shadows of the images', Plato, *The Republic*, Book VII, 360 BCE.

234 'We are in such a cave, imprisoned by the limitations on the sorts of experiments we can do...', Steven Weinberg, Nobel lecture, 8 December 1979, p. 556.

24: Deep Inelastic Scattering

238 '...like smashing two pocket watches together to see how they are put together', Richard Feynman, interview with Michael Riordan, 14 and 15 March 1984. Quoted by Riordan, p. 152.

240 'I brought it in mainly as a desperate attempt to interpret the rather striking phenomenon of point-like behaviour', James Bjorken, *Proceedings of the 1967 International Symposium on Electron and Photon Interactions at High Energy*, Stanford, California, 5–9 September 1967, p. 126. Quoted in Riordan, p. 140.

240 'The analysis programs had to be rewritten as we continued to test the workings of the spectrometer hardware...', Richard Taylor, 'The Discovery of the Point-like Structure of Matter', invited talk presented at the Discussion Meeting on the Quark Structure of Matter, Royal Society of London, England, 24–25 May, 2000, p. 7.

241 '...wondering how Balmer may have felt when he saw, for the first time, the striking agreement of the formula that bears his name...', Henry W. Kendall, Nobel Lecture, 8 December 1990, p. 694.

243 'All my life I've looked for an experiment like this...', Richard Feynman, interview with Paul Tsai, 3 April 1984. Quoted in Riordan, p. 150.

244 'I've really got something to show youse guys...', Richard Feynman, quoted by Jerome Friedman in an interview with Michael Riordan, 24 October 1985. Quoted in Riordan, p. 151.

25: Of Charm and Weak Neutral Currents

247 'You have to remember the flavour of the time...', James Bjorken, interview with Robert Crease and Charles Mann, 10 May 1983. Quoted in Crease and Mann, p. 291.

248 'The problem which was explicitly posed in 1961 was solved, in principle, in 1964...', Sheldon Glashow, Nobel Prize lecture, p. 499.

249 'Every day, invariably, one of us would come up with an idea...', John Iliopoulos, interview with Robert Crease and Charles Mann, 8 May 1984. Quoted in Crease and Mann, p. 316.

251 'Psychological barrier', Steven Weinberg, quoted by John Iliopoulos in an interview with Michael Riordan, 4 June 1985. Quoted in Riordan, p. 211.

251 'Of course, not everyone believed in the predicted existence of charmed hadrons', Sheldon Glashow, Nobel Prize lecture, p. 500.

251 'I do not care what and how...', Martinus Veltman, private communication to Andrew Pickering, quoted in Pickering, p. 178.

252 'It nearly works. You just have some factors of two wrong', and subsequent quotes, Gerard 't Hooft, interview with Robert Crease and Charles Mann, 26 September 1984. Quoted in Crease and Mann, pp. 325–6.

252 '... the psychological effect of a complete proof of renormalisabiity has been immense', Martinus Veltman in Hoddeson, *et al.*, p. 173.

253 'Either this guy's a total idiot...', Sheldon Glashow, as quoted by David Politzer, interview with Robert Crease and Charles Mann, 21 February 1985. Quoted in Crease and Mann, p. 326.

255 'I put them out of their misery and *told* them...', Sheldon Glashow, interview with Robert Crease and Charles Mann, 29 August 1983. Quoted in Crease and Mann, p. 358.

26: The Magic of Colour

258 'In memory of that occasion...', Murray Gell-Mann, in Hoddeson, *et al.*, p. 629.

259 'Specifically these findings suggest the presence, in protons and neutrons...', Walter Sullivan, *New York Times*, 25 April 1971. Quoted in Riordan, p. 179.

260 'We gradually saw that that [colour] variable was going to do everything for us!', W.A. Bardeen, H. Fritzsch and M. Gell-Mann, *Proceedings of the Topical Meeting on Conformal Invariance in Hadron Physics*, Frascati, May 1092. Quoted in Crease and Mann, p. 328.

260 'In preparing the written version...', Murray Gell-Mann in Hoddeson, *et al.*, p. 631.

261 'The one hole left in this thing was gauge theories...', David Gross, interview with Robert Crease and Charles Mann, 2 April 1985. Quoted in Crease and Mann, p. 332.

261 'He spoiled me. I thought they'd all be that good', David Gross, quoted in Woit, p. 89.

262 'Um, hum. That's very interesting – except for one problem...', and subsequent quotations, David Politzer, interview with Crease and Mann, 21 February 1985. Quoted in Crease and Mann, p. 334.

263 'It was a tremendous eye-opener...', Riordan, p. 236.

263 'The theory had many virtues and no known vices...', Murray Gell-Mann in Hoddeson, *et al.*, p. 633.

27: The November Revolution

266 'There are just three possibilities...', Sheldon Glashow, *Proceedings of the Fourth International Conference on Experimental Meson Spectroscopy*, Boston, 26–27 April 1974, p. 392. Quoted in Riordan, p. 295.

266 '...the broken remnant of a single unified gauge group that existed in the distant past', John Iliopoulos, in J.R. Smith (ed.), *Proceedings of the Seventeenth International Conference on High Energy Physics*, London, 1–10 July 1974, Section III, pp. 97–100. Quoted in Crease and Mann, p. 358.

267 'You don't want to say you've found a new state of matter on the basis of one event...', Robert Palmer, interview with Robert Crease and Charles Mann, 18 May 1983. Quoted in Crease and Mann, p. 361.

270 'I hear you got a bump at 3.1', and subsequent exchanges, Melvin Schwartz, interview with Crease and Mann, 19 February 1985. Quoted in Crease and Mann, p. 375.

272 'Burt, I have some interesting physics to tell you about...', and subsequent exchanges, Burton Richter, *Adventures in Experimental Physics*, **5**, 1976, p. 147. Quoted in Riordan, p. 287.

273 'Sam! It's the same thing! It has to be right!', Burton Richter, interview with Robert Crease and Charles Mann, 5 July 1983. Quoted in Crease and Mann, p. 380.

28: Intermediate Vector Bosons

278 'The pressure to discover the W and Z was so strong...', Pierre Darriulat, in Cashmore, *et al.*, p. 57.

279 'The idea seemed too far-fetched at the time...', Simon van der Meer, quoted in Brian Southworth and Gordon Fraser, *CERN Courier*, November 1983.

281 'Most likely, if CERN hadn't bought Carlo [Rubbia]'s idea...', Pierre Darriulat, in Cashmore *et al.*, p. 57.

283 'They look like Ws...', Carlo Rubbia, quoted in Brian Southworth and Gordon Fraser, *CERN Courier*, November 1983.

283 'His talk was spectacular...', Lederman, p. 357.

29: The Standard Model

289 'The Higgs particle itself has never been detected...', 't Hooft, p. 115.

290 'This is a mathematical description of all known particles...', 't Hooft, p. 114.

290 'The idea is that twenty or so numbers...', Lederman, p. 363.

292 'Throw deep,' Ronald Reagan, quoted in Lederman, p. 380.

292 'Pave the LEP tunnel with superconducting magnets,' Carlo Rubbia, quoted in Lederman, p. 381.

293 'Well, those were great days...', Steven Weinberg, in Cashmore, *et al.*, p. 20.

30: Hidden Variables

297 'May have to sit for a while', Albert Einstein, quoted by Bohm in an interview with Maurice Wilkins, late 1980s. Quoted in Peat, p. 92.

297 'Thorough American', Princeton University press release, 27 May 1949. Quoted in Peat, p. 95.

298 'I found it hard to accept Bohm's decision...', Wheeler, p. 216.

298 'Hard to concern myself...', David Bohm, letter to Miriam Yevick, 7 January 1952. Quoted in Peat, p. 105.

299 'EPR's criticism has, in fact, been shown to be unjustified...', Bohm, *Quantum Theory*, p. 611.

301 'This encounter had a strong effect on the direction of my research...', David Bohm, *Zygon*, **20**, 1985, pp. 113–114. Quoted in Jammer, *Einstein and Religion*, p. 227.

302 'The usual interpretation of the quantum theory is self-consistent...', David Bohm, *Physical Review*, **85**, 1952, p. 169.

304 'In this point we are in agreement with Bohr...', David Bohm, *Physical Review*, **85**, 1952, pp. 187–188.

305 '...if we cannot disprove Bohm, then we must agree to ignore him', J. Robert Oppenheimer, quoted by Max Dresden at a meeting of the American Physical Society, Washington, May 1989, and confirmed in a subsequent interview with, and letter to, F. David Peat. Quoted in Peat, p. 133.

305 'Have you noticed that Bohm believes...', Albert Einstein, letter to Max Born, [date?], 1952. Quoted in John S. Bell, *Proceedings of the Symposium on Frontier Problems in High Energy Physics*, Pisa, June 1976, pp. 33–45. This paper is reproduced in Bell, pp. 81–92. The quote appears on p. 91.

31: Bertlmann's Socks

307 'A theory founded in this way on arguments of manifestly approximate character...', John Bell, *Proceedings of the International School of Physics 'Enrico Fermi', Course IL: Foundations of Quantum Mechanics*, Academic Press, New York, 1971, pp. 171–181. Reproduced in Bell, pp. 29–39. This quote appears on p. 29.

307 '...no theory of mechanically determined hidden variables...', Bohm, *Quantum Theory*, p. 623.

308 'Probably I got that equation into my head and out on to paper...', John Bell, in Davies and Brown, p. 57.

313 'The philosopher in the street, who has not suffered a course in quantum mechanics...', John Bell, *Journal de Physique* Colloque C2. suppl. au numero 3. Tome 42, 1981, pp. 41–61. Reproduced in Bell, pp. 139–158. This quote appears on p. 139.

316 'If the [hidden variable] extension is local it will not agree with quantum mechanics...', John Bell, *Epistemological Letters*, November 1975, pp. 2–6. This paper is reproduced in Bell, pp. 63–66. This quote appears on p. 65.

317 'In a theory in which parameters are added to quantum mechanics...', John Bell, *Physics*, 1, 1964, pp. 195–200. This paper is reproduced in Bell, pp. 14–21. The quote appears on p. 20.

32: The Aspect Experiments

320 'Are you tenured?' and subsequent dialogue, John Bell and Alain Aspect, quoted in Aczel, p. 186.

323 'Our results, in excellent agreement with quantum mechanical predictions...', Alain Aspect, Philippe Grangier and Gérard Roger, *Physical Review Letters*, **47**, 1981, p. 463.

327 'But of course we already know that...', Alain Aspect, in Davies and Brown, p. 43.

327 'It is a very important experiment...', John Bell, in Davies and Brown, p. 52.

33: The Quantum Eraser

329 'If an apparatus is capable of determining...', Richard P. Feynman, Robert B. Leighton and Matthew Sands, *The Feynman Lectures on Physics*. Volume III. Addison-Wesley, Reading, Massachusetts, 1965, p. 1–9.

332 'Bohr would not have been distressed by the outcome of these considerations...', Marlan Scully, Berthold-Georg Englert and Herbert Walther, *Nature*, **351**, 1991, p. 111.

335 'Realizing quantum erasure in an experiment...', Philip Yam, *Scientific American*, January 1996, p. 30.

337 'The use of mutually exclusive settings of the experimental apparatus...', Thomas Herzog, Paul G. Kwiat, Harald Weinfurter and Anton Zeilinger, *Physical Review Letters*, **75**, 1995, p. 3037.

34: Lab Cats

339 'What exactly qualifies some physical systems to play the role of "measurer"...', John Bell, *Physics World*, **3**, 1990, p. 34.

340 'The idea that elimination of coherence...', John Bell, *Physics World*, **3**, 1990, p. 36.

345 'The quantum dynamics of the SQUID is determined...' Jonathan R. Friedman, Vijay Patel, W. Chen, S.K. Tolpygo and J.E. Lukens, *Nature*, **406**, 2000, p. 45.

346 '...symmetric [anticlockwise + clockwise] and anti-symmetric [anticlockwise – clockwise] quantum superpositions of macroscopic states...', Caspar H. van der Wal, A.C.J. ter Haar, F.K. Wilhelm, R.N. Schouten, C.J.P.M. Harmans, T.P. Orlando, Seth Lloyd and J.E. Moonij, *Science*, **290**, 2000, p. 773.

347 '...we can say that on a logarithmic scale we have come about 40% of the way...', A.J. Leggett, *Journal of Physics: Condensed Matter*, **14**, 2002, p. R447.

347 'There are many open questions regarding entanglement...', Vlatko Vedral, *Nature*, **453**, 2008, p. 1007.

35: The Persistent Illusion

351 'But in physics we are normally accustomed to require...', A.J. Leggett, *Foundations of Physics*, **33**, 2003, pp. 1474–5.

352 'There is in a case like that just considered no question of a mechanical disturbance...', Niels Bohr, *Physical Review*. **48**, 1935, 696–702. Reproduced in Wheeler and Zurek, p. 145. This quote appears on p. 148

354 'We believe that our results lend strong support to the view...', Simon Gröblacher, Tomasz Paterek, Rainer Kaltenbaek, Caslav Brukner, Marek Zukowski, Markus Aspelmeyer and Anton Zeilinger, *Nature*, **446**, 2007, p. 875.

354 'Quantum mechanics is very fundamental...', Anton Zeilinger, quoted in *SEED*, May/June 2008, p. 57.

356 'We have to remember that what we observe is not nature in itself...', Heisenberg, *Physics and Philosophy*, p. 46.

357 'Reality is merely an illusion...', quotation attributed to Einstein. See, for example, Baggott, *A Beginner's Guide to Reality*, p. 228.

357 'I'm in a small minority with that point of view...', Anthony Leggett, quoted in *SEED*, May/June 2008, p. 58.

36: The Wavefunction of the Universe

361 'Due to electron motion inside the atom...', Albert Einstein, *Preussische Akademie der Wissenschaften (Berlin) Sitzungsberichte*, 1916, p. 688. Quoted in Gorelik and Frenkel, p. 86.

363 'That fall, fifteen graduate students enrolled in my course...', Wheeler, p. 228.

366 'This result has led me to doubt...', Paul Dirac, quoted in Barbour, p. 2.

369 '...one therefore comes to the conclusion that nothing ever happens...', Bryce DeWitt, *Physical Review*, **160**, 1967, p. 1119.

370 'The Copenhagen view depends on the assumed a priori existence...', Bryce DeWitt, *Physical Review*, **160**, 1967, p. 1131.

370 'Thus with each succeeding observation (or interaction), the observer state...', Hugh Everett III, *Reviews of modern physics*. **29**, 1957, p. 454.

371 'Everett's view of the world is a very natural one to adopt...', Bryce DeWitt, *Physical Review*, **160**, 1967, p. 1141.

37: Hawking Radiation

372 'That damned equation', Barbour, p. 247.

373 'If the rays of light that form the event horizon...', Hawking, p. 100.

375 'Each black hole law, in fact, turned out to be identical...', Thorne, p. 427.

375 'I must admit that in writing this paper...', Hawking, p. 104.

375 'How can you make such a crazy claim...', Thorne, p. 429.

375 'The metal sphere will radiate...', Yakov Zeldovich, quoted by Thorne, pp. 429–430. The italics are Thorne's

376 'They convinced me that, according to the quantum mechanical uncertainty principle...', Hawking, p. 104.

377 'I didn't want particles coming out...', Dennis Overbye, *Omni*, February 1979, quoted in Larsen, p. 40.

377 'When I did the calculation, I found...', Hawking, p. 105.

378 '...Bardeen, Carter and I considered that the thermodynamical similarity...', Stephen Hawking, *Nature*, **248**, 1974, p. 31.

379 'I would rather be right than rigorous', Stephen Hawking, quoted in Thorne, p. 441.

380 'In the front row a young man in a wheelchair...', Carl Sagan, in Hawking, p. x.

38: The First Superstring Revolution

381 'Gauge group of the world', Howard Georgi and Sheldon Glashow, *Physical Review Letters*, **32**, 1974, p. 438.

381 'And then I realized that this made the proton, the basic building block of the atom, unstable...', Howard Georgi, interview with Robert Crease and Charles Mann, 29 January 1985. Quoted in Crease and Mann, p. 400.

383 'I worked on it for a long time...', Leonard Susskind, *The Landscape: A Talk with Leonard Susskind*, www.edge.org., April 2003.

384 'What do you do?', and subsequent quotation, Leonard Susskind, *The Landscape: A Talk with Leonard Susskind*, www.edge.org., April 2003.

385 'I think we were kind of struck by the mathematical beauty...', John Schwarz, interview with Sara Lippincott, 21 and 26 July, 2000, Oral History Project, California Institute of Technology Archives, 2002, p. 26.

387 'In our discussions with him it became very clear...', John Schwarz, interview with Sara Lippincott, 21 and 26 July, 2000, Oral History Project, California Institute of Technology Archives, 2002, p. 39.

388 'SO(32),' John Schwarz, interview with Sara Lippincott, 21 and 26 July, 2000, Oral History Project, California Institute of Technology Archives, 2002, p. 41.

388 'I figured out how to do everything...', John Schwarz, interview with Sara Lippincott, 21 and 26 July, 2000, Oral History Project, California Institute of Technology Archives, 2002, p. 43.

39: Quanta of Space and Time

392 'I had no desire to share the fate of Galileo...', Hawking, p. 116.

393 'One could now import into general relativity...', Abhay Ashtekar, arXiv: gr-qc/0410054v2, 19 October 2004, p. 8. Available on http://arXiv.org/

393 '...new ways of attacking a number of problems...', Abhay Ashtekar, *Physical Review Letters*, **57**, 1986, p. 2244.

393 'All of a sudden we realized that our second or third guess...' and subsequent quote, Smolin, *Three Roads to Quantum Gravity*, p. 40.

394 '...even at the start we knew that we had in our grasp...', Smolin, *Three Roads to Quantum Gravity*, p. 128.

395 'I've found the answer to all the problems,' Carlo Rovelli, quoted in Smolin, *Three Roads to Quantum Gravity*, p. 129.

396 'The loops *are* space because they are the quantum excitations...', Carlo Rovelli, *Physics World*, November 2003, p. 2.

397 'If we could draw a detailed picture of the quantum state of our universe...', Lee Smolin, *Scientific American*, January 2004, p. 72.

40: Crisis? What Crisis?

401 'Schizophrenia with a vengeance,' Bryce DeWitt, quoted in Davies and Brown, p. 36.

402 'One distinguished physicist, well versed in quantum mechanics...', Gell-Mann, p. 138.

402 'Copenhagen done right,' Robert Griffiths, *Physical Review A*, 57, 1998, p. 1604.

402 '...just as there is no single 'correct' choice of consistent [history]...', Robert Griffiths, *Physical Review A*, **57**, 1998, p. 1604.

403 'We believe Everett's work to be useful and important...', Gell-Mann, p. 138.

403 'While the 'classical' world we observe, in which particles have definite positions...', Smolin, *Three Roads to Quantum Gravity*, p. 44.

404 'So conventional quantum cosmology...', Smolin, *Three Roads to Quantum Gravity*, p. 45.

405 'As you undoubtedly know...', J. Robert Oppenheimer, letter to Frank Oppenheimer, 4 June 1934. Quoted in Kragh, *Dirac: A Scientific Biography*, p. 165–166.

405 'I don't like that they're not calculating anything...', Richard Feynman, in PCW Davies and Julian Brown, eds., *Superstrings: A Theory of Everything*, Cambridge University Press, 1988, p. 194.

405 'The fact is that this is a book about physics...', Veltman, p. 308.

406 'The failure of the superstring theory programme must be recognised...', Woit, p. 259.

406 'I don't want to discourage string theorists...', Steven Weinberg, quoted by Peter Woit, www.math.columbia.edu/ woit/wordpress/, July 2009.

Epilogue: A Quantum of Solace?

407 'It's a fantastic moment,' *CERN bulletin* 37–38/2008.

410 'I think it will be much more exciting...', Stephen Hawking, quoted in an article by Mark Henderson, *The Times*, 9 September 2008.

410 'It was a bit of a cheek...', Peter Higgs, quoted by Ian Sample in Sample, p. 211.

BIBLIOGRAPHY

Aczel, Amir D., *Entanglement*. John Wiley & Sons, London, 2003.

Bacciagaluppi, Guido and Valentini, Antony, *Quantum Theory at the Crossroads: Reconsidering the 1927 Solvay Conference*. Cambridge University Press, 2009.

Baggott, Jim, *Beyond Measure: Modern Physics, Philosophy and the Meaning of Quantum Theory*. Oxford University Press, 2003.

Baggott, Jim, *A Beginner's Guide to Reality*. Penguin, London, 2005.

Baggott, Jim, *Atomic: The First War of Physics and the Secret History of the Atom Bomb 1939–49*. Icon Books, London, 2009.

Barbour, Julian, *The End of Time*. Weidenfeld & Nicholson, London, 1999.

Barrow, John D., *Theories of Everything*. Vintage, London, 1991.

Bell, J.S., *Speakable and Unspeakable in Quantum Mechanics*. Cambridge University Press, 1987.

Beller, Mara, *Quantum Dialogue*. University of Chicago Press, 1999.

Bernstein, Jeremy, *Quantum Profiles*. Princeton University Press, 1991.

Bird, Kai and Sherwin, Martin J., *American Prometheus: the Triumph and Tragedy of J. Robert Oppenheimer*. Atlantic Books, London, 2008.

Bohm, David, *Quantum Theory*. Prentice-Hall, Englewood Cliffs, NJ., 1951.

Bohm, David, *Causality and Chance in Modern Physics*. Routledge, London, 1957.

Bohm, David, *Wholeness and the Implicate Order*. Routledge, London, 1980.

Bohm, D. and Hiley, B.J., *The Undivided Universe*. Routledge, London, 1993.

Born, Max, *Physics in My Generation* (2nd edn). Springer, New York, 1969.

Brown, Andrew, *The Neutron and the Bomb: A Biography of Sir James Chadwick*. Oxford University Press, 1997.

Cashmore, Roger, Maiani, Luciano and Revol, Jean-Pierre (eds.), *Prestigious Discoveries at CERN*. Springer, Berlin, 2004.

Cassidy, David C., *Uncertainty: The Life and Science of Werner Heisenberg*. W.H. Freeman, New York, 1992.

Crease, Robert P., *A Brief Guide to the Great Equations: The Hunt for Cosmic Beauty in Numbers*. Robinson, London, 2009.

Crease, Robert P. and Mann, Charles C., *The Second Creation: Makers of the Revolution in Twentieth-century Physics*. Rutgers University Press, 1986.

Cushing, James T., *Quantum Mechanics: Historical Contingency and the Copenhagen Hegemony*. University of Chicago Press, 1994.

Cushing, James T., *Philosophical Concepts in Physics*. Cambridge University Press, 1998.

Davies, P.C.W. and Brown, J.R. (eds), *The Ghost in the Atom*. Cambridge University Press, 1986.

de Broglie, Louis, 'Recherches sur la Théorie des Quanta', PhD Thesis, Faculty of Science, Paris University, 1924. English translation by A.F. Kracklauer.

Deutsch, David, *The Fabric of Reality*. Penguin, London, 1997.

DeWitt, Bryce. S. and Graham, Neill (eds), *The Many Worlds Interpretation of Quantum Mechanics*. Pergamon, Oxford, 1975.

Dirac, P. A. M., *The Principles of Quantum Mechanics* (4th edn). Clarendon Press, Oxford, 1958.

Dodd, J. E., *The Ideas of Particle Physics*. Cambridge University Press, 1984.

Dörries, Matthias (ed.), *Michael Frayn's Copenhagen in Debate*. University of California, 2005.

Dyson, Freeman, *Disturbing the Universe*. Basic Books, New York, 1979.

d'Espagnat, Bernard, *The Conceptual Foundations of Quantum Mechanics* (2nd edn). Addison-Wesley, New York, 1989.

d'Espagnat, Bernard, *Reality and the Physicist*. Cambridge University Press, 1989.

Enz, Charles P., *No Time to be Brief: a Scientific Biography of Wolfgang Pauli*. Oxford University Press, 2002.

Farmelo, Graham (ed.), *It Must be Beautiful: Great Equations of Modern Science*. Granta Books, London, 2002.

Farmelo, Graham, *The Strangest Man: The Hidden Life of Paul Dirac, Quantum Genius*. Faber and Faber, London, 2009.

Feynman, Richard, *The Character of Physical Law*. MIT Press, Cambridge, MA, 1967.

Feynman, Richard P., *'Surely You're Joking, Mr. Feynman!' Adventures of a Curious Character*. Unwin, London, 1985.

Feynman, Richard P., *QED: The Strange Theory of Light and Matter*. Penguin, London, 1985.

Feynman, Richard P., *Six Easy Pieces*. Perseus, Cambridge, MA, 1998.

Feynman, Richard P., *Six Not-so-easy Pieces*. Allen Lane, London, 1998.

Feynman, Richard P., Leighton, Robert B., and Sands, Matthew, *The Feynman Lectures on Physics*, Vol. III. Addison-Wesley, Reading, MA, 1965.

Fine, Arthur, *The Shaky Game: Einstein, Realism and the Quantum Theory* (2nd edn). University of Chicago Press, 1996.

French, A.P. and Kennedy, P.J. (eds), *Niels Bohr: A Centenary Volume*. Harvard University Press, Cambridge, MA, 1985.

Frisch, Otto, *What Little I Remember*. Cambridge University Press, 1979.

Gamow, George, *Mr. Tompkins in Paperback*. Cambridge University Press, 1965.

Gamow, George, *Thirty Years that Shook Physics*. Dover Publications, New York, 1966.

Gell-Mann, Murray and Ne'eman, Yuval, *The Eightfold Way*. W.A. Benjamin, New York, 1964.

Gell-Mann, Murray, *The Quark and the Jaguar*. Little, Brown & Co., London, 1994.

Gleick, James, *Genius: Richard Feynman and Modern Physics*. Little, Brown & Co., London, 1992.

Goodchild, Peter, *J. Robert Oppenheimer: 'Shatterer of Worlds'*. BBC, London, 1980.

Goodchild, Peter, *Edward Teller: The Real Dr Strangelove*. Weidenfeld & Nicholson, London, 2004.

Gorelik, Gennady E. and Frenkel, Viktor Ya., *Matvei Petrovich Bronstein and Soviet Theoretical Physics in the Thirties*. Birkhäuser Verlag, Basel, 1994.

Greene, Brian, *The Elegant Universe: Superstrings, Hidden Dimensions and the Quest for the Ultimate Theory*. Vintage Books, London, 2000.

Greene, Brian. *The Fabric of the Cosmos: Space, Time and the Texture of Reality*. Allen Lane, London, 2004.

Gregory, Bruce, *Inventing Reality: Physics as Language*. John Wiley & Sons, New York, 1988.

Gribbin, John, *Schrödinger's Kittens*. Penguin, London, 1995.

Gribbin, John, *Q is for Quantum: Particle Physics from A to Z*. Weidenfeld & Nicholson, London, 1998.

Halpern, Paul, *Collider: The Search for the World's Smallest Particles*. John Wiley & Son, Inc., New Jersey, 2009.

Hecht, Eugene and Zajac, Alfred, *Optics*. Addison-Wesley, Reading, MA, 1974.

Heilbron, J.L., *The Dilemmas of an Upright Man: Max Planck and the Fortunes of German Science*. Harvard University Press, 1996.

Heisenberg, Elisabeth, *Inner Exile: Recollections of a Life with Werner Heisenberg*. Birkhäuser, Boston, 1984.

Heisenberg, Werner, *The Physical Principles of the Quantum Theory*. University of Chicago Press, 1930. Republished in 1949 by Dover Publications, New York.

Heisenberg, Werner, *Physics and Beyond: Memories of a Life in Science*. George Allen & Unwin, London, 1971.

Heisenberg, Werner, *Encounters with Einstein*. Princeton University Press, 1983.

Heisenberg, Werner, *Physics and Philosophy: The Revolution in Modern Science*. Penguin, London, 1989 (first published 1958).

Hermann, Armin, *The Genesis of Quantum Theory (1899–1913)*. MIT Press, Cambridge, MA, 1971.

Hiley, B.J. and Peat, F.D. (eds), *Quantum Implications*. Routledge & Kegan Paul, London, 1987.

Hoddeson, Lillian, Brown, Laurie, Riordan, Michael and Dresden, Max, *The Rise of the Standard Model: Particle Physics in the 1960s and 1970s*. Cambridge University Press, 1997.

Hoffmann, Banesh, *Albert Einstein*. Paladin, St. Albans, 1975.

Holland, Peter R., *The Quantum Theory of Motion*. Cambridge University Press, 1993.

Irving, David, *The Virus House: Germany's Atomic Research and Allied Counter-measures*. Focal Point, 2002 (first published in 1968).

Isaacson, Walter, *Einstein: His Life and Universe*. Simon & Shuster, New York, 2007.

Isham, Chris J., *Lectures on Quantum Theory*. Imperial College Press, 1995.

Jammer, Max, *The Philosophy of Quantum Mechanics*. John Wiley & Sons, New York, 1974.

Jammer, Max, *Einstein and Religion: Physics and Theology*. Princeton University Press, 2002.

Johnson, George, *Strange Beauty: Murray Gell-Mann and the Revolution in Twentieth-Century Physics*. Vintage, London, 2001.

Kilmister, C.W. (ed.), *Schrödinger: Centenary Celebration of a Polymath*. Cambridge University Press, 1987.

Klein, Martin J., *Paul Ehrenfest: The Making of a Theoretical Physicist*, Vol. 1 (3rd edn). North-Holland, Amsterdam, 1985.

Kragh, Helge S., *Dirac: A Scientific Biography*. Cambridge University Press, 1990.

Kragh, Helge, *Quantum Generations: A History of Physics in the Twentieth Century*, Princeton University Press, 1999.

Kuhn, Thomas S., *Black-body Theory and the Quantum Discontinuity 1894–1912*. University of Chicago Press, 1978.

Kumar, Manjit, *Quantum: Einstein, Bohr and the Great Debate About the Nature of Reality*. Icon Books, London, 2008.

Larsen, Kristine, *Stephen Hawking: A Biography*. Greenwood Press, Westport, Connecticut, 2005.

Lederman, Leon (with Dick Teresi), *The God Particle: If the Universe is the Answer, What is the Question?* Bantam Press, London, 1993.

Lindley, David, *Where Does the Weirdness Go?* Basic Books, New York, 1996.

Liss, Tony M. and Tipton, Paul L., 'The Discovery of the Top Quark', *Scientific American*, September 1997.

Mehra, Jagdish, *The Beat of a Different Drum: The Life and Science of Richard Feynman*. Oxford University Press, 1994.

Mehra, Jagdish, *Einstein, Physics and Reality*. World Scientific, London, 1999.

Mehra, Jagdish and Rechenberg, Helmut, *The Historical Development of Quantum Theory Volume 1, Part 1: The Quantum Theory of Planck, Einstein, Bohr and Sommerfeld: Its Foundation and the Rise of Its Difficulties, 1900–1925*. Springer-Verlag, New York, 1982.

Mehra, Jagdish and Rechenberg, Helmut, *The Historical Development of Quantum Theory Volume 1, Part 2: The Quantum Theory of Planck, Einstein, Bohr and*

Sommerfeld: Its Foundation and the Rise of Its Difficulties, 1900–1925. Springer-Verlag, New York, 1982.

Mehra, Jagdish and Rechenberg, Helmut, *The Historical Development of Quantum Theory Volume 2: The Discovery of Quantum Mechanics*. Springer-Verlag, New York, 1982.

Mehra, Jagdish and Rechenberg, Helmut, *The Historical Development of Quantum Theory Volume 3: The Formulation of Matrix Mechanics and its Modifications 1925–1926*. Springer-Verlag, New York, 1982.

Moore, Walter, *Schrödinger: Life and Thought*. Cambridge University Press, 1989.

Murdoch, Dugald, *Niels Bohr's Philosophy of Physics*. Cambridge University Press, 1987.

Neumann, John von, *Mathematical Foundations of Quantum Mechanics*. Princeton University Press, 1955.

Omnès, Roland, *The Interpretation of Quantum Mechanics*. Princeton University Press, 1994.

Omnès, Roland, *Understanding Quantum Mechanics*. Princeton University Press, 1999.

Omnès, Roland, *Quantum Philosophy*. Princeton University Press, 1999.

Oppenheimer, J. Robert, *The Open Mind*. Simon & Shuster, New York, 1955.

Oppenheimer, J. Robert, *Atom and Void*. Princeton University Press, 1989.

Pais, Abraham, *Subtle is the Lord: The Science and the Life of Albert Einstein*. Oxford University Press, 1982.

Pais, Abraham, *Inward Bound: Of Matter and Forces in the Physical World*. Oxford University Press, 1986.

Pais, Abraham, *Niels Bohr's Times, in Physics, Philosophy and Polity*. Clarendon Press, Oxford, 1991.

Pais, Abraham *J. Robert Oppenheimer: A Life*. Oxford University Press, 2006.

Peat, F. David, *Infinite Potential: The Life and Times of David Bohm*. Addison-Wesley, Reading, MA, 1997.

Penrose, Roger, *The Emperor's New Mind*. Vintage, London, 1990.

Penrose, Roger, *Shadows of the Mind*. Vintage, London, 1995.

Penrose, Roger, *The Large, the Small and the Human Mind*. Cambridge University Press, 1997.

Pickering, Andrew, *Constructing Quarks: A Sociological History of Particle Physics*. University of Chicago Press, 1984.

Polkinghorne, J.C., *The Quantum World*. Penguin, London, 1984.

Popper, Karl R., *Quantum Theory and the Schism in Physics*. Unwin Hyman, London, 1982.

Rae, Alastair, *Quantum Physics: Illusion or Reality?* Cambridge University Press, 1986.

Rae, Alastair I. M., *Quantum Mechanics* (2nd edn). Adam Hilger, Bristol, 1986.

Rhodes, Richard, *The Making of the Atomic Bomb*. Simon & Shuster, New York, 1986.
Riordan, Michael, *The Hunting of the Quark: A True Story of Modern Physics*. Simon & Shuster, New York, 1987.
Rohrlich, Fritz, *From Paradox to Reality*. Cambridge University Press, 1987.
Sachs, Mendel, *Einstein versus Bohr: the Continuing Controversies in Physics*. Open Court, La Salle, IL., 1988.
Sample, Ian, *Massive: The Hunt for the God Particle*. Virgin Books, London, 2010.
Schilpp, Paul Arthur (ed.), *Albert Einstein. Philosopher-Scientist*. The Library of Living Philosophers, Volume 1, Harper & Row, New York, 1959 (first published 1949).
Schweber, Silvan S., *QED and the Men Who Made It: Dyson, Feynman, Schwinger, Tomonaga*. Princeton University Press, 1994.
Schweber, Silvan S., *Einstein & Oppenheimer: The Meaning of Genius*. Harvard University Press, 2008.
Scully, Robert J., *The Demon and the Quantum: From the Pythagorean Mystics to Maxwell's Demon and Quantum Mystery*. Wiley-VCH Verlag GmbH & Co., Weinheim, 2007.
Segrè, Emilio, *Enrico Fermi: Physicist*. University of Chicago Press, 1970.
Smolin, Lee, *Three Roads to Quantum Gravity*. Weidenfeld & Nicholson, London, 2000.
Smolin, Lee, *The Trouble with Physics: The Rise of String Theory, the Fall of a Science and What Comes Next*. Penguin, London, 2006.
Snow, C.P., *The Physicists*. Macmillan, London, 1981.
Squires, Euan, *The Mystery of the Quantum World*. Adam Hilger, Bristol, 1986.
Stachel, John (ed.), *Einstein's Miraculous Year: Five Papers that Changed the Face of Physics*. Princeton University Press, 2005.
t' Hooft, Gerard, *In Search of the Ultimate Building Blocks*. Cambridge University Press, 1997.
Ter Haar, D., *The Old Quantum Theory*. Pergamon Press, Oxford, 1967.
Tomonaga, Sin-itiro, *The Story of Spin*. University of Chicago Press, 1997.
Treiman, Sam, *The Odd Quantum*. Princeton University Press, 1999.
Veltman, Martinus, *Facts and Mysteries in Elementary Particle Physics*. World Scientific, London, 2003.
Waerden, B.L. van der, *Sources of Quantum Mechanics*. Dover Publications, New York, 1968.
Weyl, Hermann, *Symmetry*. Princeton University Press, 1952.
Wheeler, John Archibald (with Kenneth Ford), *Geons, Black Holes and Quantum Foam*. W.W. Norton & Company, New York, 2000.
Wheeler, John Archibald and Zurek, Wojciech Hubert (eds.), *Quantum Theory and Measurement*. Princeton University Press, 1983.

Wigner, Eugene, *The Recollections of Eugene P. Wigner, as Told to Andrew Szanton.* Basic Books, New York, 1992.

Woit, Peter, *Not Even Wrong*. Vintage Books, London, 2007.

Zee, A., *Quantum Field Theory in a Nutshell.* Princeton University Press, 2003.

Zee, A., *Fearful Symmetry: The Search for Beauty in Modern Physics.* Princeton University Press, 2007 (first published 1986).

PLATE ACKNOWLEDGEMENTS

1. AIP Emilio Segrè Visual Archives
2. Hebrew University of Jerusalem Albert Einstein Archives, courtesy AIP Emilio Segrè Visual Archives
3. AIP Emilio Segrè Visual Archives, Margrethe Bohr Collection
4. Deutsche Verlag, courtesy AIP Emilio Segrè Visual Archives, Brittle Books Collection
5. Max-Planck Institute, courtesy AIP Emilio Segrè Visual Archives
6. Niels Bohr Institute, courtesy AIP Emilio Segrè Visual Archives
7. Photograph by Benjamin Couprie, Institut International de Physique Solvay, courtesy AIP Emilio Segrè Visual Archives
8. Photograph by Paul Ehrenfest, courtesy AIP Emilio Segrè Visual Archives
9. Harvey of Pasadena, courtesy AIP Emilio Segrè Visual Archives
10. AIP Emilio Segrè Visual Archives
11. AIP Emilio Segrè Visual Archives
12. Harvey of Pasadena, courtesy AIP Emilio Segrè Visual Archives
13. AIP Emilio Segrè Visual Archives, Segrè Collection
14. AIP Emilio Segrè Visual Archives
15. AIP Emilio Segrè Visual Archives, Marshak Collection
16. Courtesy of the University of California, Santa Barbara
17. AIP Emilio Segrè Visual Archives, W. F. Meggers Gallery of Nobel Laureates, Gift of Frank Wilczek
18. AIP Emilio Segrè Visual Archives, W. F. Meggers Gallery of Nobel Laureates
19. © Peter Tuffy, University of Edinburgh, reproduced with permission
20. CERN
21. AIP Emilio Segrè Visual Archives, gift of Gerardus 't Hooft
22. Brookhaven National Laboratory, reproduced with permission

23. AIP Emilio Segrè Visual Archives, Physics Today Collection, W. F. Meggers Gallery of Nobel Laureates
24. AIP Emilio Segrè Visual Archives
25. Library of Congress, New York World – Telegram and Sun Collection, courtesy AIP Emilio Segrè Visual Archives
26. CERN
27. © CNRS Photolibrary/Jérôme Chatin, reproduced with permission
28. Board of Trustees of the University of Illinois
29. © Jaqueline Godany; http://godany.com
30. AIP Emilio Segrè Visual Archives, Wheeler Collection
31. AIP Emilio Segrè Visual Archives, Physics Today Collection
32. AIP Emilio Segrè Visual Archives, Physics Today Collection
33. Courtesy of Carlo Rovelli, reproduced with permission
34. CERN

INDEX

Abrikosov, Alexei 350
ADONE collider 266
Aharonov, Yakir 306, 307–8, 318, 320
alpha-particles 26, 26n, 28, 240
  stability 200
Alpher, Ralph 367
Alvarez, Luis 218
Alvarez-Gaumé, Luis 387
Ambler, Eric 209
Anderson, Carl 142, 158
Anderson, Philip 228–9
Annalen der Physik 8, 15, 65
Annals of Physics 210, 365
'anomalous' Zeeman effect 43n, 51, 52, 60, 131, 286
anti-electrons 171
anti-neutrinos 158, 205, 282, 289
  in beta-decay 212
  'right-handed' 210
anti-particles 142, 214, 290
anti-protons 158, 205 *see also* proton-anti-proton collisions
anti-Semitism 46, 142
Arab-Israeli War 219
Arnowitt, Richard 366–7
Ashtekar, Abhay 392–3, 396
Aspect, Alain xvii, 319–24, 327, 328
Aspelmeyer, Markus 347, 350, 353
Aspen Center for Physics, Colorado 258, 262, 263, 387
'associated production' 206–7
Astrophysical Journal Letters 368
asymptotic freedom 261–2, 385
atom bomb xv, 159–67, 173
atom-electron collisions 73–4
atomic emission spectra 30, 43, 46–7, 47–8, 51–2
  factor of two discrepancy 56–7, 58–9
atomic nuclei 28
  electron exchange model 199–200
atomic structure 24, 25–33, 128, 238, 287
  Bohr's quantum model xiv, 29–33, 34, 40
  new quantum theory needed 46–7
  Rutherford's planetary model 26, 28, 46
  Thomson's 'currant' model 27
atomist doctrine 7–9, 14, 15, 17
atoms 158
  reality of 16, 17, 21, 107, 111
Avogadro's number 17
Ayer, A. J. 110

background microwave radiation 367–8
ball-in-the-boxes thought experiment 152–3
Balmer, Johann Jakob 30
Balmer formula 30–1, 60, 65, 241–2
Balmer series 31
Bardeen, John 228, 259, 375, 379
Bardeen, William 259
'baryon numbers' 215
baryons 214, 214n, 215
  charmed 251, 267, 275
  quark compositions 222, 224, 259

baryons, spin-½ 215, 220, 224
  Eightfold Way 218, 219 Fig 12, 223 Fig 14
  and quarks 222–3
baryons, spin-3/2
  Eightfold Way 221 Fig 13
Becker, Ulrich 269, 271
Bekenstein, Jacob 374–5, 377
Bekenstein–Hawking formula 400
Bell, John 307–17, 320, 321, 327, 339, 340, 350
  Bertlmann's socks 313–16
Bell's inequality xvii, 308–17
  experimental tests proposed 319–22
  experimental violation 319, 322–7, 349, 351, 354
Bell's theorem xvii, 316, 318, 326
Bergman, Peter 365, 367
Berkeley 176, 204–5, 218, 237–8, 297, 319
Berlin University 10, 41, 72, 103
beta radioactive decay 158, 199, 205, 207–8, 210, 211
  and quarks 223, 288–9
beta-electrons 208–9
Bethe, Hans 158, 173, 179, 181, 182, 183, 188, 189
'big bang' 230, 367–8, 378, 391, 409
Birkbeck College, London 306, 373
Bjorken, James 239–40, 241, 244, 245, 246, 247
black body radiation *see* cavity radiation
black holes xvii, 364, 373
  entropy 373, 374–5, 377, 378, 400
  event horizon 364, 373, 375, 377, 378
  four laws 375, 379
  Hawking radiation 377–9, 386
  'no hair' theorem 373, 373n, 374
  Zeldovich radiation proposal 375–6
Bloch, Felix 63
Block, Richard 217
Bohm, David xvi, 173, 230, 298, 299–305, 306, 307–8, 318, 321, 327
  arrest and trial 297, 298
Bohr, Harald 27, 29, 104
Bohr, Margrethe (née Nørlund) 26, 30, 83
Bohr, Niels xiii, 26–33, 34, 35, 51, 57, 80, 95–102, 115, 128, 131, 132, 182, 184, 298, 305, 307, 331–2, 339
  anti-realism 110, 111, 147
  and the atom bomb 159–60, 163–5
  Como lecture xv, 103–6, 111
  and Copenhagen Institute 44
  debates with Einstein xv, 115–16, 118–25, 134–40, 142–3, 159, 327, 329
  debates with Heisenberg and Schrödinger xv, 81–5, 87, 95
  and electron spin 58
  engagement and marriage 26, 30
  escape to Los Alamos 166
  and Heisenberg, early relationship 44–6, 47
  and Heisenberg's uncertainty principle 97–102
  meeting with Heisenberg (Copenhagen, 1941) xv, 160–6
  Nobel Prize 34n
  and Pauli 52, 55, 101
  quantum model of the atom xiv, 29–33, 34, 40
  response to EPR argument 145–7, 157, 299, 329, 352
  and Rutherford 28, 32–3
  and Thomson 27
  and wave/particle complementarity 96–7, 401, 402
Bohr–Kramers–Slater (BKS) proposal 43, 48
Bohr–Sommerfeld theory 52
Boltzmann, Ludwig 7, 13–14, 77, 107, 108, 158
Boltzmann's constant 16, 17
Born, Max 42, 45, 46, 49, 50, 52, 72–8, 79n, 85, 95, 104, 131, 304, 305, 370
  career 72
  and Einstein 72, 76, 78, 87
  at fifth Solvay conference 115, 117
  leaves Germany 142

opposition to wave mechanics 85, 87
probability waves xv, 73–8
bosons (force particles) xvi, 214, 215, 215n, 342, 382, 396
baryon and lepton numbers 215
see also Higgs boson; Nambu–Goldstone bosons; Yang–Mills bosons
bottom quark 276, 285, 286, 288, 289, 291
Bragg, William L. 115, 116
Bronstein, Matvei 361
Bronstein cube 361–2, 362 Fig 26
Brookhaven National Laboratory xvi, 200, 205, 208, 210, 224, 237, 238, 249, 266–70, 272, 274
Alternating Gradient Synchrotron (AGS) 267–9
Brout, Robert 229
buckminsterfullerene 342

Cabibbo, Nicola 289
Cabibbo angle 289
Cabibbo–Kobayashi–Maskawa (CKM) mixing 289
Calabi, Eugenio 389
Calabi-Yau shapes 389, 392
calcium atoms 319, 320–1, 324–5
Caltech 210, 212, 217, 258, 262, 385, 388
Cambridge Electron Accelerator 266
Cambridge University 50, 52, 108, 128, 141, 188, 365, 373, 374, 382
J. J. Thomson laboratory 26–7, 28, 30
Candelas, Philip 389
Carlsberg Foundation 44
Carnap, Rudolf 108
Cartan, Élie 392
Carter, Brandon 373, 375, 379
causality 32, 42, 74–5, 77, 117, 118, 158, 298, 301, 308, 374
Bohm's reintroduction 304
complementarity with space-time 104–5, 134
death knell 93–4
Causality and Chance in Modern Physics (Bohm) 306
cavity (black-body) radiation 7, 9–16, 20–1, 24, 375, 377
CERN xvi, 220, 224, 230, 238, 246, 248, 252, 254–5, 259, 277–84, 285, 292–3, 307, 320, 328, 382, 383, 399, 406
'Gargamelle' bubble chamber 254
Intersecting Storage Rings (ISR) 278, 280
Large Electron Positron (LEP) collider 278, 281, 288
Large Hadron Collider (LHC) xiv, xviii, 292, 407–10
LHC ATLAS 409, 409n
LHC Compact Muon Solenoid (CMS) 409
Super Proton Synchrotron (SPS) 277, 278–80, 282, 283
UA_1 team 281, 282, 283
UA_2 team 281, 282, 283
Chadwick, James 141, 158
Chambers, Whitaker 297
charge-conjugation parity (CP) symmetry 249
violation in weak-force decays 249, 289
charm quarks xvi, 247, 250, 251, 255, 264, 275, 288, 289, 291
hunt for 266–74, 268 Fig 17
charmonium 273
Chen, Min 269, 271
Chen, W. 345
Chew, Geoffrey 215n
Chicago University 166, 198, 206, 383, 392
Churchill, Winston 160
classical model 1–4, 20–1, 29, 49, 66, 77, 93–4, 95, 101, 105, 106, 132, 177, 303
Clauser, John 319
Clinton, Bill 292
cloud chambers 83n, 85–6, 88, 89–90, 90–1

cobalt-60 209
Cockcroft, John 141
Cold War 190n, 237–8
Coleman, Sidney 262
colour force xvi, 259–61
Columbia University 56, 171, 176, 208, 220–1
  Radiation Laboratory 173–4
Commins, Eugene 319
Communist Party 297–8
complementarity 104–6, 116, 124–5, 146, 328, 401, 402
  parallels with relativity 134
  and quantum non-locality 338
Compton, Arthur 35, 37, 104, 116
Compton effect 92, 98, 116
Comte, Auguste 107, 108
Condon, Edward 145
conservation of energy 43, 48
'consistent histories' 402–4
constrained Hamilton formulation 365, 366–7, 393–3
constructive interference 2, 3 Fig 1
Cooper, Leon 228
Copenhagen Institute *see* Niels Bohr Institute for Theoretical Physics, Copenhagen
Copenhagen interpretation xv, 102, 103, 106, 109–11, 182, 298, 300–1, 305, 307, 327, 331, 346, 348, 370, 401, 402
  Bell's distrust 350
  Bohm's unease xvi, 301–2
  Bohr-Einstein debates 121–5, 135–40
  Einstein 'Kopenhagener Geist' jibe 152–3
  EPR challenge 144–8, 150–1
  Leggett's scepticism 357
  and Schrödinger's cat paradox 154, 156–7
Copernicus, Nicolaus 376
Cornell University 173, 176, 180, 184, 188, 211, 229
correspondence principle 102
cosmic censorship hypotheses 373
cosmic rays 142, 158, 204, 216, 405
Cronin, James 249
current algebra 239
cyclotron 204–5

Dalibard, Jean 319, 324
dark energy 410
dark matter 410
Darriulat, Pierre 278, 281
Darwin, Charles 28–9, 131
De Broglie, Louis 99, 104, 115, 306
  wave/particle 'double solution' xiv, 37–42, 62–3, 64, 71, 116, 119, 298, 304–5
Debye, Pieter 35, 62, 63, 103, 104
decoherence theory 340
decoherent histories 403
deep inelastic scattering 239–42, 243–6, 247, 258–9, 263, 319
  absence of quarks 256
delta particles 205, 218, 220, 239
Deser, Stanley 366–7
destructive interference 2, 2 Fig 1, 40, 177
determinism 75, 77, 118, 158, 298, 301, 304, 306, 308
deuterium isotopes 160
Deutsches Elektronen Synchroton (DESY), Hamburg 276
DeWitt, Bryce 368–71, 372, 401 *see also* Wheeler–De Witt equation
Di Lella, Luigi 283
dialectical materialism 216, 301
Dicke, Robert 367
diffeomorphism constraints 395
Dirac, Paul 50, 59, 60, 115, 126–32, 146, 182, 382, 392
  anti-proton 158
  canonical approach to quantum gravity 365, 366, 367
  childhood and career 127–8
  discovery of positrons 141–2
  and 'energy holes' 133–4
  Nobel Prize 142
  relativistic quantum theory of the electron xv, 128–32, 171, 172, 174
  on Weyl 193

Dirac, Charles 127
Döpel, Robert 160
double-slit experiment
  Bohr's hypothetical measurement apparatus 122–4, 123 Fig 6
  Einstein 121–2, 122 Fig. 5
  Feynman 328–9, 331
  real-life 338
Dowker, Fay 403–4
down quarks 222, 223, 224, 245, 265, 275, 287, 288, 289
Drühl, Kai xvii
  photon interference thought experiment 330–3, 331 Fig 24
Dubna accelerator 205, 237–8
Dyan, Moshe 218
Dyson, Freeman xv, 184, 188–91

$E=mc^2$ 20, 36–7, 135
Eddington, Arthur 127
efficiency loophole 325
Ehrenfest, Paul 21n, 58, 140
eigenfunctions 66, 76
eigenstates 397
eigenvalues 66, 102
Eightfold Way 216–18, 219 Fig 12, 220, 260
  Ne'eman version 220
  quark explanation 220–4, 223 Fig 14
Einstein, Albert xiii, 17–24, 51, 52, 58, 103, 108, 111, 115, 173n, 193, 326, 348, 350, 389
  annus mirabilis 19–20, 289
  and anti-Semitism 46
  and Bohm xvi, 297, 298, 301, 305
  and Born 72, 76, 78, 87
  and causality 42, 74–5, 78, 117
  correspondence with Planck 24
  correspondence with Schrödinger 150–4, 157
  and de Broglie 'double solution' 41, 62, 63, 116–17
  debates with Bohr xv, 115–16, 118–25, 134–40, 142–3, 159, 327, 329
  $E=mc^2$ 20, 36–7, 135
  fury at BKS proposal 43
  general theory of relativity 41, 52, 127–8, 138, 139, 198, 361, 364
  'ghost field' suggestion 74, 78
  gravitational field equations 362–3, 363–4, 367, 369, 371
  'heuristic principle' 22
  hidden variable theory 298–9
  light-quantum hypothesis xiv, 20, 21–4, 34–5, 37, 43, 74
  marriage and children 18–19
  Nernst visit 25
  Nobel Prize 34n
  on observation and measurement 9
  on Pauli 52
  special theory of relativity 20, 24, 35–6, 57, 60, 64, 119–20, 126, 129, 135, 144, 148, 243, 349
  Swiss Patent Office 17, 18, 19
  Swiss Patent Office promotion 24
  warning to Roosevelt 160
Einstein-Podolsky-Rosen (EPR) argument xv, xvi, 142–8, 150–1, 153, 175, 298, 317, 319, 321, 326, 349, 329
  Bohm's reimagining 300–1, 306
  Bohr's response 145–7, 157, 299, 329, 352
'Einstein separability' 151
Einstein, Hans Albert 19
Einstein, Lieserl 18–19, 19n
Einstein, Hermann 18
Einstein, Mileva 18–19
elastic scattering 239
electric charge conservation 194, 195, 197, 198, 199
electromagnetism 2–3, 208, 261, 382
  force carriers 275, 289
  weak force compared 207, 210
  see also unified field theories
electron accelerators 238–45
'electron holes' 133–4
electron neutrino 275
electron/photon diffraction experiment 118–20, 119 Fig 4
electron-positron collisions 265–6

electron-positron pairs 171, 181, 186, 265, 267–9, 283
electron-proton collisions 239
electrons 12n, 26, 110, 214, 275, 287, 289, 290
  baryon and lepton numbers 215
  in beta-decay 212
  charge estimated 17
  as 'currants' 27
  discovery 25
  empirical reality 356
  in Feynman diagrams 184–7
  fixed stable orbits 29, 30, 31–2, 34, 35
  g-factor anomaly 174–5, 181, 186
  g-factor anomaly solved 191
  high-energy 282
  mass in an electromagnetic field 175
  negative-energy state 132, 133
  particle-like properties 88
  Pauli exclusion principle 54–5, 56, 60, 126, 228, 342
  photon exchange 207
  in planetary model, problems 28
  'shells' 53–5
  spin xiv, 56–9, 60, 126, 127, 129–30, 130–1, 171
  standing waves 40
  in superconductors 228, 342–6
  'two-valuedness' 54, 56, 131
  wave-like properties 38, 41–2
  wave/particle duality 95–7
  wavefunction 64–5, 71–2, 73–8, 88–9, 198, 199, 256, 370
  see also photoelectric effect; quantum jumps; relativistic quantum theory of the electron
empirical reality xvii, 108–9, 156, 356
empiricist tradition 107–9
energy 'buckets' 13–14
energy conservation 50, 186, 194
energy elements (ε) 14–15
energy-time uncertainty 93, 98, 99, 100, 135–40, 186, 378
Englert, Berthold-Georg 332
Englert, François 229
Enrico Fermi Institute 391, 393
entropy 8–9, 13, 279–80
  black holes 373, 374–5, 377, 378
  and probability 13–14
eta particles 218
ETH (formerly Zurich Polytechnic) 18, 20n, 25, 25n, 62, 63, 103
ether 2, 3–4, 35
Euler, Leonhard 383
Euler's beta function 383
Euler's formula 195
Evans, Lyn 407
event horizon *see* black holes
Everett III, Hugh 370–1, 401–2, 404
Everhart, Glen 269
expanding universe 367, 392
extended supersymmetric theories 382

Facts and Mysteries in Elementary Particle Physics (Veltman) 405
Faraday, Michael 2, 3
Feinberg, Gerald 211
Fermat, Pierre de 177
Fermi, Enrico 104, 158, 159, 182, 199, 206, 208, 210, 214, 214n, 216
Fermilab (formerly National Accelerator Laboratory) 255, 255n, 260, 276, 281, 292, 372, 399
  proton accelerator 277
  Tevatron 285–6, 408
fermions (matter particles) 214, 246, 287, 290, 382, 386, 396
  Pauli exclusion principle 224, 256, 228
  string vibrational patterns 384
ferromagnets 225–6
Feynman, Arline 176, 189
Feynman, Richard xiii, 166, 180, 210, 246, 259, 305, 405
  and Dyson 189, 191
  Nobel Prize 242
  parton model 242–5
  path-integral approach 178–9, 181, 182, 185, 365, 391
  personality and approach 176–7, 181, 188–9
  Pocono conference 182

and quantum electrodynamics xiv, 184–8, 191, 242
Shelter Island conference 173, 177–9
two-slit experiment 328–9, 331
Feynman diagrams 184–8, 185 Fig 9, 187 Fig 10, 191, 192, 365, 388
Feynman histories 178, 402
fields 171n, 172
Fitch, Val 249
Foley, H. M. 174
force particles *see* bosons
Foundations of Physics 350
Fourier series 47
Fowler, Ralph 50, 128
Franck, James 76, 88, 95, 104, 142
Frank, Philipp 135
Franklin, Benjamin 199
Freedman, Stuart 319
Fresnel, Augustin 118
Friedman, Jerome 240–1, 244
Friedman, Jonathan 345
Friedmann, Alexander 369
Frisch, Otto 159
Fritzsch, Harald xvi, 257–60, 262–3
Fuchs, Klaus 176

Galileo 392
gamma-ray photons 171, 400
microscope experiment 92–3, 97–8 101–2
Gamow, George 367
'gauge factor' 62, 198
gauge symmetry 197–8, 365 *see also* global gauge symmetry; local gauge symmetry; symmetry fields
Geiger, Hans 26
Geiger–Marsden experiments 26, 238, 240
Gell–Mann, Murray 192, 210, 215, 219, 220, 243, 262
cabaret performance 388
career 206
decoherent histories 402,
Eightfold Way 216–18, 260
and Fritzsch 258–9
and Glashow 212–13
on 'many worlds' 402
Nobel Prize 258, 258n
and quantum chromodynamics xvi, 260–3
and quark colours 259–60
quark theory xvi, 220–4, 257, 306
'squalid-state' physics 226
and 'strangeness' 206–7
and superstring theory 384, 387
general relativity 41, 52, 127–8, 138, 139, 198, 361, 377
equivalent coordinate systems 395
exotic objects predicted 364, 372–3
recasting as 'spin system' 392–3
and superstring theory 391
unification with quantum theory xvii, 361–71, 401
Gerlach, Walther 52
Ginzburg, Vitaly 350
Glashow, Sheldon 381
and charm quark hunt 266–8, 271, 274
charm quark proposal 250, 255
and 'charmonium' 273
and grand unified theory 381
Nobel Prize 277
weak/electric force unification 217
and unified electro-weak theory xvi, 211–13, 217, 232, 233, 247–8, 249–51, 253, 284, 406
Glashow-Iliopoulos-Maiani (GIM) mechanism 249–51, 253
global symmetry 195, 195n
global symmetry groups 216–18, 260
gluon 'jets' 276
gluons xvi, 246, 260, 263, 275, 275n, 288, 290
'God particle' *see* Higgs boson
Goldhaber, Gerson 270, 271, 272, 274
Goldhaber, Maurice 210
Goldstone, Geoffrey 226–7
Goldstone theorem 227, 228, 233
Gordon, George 149
Gordon, Walter 126
Göttingen University 30, 44, 45, 46, 52, 58, 72, 76, 77, 87, 88, 128, 149

Goudsmit, Samuel 57, 130
Grangier, Philippe 319, 322
gravitino 382
graviton xvii–xix, 292, 366, 382, 386, 398, 400
gravity xvi, 261, 291–2, 373–3, 389
  action-at-a-distance 1–2, 4, 41
  superstring theory 386
  see also quantum gravity theory
Green, Michael 386, 387–8
Greenberg, Oscar 257
Greenberger, Daniel 326
Greenberger–Horne–Zeilinger states 326
Griffiths, Robert 402
Gröblacher, Simon 353
Gross, David 261–2, 383, 387, 388
Grossmann, Marcel 18
group theory 194, 216
gunpowder thought experiment 153
Guralnik, Gerald 229

Habicht, Conrad 18, 19, 22
hadron collisions 238, 243
hadron to muon production ratio (R) 265–6
hadrons 215, 239
  parton model 243
  quark composition 246
  and unified field models 233, 234, 251
Haganah 219
Hagen, Carl 229
Hahn, Otto 159
Hamburg University 52, 103, 126
Hamlet 45
Han, Moo-Young 257
Hansen, Hans 30
Harish-Chandra 188
Hartle, James 391, 402
Harvard University 176, 211, 232, 248, 255, 262, 368, 381, 393
Harvey, Jeffrey 388
Hawking, Stephen xvii, 373–5, 376–80, 382, 399, 403, 410
  amytrophic lateral sclerosis (ALS) 374, 379
  'no boundary' assumption 391–2
  Royal Society induction 379–80
Hawking radiation 377–9, 386
Hegel, Georg 110
Heisenberg, Werner 44–50, 57, 58, 77, 104, 121, 183, 258, 356, 405
  and atomic bomb 160, 163–6
  and Bohr, early relationship 44–5, 45–6, 47
  Cambridgeshire internment 166–7
  career 45, 103
  on classical/quantum paradox 106
  debates with Bohr and Schrödinger xv, 81–5, 87, 95
  and Dirac probability waves 127, 129, 132
  electron exchange model 199–200
  at fifth Solvay conference 115, 117
  fury over wave mechanics 79–80
  matrix mechanics xiv, 47–50, 60, 73, 128
  meeting with Bohr (Copenhagen, 1941) xv, 160–6
  Nobel Prize 142
  on Pauli's exclusion principle 55
  positivism 109
  quantum electrodynamics 172
  repulsed by Schrödinger theory 67
  and S-matrix 190
  uncertainty principle conflict with Bohr 97–102
  uncertainty principle discovered xv, 87–94, 96
Heisenberg, Elisabeth 161
helium atoms 26n, 31–2
  emission spectrum 43
  see also alpha particles
Herman, Robert 367
Herzog, Thomas 335–7
Hess, Rudolph 46n
hidden variable theories xvi, 307, 317
  Bell 308–13, 310 Fig 20, 311 Fig 21, 312 Fig 22

Bohm 302–5
Einstein 298–9
experimental refutation 322–6, 353–4
non-local 350–4
von Neumann's impossibility proof 158, 307
Higgs, Peter 229, 232, 410
Higgs boson xviii, 231, 252, 289, 290, 399, 409–10
mass 292
Higgs field 230, 231, 277, 289, 291
Higgs mechanism xvi, 229–34, 248, 251, 253, 284, 410
Hilbert, David 72
Hiroshima bombing 167
Hiss, Alger 297
Hitler, Adolf 45, 46n, 142, 158
Hooke, Robert 2
Horne, Michael 326
Horowitz, Gary 389
House Un-American Activities Committee (HUAC) 297, 298
Hubble, Edwin 367
human sense organs 355–6
Hume, David 107, 108
Husserl, Edmund 62
Huygens, Christiaan 2
hydrogen atom 133, 199, 199n, 370
Bell's thought experiment 308–12
Bohm's thought experiment 299–302, 305
electron wavefunction 64–5
emission spectrum 30, 34, 43, 50, 51, 60, 174

Iliopoulos, John 248, 249–51, 266
imaginary numbers 76, 195
Imbert, Christian 320
inelastic scattering 239
Institute for Theoretical and Applied Optics, Paris 319
intermediate vector bosons *see* W particles; Z particles
International Education Board 44
interpretation xviii, 401–4
isospin 199
and quarks 223
isospin symmetry 201
isotopes, stability 200
Israel, Werner 373

J/ψ particle xvi, 271, 272, 273 Fig 18, 275
Jacobson, Theodore 393–4
Jeans, James 20
Jenson, Peter 161
John Paul II 392
Jordan, Pascual 49, 50, 87n, 88, 103
Josephson junctions 343, 344–5, 346

Kaluza, Theodor 389
Kaluza-Klein theory 389
kaons (tau and theta) 204, 206, 209–10, 215, 259, 271, 345n
decay 248, 249, 250, 253
'oddness' 248–9
quantum eraser experiments 338
Kelvin, Lord (William Thomson) 4, 290
Kendall, Henry 240–1, 244
Kendall graph 241–2, 242 Fig 15
Kent, Adrian 403
Kibble, Tom 229
Kim, Yoon-Ho 337
Kirchhoff, Gustav 10
Klein, Felix 72
Klein, Oskar 100, 102, 103, 104, 126, 284n, 389
Klein–Gordon equation 126, 128, 129
Kobayashi, Makoto 289
Kocher, Carl 319
Kossel, Walther 53
Kramers, Hendrik 43, 46, 57, 79, 173, 175
Kronig, Ralph de Laer 56, 58, 131
Kudar, Johann 129
Kulik, Sergei 337
Kurlbaum, Ferdinand 12
Kusch, Polykarp 174
Kwait, Paul 335–7

Lamb, Willis 173
Lamb Shift 174, 175, 179, 180, 181, 183, 188, 191
lambda particles 204, 206, 216, 267
  and Eightfold Way 218
Landé, Alfred 51, 56–7
Landé 'g-factor' 56, 131
Langevin, Paul 41, 134
Large Hadron Collider *see* CERN
LaSauce, Helen 267
Lawrence, Ernest 141, 204
Lawrence Berkeley National Laboratory 270, 270n
Lecture on Physics (Feynman) 328
Lederman, Leon 276, 283, 290
Lee, Benjamin 253
Lee, Tsung-Dao 209
  Nobel Prize 210n
Leggett, Tony 341, 342, 346, 347, 350–4, 357
  Nobel Prize 350
Leggett's inequality xvii, 352–3
  experimental violation 353–4
Leipzig University 46, 79, 103, 160
Lenard, Philipp 23, 46
'lepton numbers' 215
leptons xvi, 214, 214n, 247, 274, 276, 290, 291
  electro-weak unified theory 233–4, 251
Leutwyler, Heinrich 262–3
Lewis, Gilbert N. 92n
Lie, Sophus 194
Lie Groups 194–5, 217 *see also* symmetry groups
light
  coherence 177
  corpuscular theory 2
  path of least time 177–8
  wave theory 2–4, 22–3, 24, 74, 177
  see also speed of light
light-quantum hypothesis xiv, 20, 21–4, 34–5, 37, 43, 74
Lindemann, Frederick 149–50, 160
local gauge symmetry 195–8, 228
local gauge theories xv–xvi, 193, 198–203, 253, 260
  and asymptotic freedom 261–3
local reality 151, 308, 316, 317
locality loophole 324, 351
logical positivism 108, 110
loop quantum gravity xviii, 392, 394–8, 400
Lorentz, Hendrik 41, 57, 71, 104, 115, 117, 118, 134
Los Alamos 166, 173, 176, 243
Low, Francis 250–1
Lukens, J. E. 345
Lummer, Otto 10

M-theory 399–400
Mach, Ernst 52, 107, 108
MacInnes, Duncan 172–3
macroscopic quantum objects xvii, 341–8
macroscopic realism 341–2
  directness 341
  disconnectivity 341, 342, 346
  distinctness 341
  extensive difference 341, 342, 346
Maiani, Luciano 149–51, 266, 293, 248
Manhattan Project 166, 173, 188, 208n
'many worlds' theory 370–1, 401–2
March, Arthur 149
March, Hildegunde 149
Marsden, Ernest 26
Marshak, Robert 210
Martinec, Emil 388
Marxism 301
Maskawa, Toshihide 289
Massachusetts Institute of Technology (MIT) 72, 176, 206, 233, 240, 242, 243, 250–1, 263, 266, 269, 346, 367
Mathematical Foundation of Quantum Mechanics (von Neumann) 158
matrices 49
matrix mechanics xiv, 50, 60, 72–3, 88, 109, 117, 128
  and observable kinematic data 88–90
  and wave mechanics 66, 79–86, 87–8
matter particles *see* fermions

Max Planck Institute for Quantum Optics, Munich 330, 332
Max Planck Institute for Physics, Munich 258
Maxwell, James Clerk 2–3, 4, 279
Maxwell's demon 279–80
Maxwell's electromagnetic field theory 2–3, 3n, 9, 21–2, 28, 35, 171, 195, 197
McCarthy, Joseph 298
McGovern, George 387
membranes 400
Mendeleev, Dmitri 215
Mercury 364
mesons 215
  charmed 266
  quark composition 222, 224, 259
mesons, spin-0 215
  Eightfold Way 218
  quark composition 223
metaphysics 107, 109, 110
Meitner, Lise 159
Michelson, Albert 4, 35
microscope resolution 98, 99
Millikan, Robert 34, 104
Mills, Robert xvi, 194, 200–3, 406
Minkowski, Hermann 72
Misner, Charles 364, 365, 366–7
mixing angles 291
mole 17, 17n
molecular motion 20
molecules 158
  reality of 7, 16, 17, 21, 107
Møller, Christian 163
momentum conservation 194
Morette, Cecile 368
Morley, Edward 4, 35
Munich University 7, 44, 45, 46, 52, 80, 81, 99
muon anti-neutrinos 286
muon neutrinos 275, 288
muon neutrino-proton collisions 253
  muonless events 253–5
muons (mu-mesons) 158, 204, 208, 214, 215n, 248, 250, 275, 286, 288, 290
musical notes 39 Fig 3, 39–4
Musset, Paul 254

Nafe, John 174
Nagasaki bombing 167
Nambu, Yoichiro 226, 257, 383
Nambu–Goldstone bosons 227, 228, 229–30, 232–3
National Accelerator Laboratory (NAL), Chicago *see* Fermilab
National Socialist Party 45–6, 142
Nature 76, 104, 146, 158, 345, 347, 353, 378
Naturwissenschaften, Die 154
Nazis 161, 163, 165
Neddermeyer, Seth 158
Ne'eman, Yuval 215, 218–20
Nelson, Edward 174
Nelson, Steve 297
Nernst, Walther 25, 115
Neurath, Otto 108
neutrino-proton collisions 267
neutrinos 158, 205, 214, 233–4, 282, 290
  'left-handed' 210
  limited to three 288
neutrons 214, 216, 287
  baryon and lepton numbers 215
  conversion into protons in beta-decay 207, 212, 223, 289
  discovery 141, 159
  Eightfold Way 218, 219 Fig 12
  instability 208
  quark composition 245, 259
  structure 245
Neveu, André 385, 386
New York Times 145, 258–9
Newton, Isaac 1–2, 4, 12, 134, 302, 380, 382, 400
Newton's gravitation constant (G) 12, 361
Niels Bohr Institute for Theoretical Physics, Copenhagen 44, 45, 46, 72, 77, 79, 85, 88, 100, 103, 128, 160, 163, 211, 239, 247, 383

Nielson, Holger 383
Nishina, Yoshio 182
Nixon, Richard 387
noble gases 53
Noether, Amalie Emmy 194
Noether's theorem 194, 195
'normal' Zeeman effect 51
'November revolution' xvi, 265–319
nuclear chain reaction 141, 160
nuclear fission 159–60
nuclear reactors 166

operators 66–7, 66n
Oppenheimer, J. Robert 76, 166, 167, 173, 174, 176, 180, 182, 183, 189, 190–1, 190n, 202, 202n, 207, 297, 298, 305, 404–5
Oppenheimer, Frank 404
Oxford University 149, 150, 377

Page, Don 375
Pagels, Heinz 263
Pais, Abraham 173, 179, 205–6
Palmer, Robert 267
parity conservation 208–9
particle accelerators xvi, 141, 205, 216, 237–8
particle colliders 265
particles *see* quantum particles
partons 243–5, 244 Fig 16
Paschen, Friedrich 10, 31, 104
Paschos, Emmanuel 243, 245
Patel, Vijay 345
Pauli, Wolfgang 52–7, 67, 80, 87n, 102, 104, 115, 121, 126, 158, 173, 194, 209, 284, 298, 305, 392, 406
  and anomalous Zeeman effect 51, 52, 286
  and Bohr 52, 59, 101
  career 52, 103
  and electron spin 58, 129–30
  fury over EPR paper 146
  and Heisenberg 55, 88–9, 94
  matrix mechanics paper 60, 65, 89
  and parity conservation 209
  and position-momentum commutation relation 89
  and quantum electrodynamics 172
  and wavefunction as probability 88–9
  and Yang–Mills field theory 202
Pauli exclusion principle xiv, 54–5, 56, 60, 126, 228, 342
  quark problem 224, 256, 257, 260
Pauli spin matrices 129–30, 392
Pearl Harbor 166
Peebles, Jim 367, 368
Peierls, Rudolf 158, 161
Penrose, Roger 373, 396–7
Penzias, Arno 367–8
periodic table 53, 54, 54n, 55, 215
Perl, Martin 275–6
perturbation expansion 172, 172n, 179, 181, 185, 201–2
perturbation theory 183, 365, 399n
phase (resonance) condition 40, 42, 63
phase factor 62, 198
phase waves 38–9
Philosophical Magazine 53
phonons 229, 400
photoelectric effect 22–3, 34, 72
photon box 135–40, 137 Fig 7
photons xvi, 92n, 172, 215, 275, 289, 290, 400
  decay routes 265–6
  entangled triplets 326
  experimental studies of entangled 318–27, 335–7, 353–4
  in Feynman diagrams 184–7, 191
  non-local hidden variables theory 350–3
  quantum eraser thought experiments 330–5, 331 Fig 24
  weak force carrier speculation 210, 211, 212
Physical Review 145, 146, 150, 183, 203, 224, 227, 305, 320–1, 369, 384
Physical Review Letters 261, 322, 336, 337, 383, 388, 393, 432
Physics Letters 224, 283, 388, 392
Pickering, Edward Charles 31

Pickering series 31
Pierce, Charles Sanders 110
pion-kaon pairs 268
pions (pi-mesons) 204, 205, 206, 209, 215, 239
decay of neutral 257, 259, 260
and kaon decay 249
spin-0 288
Planck, Max 7–16, 17, 35, 104, 115
correspondence with Einstein 24
discovery of photons 289
Nobel Prize 34n
personality 10–12
plea to Hitler 142
'quantum of action' xiv, 15–16, 29
Planck length 364, 389, 397
Planck scales 394, 394
Planck time 365, 397
Planck's constant (*h*) 12, 15, 17, 23n, 29, 31, 38n, 49, 56, 91, 124, 134, 138, 361
Planck's equation (ε=hν) 15–16, 23, 99
Planck's radiation law 11 Fig 2, 12–13, 14–16, 17, 20, 21, 24, 29, 82–3, 286, 375
Plato's cave allegory 234, 237, 355
Pocono conference 181–2, 183
Podolsky, Boris 142, 145, 151
Poisson, Siméon 50
Politzer, David 262
position-momentum commutation relation 49, 50, 89, 91–3
position-momentum uncertainty 98–9, 100, 143–4, 147
positivism 107–8, 110
positrons 141–2
high-energy 282
Powell, Cecil 204
pragmatism 110
Princeton Institute for Advanced Study 142, 149, 176, 189, 190, 199, 201, 206, 257, 298, 365, 368, 369, 393
Princeton University 261, 297, 318, 363, 370, 374, 385, 387, 389
Principia Mathematica (Newton) 1–2
Pringsheim, Ernst 10
probability
Born's probability waves 73–8, 102, 116, 129
and entropy 13–14
Feynman and 178, 185
Pauli and 88–9
quantum vs. classical 77, 303–4
Proceedings of the Royal Society 60, 132
'projection postulate' 148
proton accelerators 238
proton-anti-proton collision 267–9, 278–9, 281–3, 285–6
anti-proton beams 278–9, 280
proton-neutron interactions 199–200
proton-proton collisions 408
proton structure xvi, 239–40, 241–2
parton model 243–5
quark model 240, 245–6
protons 205, 206, 214, 216, 287
baryon and lepton numbers 215
conversion from neutrons in beta-decay 207, 212, 223, 289
Eightfold Way 218, 219 Fig 12
and 'electron holes' 133–4
quark composition 222, 245, 256, 259
stability 381
structure 239–42
Prussian Academy of Science 24, 361

quantum chromodynamics (QCD) xvi, 260–3, 265, 385, 394
quantum computers 384, 384n
quantum electrodynamics (QED) xv, 172, 175, 179–80, 181–92, 198, 201, 242–3, 306, 376
quantum entanglement 151, 300–1, 332
and 'disconnectivity' 341
experimental studies 318–27, 335–8, 353–4
macroscopic 347
quantum erasers xvii
delayed choice 333–5, 337–8
experimental realization 335–8
thought experiments 330–5, 335 Fig 25

quantum field theory xv–xvi, 194, 213, 229, 406
 and asymptotic freedom 261–2
 of gravity 299, 365, 372, 379, 391
 and string theory 399
 welding with general relativity 377–9
 see also Yang–Mills field theories; unified field theories
quantum geometrodynamics 364
quantum jumps xv, 32, 34, 42, 47, 66, 72, 73, 78, 80–1, 81–3, 84, 88, 93
quantum gravity theory xvii, xviii
 canonical approach 365–7, 369, 371, 391–2, 392–3
 covariant approach 365–6, 372, 379, 391
 early frustrations 361–3
 supergravity 382
 Wheeler–DeWitt equation xvii, 369–71, 372, 391, 393–4, 404
 see also loop quantum gravity; superstring theory
quantum numbers 32, 35, 40, 42, 174
 'colour' 259
 j (fourth) 51, 54, 56, 130
 k 34, 53, 54, 65, 130–1
 m 34, 47, 51, 53, 54, 65
 N 205–6
 n 34, 47, 53, 54, 65
'quantum of action' xiv, 15–16, 29
quantum particles
 black hole emission 377–8
 masses 290–1
 proliferation 214–15, 382
 as strings 384–5
 'vanilla' properties 356
 see also Standard Model; strange particles *and* individual particles
quantum potential 303–4
Quantum Theory (Bohm) 298, 299–302
quantum tunnelling 343, 343n
quark-anti-quark pairs 222, 223, 245, 246, 259, 264, 265, 273, 276, 288
quark 'jets' 276, 286
quarks xvi, 220–4, 247, 275, 276, 306, 319
 asymptotic freedom problem 256, 261–3
 colour 259–61, 287–8, 288, 290
 confinement 256, 263–4, 287–8
 18 types 290
 exclusion principle problem 256, 257
 flavour 287, 288, 290
 Ham–Nambu model 257, 259
 mass 290–1
 naked/dressed 94, 290–1
 and nucleon structure 240, 245–6
 obstacles to 256–7
 and partons 245
 and strong-force field theory 258
 see also bottom quarks; charm quarks; down quarks; strange quarks; top quarks; up quarks
qubits 348

Rabi, Isidor 158, 171, 173, 174–5, 175–6, 405
radioactivity 28
Ramond, Pierre 384
Rayleigh, Lord (William Strutt) 20
Rayleigh–Jeans law 11 Fig 2, 20–1, 22
Reagan, Ronald 292
realism, assumption of 349–40
reality
 empirical xvii, 108–9, 156, 356
 illusion of 355–7
 local 151, 308, 316, 317
 non-local 318, 338
 non-local, experimental validation xvii, 323–7, 328
relativistic quantum theory of the electron xv, 128–32, 171, 172, 174, 362
relativity 127–8
 parallels with complementary 134
 see also general relativity; special relativity
Renteln, Paul 393
Reviews of Modern Physics 365
Rhoades, Terence 269
Richter, Burton 266, 271, 272–3
 Nobel Prize 274

Riordan, Michael 263
Robinson, David 373
Rochester conferences 213, 220, 260, 266
Roger, Gérard 319, 322, 324
Rohm, Ryan 388
Roll, Peter 367
Roosevelt, Franklin 160, 166
Rosen, Nathan 142
Rosenfeld, Léon 105, 131–2, 135, 145–6, 361
Rovelli, Carlo 395–6, 397
Royal Society 2, 379–80
Rozental, Stefan 163
Rubbia, Carlo 255, 280–1, 283, 292
  Nobel Prize 284
Rubens, Heinrich 12, 13
Russell, Bertrand 108
Rutherford, Ernest 26, 28, 30, 32–3, 104, 133, 174
Rutherford laboratory, Manchester, 28, 30, 33
Rydberg, Johannes 31
Rydberg constant 31
Rydberg formula 31

S-matrix (scattering matrix) 190, 215n
Sagan, Carl 380
Sakata, Soichi 216, 217
Salam, Abdus xvi, 212, 219, 226, 227, 234, 406
  Nobel Prize 277
Samios, Nicholas 220, 224, 267, 274
Savi, Helene 10n
scaling 241–2, 242 Fig 15, 243, 245, 246, 256, 261
  violations 263
scattering amplitudes 383
Scherk, Joël 385, 386
Scherrer, Paul 62
Schlick, Moritz 108, 109n
Schrieffer, John 228
Schrödinger Annemarie 61–2, 63–4, 81, 149
Schrödinger, Erwin xv, 61–7, 99, 115, 149–58, 194, 348
  career 61, 149
  cat paradox xv, 154–7, 155 Fig 8, 339, 340–1, 371,
  and complex wavefunctions 76
  correspondence with Einstein 150–4, 157
  criticism of Born 78
  and de Broglie paper 62–3
  debates with Heisenberg and Bohr xv, 80–6, 87, 95
  leaves Germany 142
  marriage difficulties 61–2, 63–4, 149–50
  Nobel Prize 142, 149
  rejection of Copenhagen interpretation 103, 111
  repulsed by Heisenberg theory 67
  wave mechanics xiv, 64–7, 71, 72, 73, 74, 79, 116, 117, 126, 129, 198, 298, 370, 371
Schwartz, Melvin 269, 270, 274
Schwarz, John 384–5, 386–8
Schwarzchild, Karl 363–4
Schwinger, Julian 368
  Nobel Prize 242
  personality and approach 176, 181
  Pocono conference 182
  precocity 175–6
  and quantum electrodynamics xv, 181, 182, 189, 192
  Shelter Island conference 173, 175, 177
  weak force carrier speculations 210–11, 284
Schwitters, Roy 270, 292
Science 274, 346
Scully, Marlan xvii, 337
  photon interference thought experiment 330–3, 331 Fig. 24
Segrè, Emilio 207
Sen, Amitabha 392
Serber, Robert 173, 220–2, 223
Shaknov, Irving 318
Shelter Island conference 173–9
Shih, Yanhua 337
sigma particles 205, 214

sigma-star particles 220
singularity 373, 364
Slater, John C. 43
Smolin, Lee 393–8, 403–4, 406
Smyth, Henry D. 188
solar eclipse (1919) 41
solid-state physics 225–6, 227–9
Solovine, Maurice 18, 36
Solvay conferences
  fifth xv, 111, 115–25, 128, 299
  first 115
  sixth 134–9
Solvay, Ernest 61, 115
Sommerfeld, Arnold 34, 51, 52, 56, 80, 81, 99, 103, 104
space-time
  curvature 364, 394, 400
  in general relativity (four-dimensional) 362–3, 366
  expanding 367, 369
  extremely small distances 364–5
  'foam' 364
  as a lattice 394
  loop quantum theory 394–8
  'no boundary' assumption 391–2
  particle displacement in 382
  seventh dimension 400
  ten dimensions 385, 389
  three-space 366, 369
space-time intervals 366, 366n
special relativity 20, 24, 35–6, 57, 60, 64, 119–20, 126, 129, 135, 144, 148, 243, 349
spectroscopy 30–1
speed of light 3, 4, 12, 35, 36, 134, 144, 349, 361, 362, 366
  variable 400
Speer, Albert 166
spin networks 396–8, 400
spinors 392–3
Spinoza, Baruch 19
Sputnik I 237
Standard Model xvi, xviii, 275, 276, 285–93, 319, 328, 381, 386, 398, 399, 404, 408, 410
standing waves 39–40
Stanford Linear Accelerator Center (SLAC) xvi, 238–45, 247, 258–9, 263, 265, 273–4, 307
Stanford Positron Electron Asymmetric Rings (SPEAR) 266, 270, 271–2, 274, 275–6
Stark, Johannes 60
Stark effect 60
Starobinsky, Alexei 376
Stern, Otto 52, 58, 104, 121
stochastic cooling 280
Stoner, Edmund 42–4
strange particles 204–7, 208, 222–3
strange quarks 222, 266, 267, 275, 288, 289, 291
strangeness 206–7, 248
  and Eightfold Way 218, 219 Fig 12, 220
Strassman, Fritz 159
Strominger, Andrew 389
strong nuclear force 243
  and asymptotic freedom 246, 261, 256, 262
  electron-exchange model 199–200
  gluon carriers 246, 275
  meson carriers 215
  quantum chromodynamics 385
  and quark colours 288
  strange particle production 205
  strangeness conservation 206
  string theory 384–5
  three particles predicted 201–2, 284
  unification with electro-weak force 381
  Yang–Mills quantum field theory 201–3, 284
  Weinberg's symmetry-breaking application attempt 232–3
'structure function' 241
Sudarshan, George 210
Sullivan, Walter 258–9
superconducting quantum interference devices (SQUIDS) 344, 348
  macroscopic quantum state experiments 345–6, 347

Superconducting Supercollider (SSC), Texas 292
superconductivity 226
  Barleen–Cooper–Schrieffer theory 228, 259
  macroscopic quantum states 342–6
supergravity 382, 386, 399
superstring theory xvii–xviii, 384–90, 391, 398, 399–400, 405–6
  early string theory 383–4
  gauge anomalies 387–8
  heterotic 388, 399
  open and closed superstrings 386
  ten-dimensional space-time 389, 392
  Type I 386, 387–8, 399
  Type IIA 386, 387, 399
  Type IIB 386, 387, 399
supersymmetry 372, 382, 399, 405
Susskind, Leonard 383–4
symmetry-breaking xvi, xviii, 225–34, 248, 251, 252, 284, 291
symmetry groups
  SO(32) 388, 399
  SU(2) 201
  SU(3) 217–18, 220–3, 260
  SU(5) 381
  U(1) 195, 196–7 Fig 11, 198, 200
symmetry transformation 193–8, 193n
't Hooft, Gerard 251–3, 261, 289–90, 372
  Nobel Prize 372

tachyons 374
tau particles 275–6, 288, 290
tau neutrino 276, 288
  discovery 286
Taylor, John 379
Taylor, Richard 240–1
Teller, Edward 173, 198
thermodynamics 7–8, 375, 379
  second law 8–9, 13, 279–80, 373, 374, 378
Thomas, Llewellyn Hilleth 59
Thomson, Joseph John 25, 26–7, 28
Thorne, Kip 375–6, 379
time
  'disappearance' in three-space 366, 369
  'phenomenological' 369
  nature of 366n
time, travel backwards 181, 182
Ting, Samuel Chao Chung 266–7, 267–70, 271, 272–3, 274
  Nobel Prize 274
Tolpygo, S. K. 345
Tomonaga, Sin-Itiro xv, 182–4, 189
  Nobel Prize 242
top quarks 276, 285–6, 288, 289, 291, 328
  discovery 286
Trouble with Physics, The (Smolin) 406
Truman, Harry 297
two-slit experiments *see* double-slit experiments

Uhlenbeck, George 57, 130
ultraviolet catastrophe 21n
uncertainty principle xv, 124, 138, 182, 186, 344, 363, 376
  'beating' 328–9
  Bohr–Heisenberg conflict 97–102, 329
  Heisenberg's discovery xv, 89–94, 96, 143, 144, 378
  and space-time 364–5
  see also energy-time uncertainty; position-momentum uncertainty
unified field theories
  SU(2) x U(1) electro-weak xv, 212–13, 217, 232, 233–4, 247–8, 249–51, 252–3
  SU(3) x S(2) x U(1) strong-electro-weak 263, 266, 286, 381
  SU(5) grand 381
Universal Fermi Interactions 208, 210
Unruh, William 376
up quarks 222, 223, 224, 245, 265, 275, 287, 288, 289
upsilon particle 276
uranium-235 159, 166
Uranverein 160, 161, 166
US National Bureau of Standards 209

van der Meer, Simon 279–80
  Nobel Prize 284
Van Vleck, John 173
Vedral, Vlatko 347
Veltman, Martinus 251–2, 372
  Nobel Prize 372
Veneziano, Gabriele 383, 385
Vienna Circle 108, 109n
Volta, Alessandro 104
von Hevesy, George 28
von Laue, Max 104
von Neumann, John 158, 173, 194, 307
  collapse/projection postulate 148, 156, 339

W particles 250, 267, 275, 288, 290, 408
  acquisition of mass 234, 253
  decay 286, 289
  emission in beta-decay 211, 223
  observation xvi, 283, 293, 328
  predicted xvi, 210, 212, 213, 232, 233
  search for 276–82
Walther, Herbert 332
Walton, Ernest 141
Ward, John 212
wave mechanics xiv, 64–7, 71, 72–3, 88, 99, 116, 117, 198, 298, 370, 371
  and matrix mechanics 66, 79–86, 87–8
'wave packet' states 71, 88, 98–9, 369
wave/particle duality 21, 35, 100, 101, 102, 329, 332–3
  de Broglie 'double solution' xiv, 37–42, 62–3, 64, 71, 116, 119, 298, 304–5
  Bohm's hidden variables theory 302–5
  consistent histories interpretation 402, 403
  Einstein's hidden variable approach 298–9
wavefunction
  Bohm's interpretation 302–4
  Born's probability interpretation 72, 73–8, 89, 129
  Copenhagen interpretation 102
  'decoherence' 340
  in entangled atoms 300
  gauge invariance 199
  as 'guiding field' 298–9
  kaons 248
  in macroscopic states 343
  parity 208
  Pauli's interpretation 88–9
  and physical reality 153–4
  Schrödinger's interpretation xv, 64–7, 71, 76, 89, 116
  symmetry properties 198, 256
  of the universe 369–71, 391–2, 401
wavefunction, collapse 94
  Bell's rejection 307, 308
  Bohm's elimination 304, 308
  Einstein's objection 118–20
  ensemble alternative 120
  experimental validation 323
  and Schrödinger's cat paradox 154–7, 339
  two-particle 147–8, 150–1, 300
weak nuclear force 205, 206, 207–10, 289–90
  analogy with electromagnetism 207, 210
  carriers *see* W particles; Z particles
  parity violation in decay 209, 249, 289
  and quark flavours 288–9
  see also unified field theories
weak neutral currents 212, 213, 233–4, 247–8, 250, 251, 253
  discovery 253–5, 277, 281, 292–3, 319
Weinberg, Joseph 297, 298
Weinberg, Steven xvi, 211, 226, 227, 237, 253, 254, 293
  Nobel Prize 277
  and SU(2) x U(1) field theory 232–4, 248, 251, 252, 277, 406
Weinstein, Roy 274
Weisskopf, Victor 173, 174, 175, 209, 224
Weizsäcker, Carl Friedrich von 160, 161, 162, 165
Weizsäcker, Ernst von 162

Wess, Julius 382
Weyl, Helene Joseph 62
Weyl, Hermann 62, 65, 66, 149, 193–8
Wheeler, John 159, 173, 181, 190, 298, 318, 339–40, 363–5, 368–9, 370, 376, 401–2
Wheeler–DeWitt equation xvii, 369–71, 372, 391, 393–4, 401
White, Harry Dexter 297
Wien, Wilhelm 45, 65, 78, 80, 81, 84, 99
Wien's law 10, 11 Fig 2, 12, 21
Wienfurter, Harald 335–7
Wigner, Eugene 105, 194
Wilczek, Frank 261–2
Wilde, Jane 374, 376, 379
Wilkinson, David 367
Wilson, Charles 83n
Wilson, Kenneth 394
Wilson, Robert 367–8
'Wilson loops' 394
Witten, Edward 387, 388, 389, 399–400
Wittgenstein, Ludwig 108
Woit, Peter 406
work function 23
World War I 37, 61
World War II 160–7, 188
wormholes 364
Wright, Courtney 207
Wu, Chien-Shiung 208, 208n, 318

xi particles 205, 214
xi-star particles 220

Yale University 206, 219, 375, 393, 395
Yang, Chen Ning xvi, 194, 198–9, 200–3, 209, 216, 406
  Nobel Prize 210n
Yang-Mills bosons 226, 227
  acquisition of mass 229–31
Yang-Mills theories
  renormalizability 251–3, 372
  SU(2) 200–3, 210, 211, 212, 284, 306
  supersymmetric 386
  see also quantum chromodynamics; unified field theories
Yau, Shing-tung 389
Young, Thomas 2, 3 Fig 1
Yu, Rong 337
Yukawa, Hideki 182, 202, 204

Z particles 250, 275, 288, 290, 408
  acquisition of mass 234, 253
  decays 288
  predicted xvi, 212, 213, 232, 233, 234, 247, 248
  observation xvi, 283, 293, 328
  search for 276–82
Zeeman, Pieter 43n, 104
Zeh, Dieter 340
Zeilinger, Anton xvii, 326, 335–7, 342, 347, 350, 353, 354–5
Zeitschrift für Physik 60, 65, 97
Zeldovich, Yakov 375–6, 377, 379
Zermelo, Ernst 7–8
Zumino, Julian 382
Zurich University 61, 63, 65, 72
Zweig, George 223n, 306